D'Arcy Wentworth Thompson's Generative Influences in Art, Design, and Architecture

D'Arcy Wentworth Thompson's Generative Influences in Art, Design, and Architecture

From Forces to Forms

Edited by
Ellen K. Levy and Charissa N. Terranova

BLOOMSBURY VISUAL ARTS
LONDON • NEW YORK • OXFORD • NEW DELHI • SYDNEY

BLOOMSBURY VISUAL ARTS
Bloomsbury Publishing Plc
50 Bedford Square, London, WC1B 3DP, UK
1385 Broadway, New York, NY 10018, USA
29 Earlsfort Terrace, Dublin 2, Ireland

BLOOMSBURY, BLOOMSBURY VISUAL ARTS and the Diana logo are trademarks of Bloomsbury Publishing Plc

First published in Great Britain 2021

Series design by Toby Way
Cover image: *Skies Painted with Unnumbered Sparks* sculpture by Janet Echelman, installed in Vancouver, Canada (2014). Photo by Ema Peter, courtesy Studio Echelman.

A catalogue record for this book is available from the British Library.

Library of Congress Cataloging-in-Publication Data
Names: Levy, Ellen K., editor. | Terranova, Charissa N., editor.
Title: D'Arcy Wentworth Thompson's generative influences in art, design, and architecture : from forces to forms / edited by Ellen K. Levy and Charissa N. Terranova.
Description: London ; New York : Bloomsbury Visual Arts, 2021. | Series: Biotechne : interthinking art, science and design | Includes bibliographical references and index.
Identifiers: LCCN 2020038564 (print) | LCCN 2020038565 (ebook) | ISBN 9781350191112 (hardback) | ISBN 9781350191129 (pdf) | ISBN 9781350191136 (epub)
Subjects: LCSH: Thompson, D'Arcy Wentworth, 1860-1948. | Thompson, D'Arcy Wentworth, 1860-1948--Influence. | Thompson, D'Arcy Wentworth, 1860-1948. On growth and form. | Zoologists—Scotland–Biography. | Growth. | Form (Aesthetics) | Morphology (Animals) | Influence (Literary, artistic, etc.) | Art and science.
Classification: LCC QP84 .D35 2021 (print) | LCC QP84 (ebook) | DDC 590.92 [B]–dc23
LC record available at https://lccn.loc.gov/2020038564
LC ebook record available at https://lccn.loc.gov/2020038565

ISBN: HB: 978-1-3501-9111-2
ePDF: 978-1-3501-9112-9
eBook: 978-1-3501-9113-6

Series: Biotechne: Interthinking Art, Science and Design

Typeset by Integra Software Services Pvt. Ltd.
Printed and bound in Great Britain

To find out more about our authors and books visit www.bloomsbury.com and sign up for our newsletters.

Contents

Illustrations

Plates

Figures

Contributors

Philip Ball is a freelance writer and broadcaster and worked previously for over twenty years as an editor for *Nature*. He writes regularly in the scientific and popular media and has authored many books on the interactions of the sciences, the arts, and the wider culture, including *H_2O: A Biography of Water*, *Bright Earth: The Invention of Colour*, *The Music Instinct*, and *Curiosity: How Science Became Interested in Everything*. His book *Critical Mass* won the 2005 Aventis Prize for Science Books. Philip is a presenter of Science Stories, the BBC Radio 4 series on the history of science. He was trained as a chemist at the University of Oxford, and as a physicist at the University of Bristol. His latest book is *Beyond Weird* (2018), a survey of what quantum mechanics means.

Philip Beesley is a professor in the School of Architecture at the University of Waterloo. A practitioner of architecture and digital media art, he was educated in visual art at Queen's University, in technology at Humber College, and in architecture at the University of Toronto. His Toronto-based practice PBAI is an interdisciplinary design firm that combines public buildings with exhibition design, stage, and lighting projects. His work was selected to represent Canada at the 2010 Venice Biennale for Architecture, and he has been recognized by the Prix de Rome in Architecture, VIDA 11.0, FEIDAD, two Governor General's Awards and as a Katerva finalist. Beesley's funding includes core CFI, SSHRC, NSERC and Canada Council grants.

Sarah Bonnemaison is a professor of architecture at Dalhousie University. Her practice specializes in tensile structures and festival architecture. Sarah is also a writer. Her books include *Architecture and Nature; Festival Architecture;* and *Installations by Architects* as well as numerous book chapters and essays. Her passion lies on bringing history and theory to life through interactive exhibitions. She is currently writing a book about modern female architects who searched for the organic and whose projects contributed to the philosophy of *Organicism*.

Matthew Jarron is Curator of the University of Dundee Museum Collections, which comprise art, science, and medical history as well as the D'Arcy Thompson Zoology Museum. He is the author or co-author of several books, including *The Artist & the Thinker: John Duncan & Patrick Geddes in Dundee* (2004), *D'Arcy Thompson and His Zoology Museum in Dundee* (2010), and *Independent & Individualist—Art in Dundee 1867–1924* (2015). He has also guest-edited issues of the *Journal of the Scottish Society for Art History* and *Interdisciplinary Science Reviews*. He was Chair of the Scottish Society for Art History 2005–12 and continues to be an active committee member of the society.

Bart Kahr was born in New York City in 1961. He studied chemistry with I. D. Reingold at Middlebury College, with Kurt Mislow at Princeton University (PhD, 1988), and with J. M. McBride at Yale University. He has been a faculty member at Purdue University, the University of Washington, and New York University (since 2009). Kahr's research group studies complex organized media. Projects emphasize crystal growth mechanisms, polycrystalline pattern formation, new methods of metrology using polarized light, and the analysis of chiroptical anisotropy. He practices the experimental history of chemistry and crystallography, that is, those aspects of the development of a science that can only be informed by contemporary laboratory experiments. More recently, he has been advocating for changes in the way that universities and government agencies manage scientific misconduct. In 2014 he was a National Science Foundation distinguished lecturer in the mathematical and physical sciences.

Assimina Kaniari is Assistant Professor in the Department of Art History and Theory, Athens School of Fine Arts. Kaniari received her doctorate from the Department of Art History, University of Oxford, working under Martin Kemp on the location of the ornament in nineteenth-century science and aesthetic theory. Her research focuses on the relations of art and science in aesthetic theories and art practices from the long nineteenth century to now, the history of collections and exhibitions, and the historiography of art.

Evelyn Fox Keller is Professor Emerita of History and Philosophy of Science in the Program in Science, Technology and Society at MIT. She received her PhD in theoretical physics at Harvard University, worked for a number of years at the interface of physics and biology, and then turned to the study of gender and science, and, more generally, to the history and philosophy of science. Author of many books, Keller is the recipient of many awards and honorary degrees; a member of the American Philosophical Society, the American Academy of Arts and Sciences, a MacArthur Fellow, a recipient of the Chaire Blaise Pascal in Paris and, in 2018, of the Dan David Prize for her work in the history of science.

Justine Kupferman is a scientist in New York City. She currently works at Kallyope, Inc., a biotechnology company that studies the gut-brain axis to develop transformative therapeutics for the diseases of the gastrointestinal tract and the brain, including obesity, NASH, depression, and Parkinson's disease. She received her undergraduate degree from Reed College. Her PhD in biology and her postdoctoral training in neuroscience were both completed at Columbia University. In addition to her research, she has taught for Columbia's Science Honors Program and the American Museum of Natural History. She is also an avid bird watcher.

Ellen K. Levy, a past president of the College Art Association, publishes widely on art, evolution, and complex systems, including in Leonardo/ISAST. Before earning her doctorate in art and neuroscience (University of Plymouth, UK, 2012), she was a Distinguished Visiting Fellow in Arts and Sciences, a Henry Luce Foundation-supported position at Skidmore College (1999). She guest-edited *Art Journal's* special issue, "Contemporary Art and the Genetic Code" (1996), the first in-depth academic publication about genomics and art. Levy has taught at The New School, Cooper Union, and Brooklyn College and was Special Advisor on the Arts and Sciences to the Institute for Doctoral Studies in the Visual Arts. She has exhibited her art internationally and at NASA and in such landmark exhibitions as *Weather Report* (Boulder Museum) and *Gregor Mendel: Planting the Seeds of Genetics* (Field Museum, Chicago). She was represented by Associated American Artists and Michael Steinberg Fine Arts (NYC).

Caroline O'Donnell is an Irish designer and theorist, based in New York. She is principal of the design firm CODA, and author of the book *Niche Tactics* (Routledge, 2015). O'Donnell holds the Edgar A. Tafel Professorship in the Architecture Department at Cornell University, where she is also director of the MArch program and editor-in-chief of the *Cornell Journal of Architecture*.

Henry Petroski is Aleksandar S. Vesic Professor of Civil Engineering and a professor of history at Duke University. He has written and lectured broadly on the topics of design, success, and failure, and the history of engineering and technology. His nineteen books on these subjects and others include *The Pencil, The Evolution of Useful Things, To Engineer Is Human,* and *Engineers of Dreams.* His most recent book is *The Road Taken: The History and Future of America's Infrastructure*. He is an elected member of the American Academy of Arts and Sciences, the American Philosophical Society, and the US National Academy of Engineering.

Carolee Schneemann (1939–2019) was an artist who pioneered investigations into subjectivity, the social construction of the female body, and the cultural biases of art history, which have had significant influence on subsequent generations of artists. She was awarded the Golden Lion Award for Lifetime Achievement at the Venice Biennale in 2017. In the same year, her first comprehensive retrospective traveled from the Museum of Modern art in Salzburg, to the Frankfurt Museum of Modern Art—and then to MoMA PS1 in New York, where it opened in October. *Uncollected Texts*, a compilation of her writings, was released in February 2018.

Marjorie Senechal is Louise Wolff Kahn Professor Emerita in Mathematics and History of Science and Technology at Smith College, holds a BS from the University of Chicago and a PhD from the Illinois Institute of Technology. *I Died for Beauty: Dorothy Wrinch and the Cultures of Science* (OUP, 2013) is her most recent book.

Todd Siler is a visual artist who received a PhD Interdisciplinary Studies in Psychology and Art from MIT and the 2011 Leonardo da Vinci World Award of Arts. His books include *Breaking the Mind Barrier* and *Think Like a Genius*. He was a Research Fellow at MIT's Center for Advanced Visual Studies and a Visiting Artist/Scientist at the Computer-Aided Design Lab in the Department of Mechanical Engineering. He was awarded an IBM Thomas J. Watson Fellowship to Paris (1975–1976) and a Fulbright Fellowship to India (1985–1986). Siler is represented by Ronald Feldman Fine Arts, and his artworks are in numerous public collections, including: The Solomon R. Guggenheim Museum, The Metropolitan Museum of Art, The Museum of Modern Art, The Whitney Museum of American Art, The Israel Museum in Jerusalem, and The Pushkin Fine Arts Museum in Moscow.

Hadas A. Steiner is Associate Professor at the University at Buffalo, SUNY, who researches cross-pollinations of technological, scientific, and cultural aspects of architectural fabrication. She is at work on a manuscript, *The Accidental Visitant*, which studies the influence of the modern field of ornithology on architecture and the conceptualization of the built environment as an ecosystem. Steiner is the author of *Beyond Archigram: The Technology of Circulation* (Routledge), and her scholarship and reviews have been published in *OCTOBER*, *Grey Room*, *New Geographies*, *Journal of the Society of Architectural Historians*, *Journal of Architectural Education*, *Journal of Architecture*, and *arq*.

Brandon Taylor is Professor Emeritus of History of Art, University of Southampton, England, and Tutor in History and Theory of Art, Ruskin School of Art, Oxford University. His research interests include modern and contemporary art, artists' writings, the history of art institutions, and East European art. His most recent books include *Collage: The Making of Modern Art* (Thames and Hudson, London 2004), *After Constructivism* (Yale University Press 2014), and *St Ives and British Modernism* (Pallant House Gallery, Chichester 2015). He exhibits occasionally as a painter.

Charissa N. Terranova is a writer and educator who researches complex biological systems from a cultural purview, focusing on the history of evolutionary theory, biology, and biocentrism in art, architecture, and design. Associate Professor of Aesthetic Studies at the University of Texas at Dallas, she lectures and teaches seminars on modern and contemporary art and architectural history and theory, the history of biology in art and architecture, and media and new media art and theory. She is the author of *Biology in the British Bauhaus: Morphogenic Modernism in Art, Science, and Design* (2019), *Art as Organism: Biology and the Evolution of the Digital Image* (2016), *Automotive Prosthetic: Technological Mediation and the Car in Conceptual Art* (2014), and coeditor with Meredith Tromble of *The Routledge Companion to Biology in Art and Architecture* (2016). She also co-edits with Tromble *Biotechne: Interthinking Art, Science, and Design*, a book series published by Bloomsbury Visual Arts.

Meredith Tromble is an artist, writer, and educator whose practice and research centered on collaboration, cognition, and creativity have been informed by working with scientists at several University of California campuses, including a long-term residency at the Complexity Sciences Center (CSC) at the University of California, Davis (UCD). In addition to authoring a number of book chapters and a blog on art and science funded by the Andy Warhol Writers Grant 2012–2013, Tromble served as editor of *Artweek*, the journal of record for art on the West Coast of the United States in the mid-1990s, edited *The Art and Films of Lynn Hershman Leeson* (University of California Press, 2005) and co-edited with Charissa Terranova *The Routledge Companion to Biology in Art and Architecture* (2016).

Stephen Wolfram is the creator of Mathematica, Wolfram|Alpha, and the Wolfram Language; the author of *A New Kind of Science*; and the founder and CEO of Wolfram Research. Over the course of nearly four decades, he has been a pioneer in the development and application of computational thinking—and has been responsible for many discoveries, inventions, and innovations in science, technology, and business. In recognition of his early work in physics and computing, Wolfram became in 1981 the youngest recipient of a MacArthur Fellowship. Wolfram has been involved with education for many years, founding the Wolfram Summer School in 2003, and in 2015 publishing *An Elementary Introduction to the Wolfram Language* to introduce young students and others to modern computational thinking.

Preface

Ellen K. Levy and Charissa N. Terranova

Scottish zoologist D'Arcy Wentworth Thompson's *On Growth and Form* has influenced thinkers far afield—artists, designers, and architects as well as scientists. A primary goal of this anthology is to spread knowledge of Thompson's work, his influence on art, architecture, and design, and his prescience vis-à-vis contemporary theories of evolution. Throughout this anthology we explore in what ways Thompson's tome has broadened the scope of creativity and how new innovations continue to result from his ideas. We analyze the role of *On Growth and Form* in cross-disciplinary research in art and science, revealing Thompson's ideas to be living and generative in the twenty-first century, a full century after it was published. The current infusion of ideas and processes issuing from biology, material science, and complex systems within the arts dovetails in remarkable ways with Thompson's approach to science that expands upon an understanding of matter as dynamic and interactive with other matter situated in the environment. As stated by the late microbiologist Carl Woese, who echoed Thompson's holistic approach, "The time has come to replace the purely reductionist 'eyes-down' molecular perspective with a new and genuinely holistic, 'eyes-up,' view of the living world, one whose primary focus is on evolution, emergence, and biology's innate complexity."[1]

Woese's statement echoes the outlook of many artists, designers, and architects working in the twentieth and twenty-first centuries. In like terms, Thompson's enterprise thrives today; practitioners in the arts continue to look to nature and its evolutionary processes for inspiration. The range of chapters in this volume is moreover evidence of Thompson's persuasive "eyes-up" approach. For example, chapters in this volume explore how, in the mid-1950s, artist Richard Hamilton utilized *On Growth and Form's* potential for creating more credible and vivid representations of life. Other chapters elaborate how artists, designers, and architects have deployed the dynamics within nature to animate inanimate forms in accordance with principles first suggested by Thompson. Authors collectively explore art, design, and architecture, describing a taxonomy of cultural practices in which Thompson's ideas have been formative, uniquely and with respect to the diverse methodologies characteristic of each field. Artists often deploy Thompson's ideas poetically or philosophically, with attention to generating form; architects use his ideas structurally, with concern for the act of building; and a subset of designers called biodesigners utilize biology to problem-solve issues of design. Bioart and bioarchitecture are new modes of "design"—creative problem solving—building on the overlapping histories of science, art, and architecture. We have included practitioners and historians who reference Thompson in their work in all three of these creative fields.

For Thompson, the living and nonliving share the same laws of physics; the matter within both spheres has meaning and is generative. His was a monistic conception, meaning it was based on a model of life as the concrescence of the organic from the inorganic. As such, Thompson's ideas can be viewed as progenitors of the contemporary field of material studies, which encompasses matter and agency, stuff and its animation. In this framework, matter is conceived as active and creative. For Thompson, careful observation of nonliving inanimate shapes could bring greater understanding of living animate shapes. While artificial forms and contours cultivated his thinking, it was the organic that was his true focus, for the primary argument of *On Growth and Form* unfolded around the claim that living organisms take form from physical forces. A later outgrowth of this position involves physical embodiment. Building on Thompson, philosophers George Lakoff and Mark Johnson have argued that both experience and metaphor are mediated by our bodies.[2] Embodied simulation—the means by which humans reconstruct mental enactments—suggests that our thinking and viewing is never purely abstract but engages our bodily senses. We are connected to the world through our bodies, however symbolic and reductive our communication.

Thompson's work speaks to the current questioning of basic categories of the living and non-living. In the age of biotechnology, the line between the two is difficult to distinguish, with certain philosophers conceiving matter as vibrant with properties of self-organization. Writings by philosophers Bruno Latour and Manuel DeLanda express renewed interest in the relevance of the material world to social and political concerns.[3] One thinks of Latour's human and non-human "actants," which interact in unpredictable ways. In keeping with this, humans are actively inhabited by chemical, hormonal, and microbial multiplicities that lie outside the regulatory control of genetics. Thompson's views of matter are consistent with early theoretical currents in materialisms as examined by writers such as Donna Haraway and N. Katherine Hayles.[4] New feminist materialisms integrate conceptions of agency and embodiment as explored more recently by Karen Barad, Jane Bennett, Diana Coole, and Samantha Frost, among others.[5]

In addition to Thompson's prescience with respect to contemporary feminist materialisms, Thompson was forward-thinking in his own day, supporting women scientists who were typically marginalized. For example, in this compendium Marjory Senechal explores how his mentorship of Dorothy Wrinch resulted in her helping to transform crystallography into a science of structure via principles of local force. Thompson's support was a world apart from the exclusionary treatment Rosalind Franklin later received during the 1940s and 1950s. Franklin's pivotal role in creating a crystallographic image of the DNA molecule that was key to deciphering its structure was ignored for decades.

Like Thompson, many of today's art practitioners are fascinated by engagements with living forms and their processes. Thompson has functioned as a visual thinker for intellectuals from across disciplines, from art to engineering, architecture to neuroscience. This is not only evidenced by the 554 images in his seminal work but by the prose of the text. Simply put, he made pictures using language. Words do the work of lines, brushstrokes, chemical substrates activated by light, filmic montage, and pixels. Passages devoted to an array of force-based causal relations—such as the magnitude of a flea's jump, comparisons of the spongey trabeculae of bones to the cantilevering of the Forth Bridge, and the temporal logic of the shapes and silken surfaces of a cell—crystallize as images in the mind's eye through Thompson's ability to limn with words. The main thrust of his book concerns morphology and how living and nonliving shapes come into form.

Thompson represents what might be considered an early incarnation of contemporary animal studies, explorations into "interspecies communication," or variably the "post-human turn" and "non-human turn." His love of animals bears deep roots in the field of contemporary post-humanist studies, and he was an ecological activist. From this perspective, we locate an early progenitor of contemporary ecological, companionist, and symbiogenetic thinking. Many of the developments in art that we find emblematic of our human existence are informed by current scientific thought with its basis in a holism practiced by Thompson and continued by other scientists. Thompson's 1917 thesis fell partially hidden in the shadow of the rising field of genetics, but nonetheless made a sustaining impact in the arts, design world, and collaborative art-and-science initiatives.

Artists working across media, from painting to bioart, continue to benefit from Thompson's tome, in particular drawing from its fresh new resonances in current understandings of evolutionary biology, neuroscience, and nanotechnology. The latter carries many of Thompson's insights into the microbiological realm. By now, artists and art historians look for understanding of the mind and body as an interconnected system and to comprehend how we generate, simulate, mirror, and experience emotional states. Neuroscience builds on Thompson's forays into development of the human brain cortex with its gyri, sulci, and convolutions, understanding brain folding as a process of self-organization. Only recently has it been resolved that the brain folds like crumpled paper in response to the thickness and surface area of tissue at the time forces are applied.[6] The dynamism and interactivity with the local environment that we view during differentiation is echoed today by developments in new media and biomedia that allow for visualization in real time. Contemporary art elicits behavioral responses, including motor actions as it branches into performance, simulations, video, interactive new media, biomedia, artificial intelligence, and augmented and virtual reality. For artist Carolee Schneemann, close readings of Thompson and others, including Gaston Bachelard and Antonin Artaud, elicited kinetic visual properties that activated her body.

The design field has gained immensely from Thompson's exemplary text. Biomimetics is a case-in-point. It utilizes principles from engineering, chemistry, and biology to synthesize materials that have functions that mimic biological processes. As another example, during the 1980s, the merging of disciplines such as textiles with biotechnology, materials science, and chemistry realized some of the potential for applying dynamical principles of growth to objects. The deployment of such principles resulted in the production and wide use of smart materials, self-organizing colloids, and high-performance fibers and aramids. In 2005, two extraordinary exhibitions, "Extreme Textiles" at the Cooper-Hewitt Museum (NYC) and "Design Takes on Risk" at the Museum of Modern Art (NYC), displayed a synthesis of art, materials, and processes of evolutionary biology that were intended to assist human survival.[7]

Recent exhibitions have continued to signal that perspectives from materials science, evolutionary biology, and cognitive science are pertinent to artists and designers. They matter with regard not only to traditional artistic concerns of embodiment, memory, and emotion but to artistic conception, production, and communication.

In many ways, architecture is the most logical destination for the application of Thompsonian principles, given the necessity of architects to consider the effects of external forces on a body through tension and compression analyses. Problems of structure hold ready analogies to the morphological problems faced by animals. As contributor Philip Beesley noted, pluripotency is now positioned as a key in architectural design. In discussing his collaborations within the Living Architecture Systems Group, Beesley stated that distributed kinetic architectural systems are starting to integrate hybrid circulation systems and artificial metabolisms. He found that Jeremy England's recent writing on dissipative adaption offers new interpretations of how physical laws shape forms of life.

Many practitioners in art, design, and architecture are active environmentalists, which is also in keeping with Thompson's practice as a researcher and writer. Recalling Thompson's exploration of environmental and species sustainability, the works of these artists call for a re-engagement with science at a time when efforts and ingenuity need to be devoted to the containment of global warming and when science deniers need to be confronted with evidence that its cause is man-made. We are explicit in placing Thompson's work within contemporary hybrid practices of art-science-design as a counterforce to the anti-science movement.

While Thompson continues to influence the art, design, and architecture fields, it is counterintuitive that Thompson's work would still speak productively to contemporary science. One reason is that key developments, particularly in genetics, cognitive science, and epigenetics, were not yet discovered in 1917. *On Growth and Form* was published thirty-six years before the discovery of the structure of DNA in 1953. Moreover, until documented in 1998, it was believed that the brain's ability to grow new neurons, a process called neurogenesis, was limited to infants.[8] Simply put, the landscapes of biology, genetics, and neuroscience have dramatically changed since 1917.

Working at the turn of the last century, both the recently identified "gene" and long-held theories of inheritance were important to the grounding of Thompson's thinking, but by no means central to his argument in the book. Claims to "natural selection," the engine of evolution for absolutist Darwinism, were rather anathema to Thompson's thinking. Natural selection was something of a conceptual trap, equal parts monological and tautological. At its worst, it was a veiled monotheism. "To buttress the theory of natural selection the same instances of adaptation are used," Thompson argued, is to fall prey to what "in an earlier but not distant age testified to the wisdom of the Creator."[9] Natural selection, though an undeniable force within a coterie of evolutionary determinants, would come to dominate biology and the greater dialogue of Darwinism partially in the form of genetics and inheritance during the twentieth century.

The compendium reinforces the importance of Thompson's thinking in light of contemporary postgenomic sciences, or the "omics" fields of complex biological systems that have flourished since the mapping of the human genome in 2003. These include proteomics, nutrigenomics, metabolomics, and transcriptomics.[10] With essays on art, architecture, and art-and-science hybrids, the anthology reveals how Thompson's thesis has been more recently confirmed by knowledge systems that have emerged since this mapping. For example, the current evidence that certain processes can impact gene expression by altering chromatin structure or recruiting *histone* modifiers has led to reconsideration of factors apart from the gene that can be inherited. This has developed into a broad field of research known as epigenetics.[11] It intimates that some of Thompson's physico-chemical forces may be operating at the cellular level. The book situates Thompson's non-gene centric take on morphogenesis within the history of evolutionary theory, showing that it validates non-genetic drives and impulses—ranging from physics to culture—within the formation of life.

In the Postscript to Thompson's tome, British biologist Peter Medawar captures the book's originality, through its impact, stylistics, and peculiar contents. "The influence of Growth and Form in this country and in America has been very great, but it has been intangible and indirect," Medawar says, reinforcing the book's effective and broad reach. The sometimes knotty, sometimes fluid writing style of the book is part of an organic process itself. That to-and-fro, Medawar writes, reminds us that "science cannot be divided into what is up to date and what is merely of antiquarian interest, but is to be regarded as the product of a growth of thought."[12] Science writers have contemplated the singularity of Thompson's book. They view it, on the one hand, as avant-garde and out-of-time, with its transformation grids predicting digital imagery fifty years before the technology was available as Stephen Jay Gould would have it, and, on the other, fundamentally in tune with its moment and in-time with its counterstrategy to Darwinian natural selection for explaining morphology and morphogenesis, as Maurizio Esposito suggests.[13] For Esposito, Thompson's book was framed by contemporary discourses and as "part of a vibrant community of international scholars who … shared not only his anti-Darwinian rhetoric and skepticism, but, above all, his ideas on morphology, mathematics, and evolution."[14] Thompson's intellectual context bore geographical tentacles, connecting his work in Scotland to scientists located in Central Europe.

Part of the uniqueness of Thompson's approach stems from his experiences working as a scholar alongside of scientist-brothers, Hans and Karl Przibram, a zoologist and physicist respectively, in Vienna at the Vivarium during the early decades of the twentieth century. The Vivarium was a state-of-the-art laboratory, "a daring intellectual environment," that opened in 1903 in a rehabilitated aquarium located in an urban garden, the Prater in the city of Vienna.[15] Scientists such as Paul Kammerer, Paul Weiss, Eugen Steinach, Karl Frisch, and Thompson met there in pursuit of developing mathematical systems that could account for the dynamical topologies of developmental biology. They created a holistic framework and collectively pursued what we call "complex reductionisms" in the combined fields of biology and math. They sought a form of math that maintained and accounted for the nuances of multi-vectored and changing models of organic life. As Deborah R. Coen explains, the Vivarium's scientists "converged in their search for a 'third way' between mechanical determinism and pure spontaneity, a framework that would do justice to the complex interaction between organism and environment."[16]

In this volume, we strive for a similar justice for Thompson's *On Growth and Form*: an impartiality facilitated by today's knowledge systems. It is an exploration of his work by artists, architects, scientists, and hybrid art-and-science practitioners after the "century of the gene" within our contemporary moment where the complex, holistic approach of Thompson's is not merely plausible but correct and welcomed anew.[17]

Notes

1 Woese, C. R., "A New Biology for a New Century," *Microbiol Mol Biol Rev*, Vol. 68 (2004), 175.
2 Lakoff, G. and Johnson, M., *Metaphors We Live By*, University of Chicago Press, Chicago; also see their 1999, *Philosophy in the Flesh: The Embodied Mind and Its Challenge to Western Thought* (New York: Perseus Books, 1980).
3 DeLanda, M., *Intensive Science and Virtual Philosophy* (London: Continuum, 2002); Latour, B., *An Inquiry into Modes of Existence: An Anthropology of the Moderns*, trans. Catherine Porter. (Cambridge, MA: Harvard University Press, 2013).
4 Haraway, D., *Simians, Cyborgs, and Women: The Reinvention of Nature* (New York: Routledge, 1991); Hayles, N. K., *How We Became Posthuman: Virtual Bodies in Cybernetics, Literature and Informatics* (Chicago, IL: University of Chicago Press, 1999).
5 Coole, D. and Frost, S., "Introducing the New Materialisms," in *New Materialisms: Ontology, Agency, and Politics*, edited by D. Coole and S. Frost (Durham, NC & London: Duke University Press, 2010), 1–46; Dolphijn, R. and van der Tuin, I., *New Materialism: Interviews & Cartographies* (University of Michigan: Open Humanities Press, 2012); Barad, K. and Bennett, J., *Vibrant Matter: A Political Ecology of Things* (Durham, NC: Duke University Press, 2010).
6 Mota, B. and Herculano-Houzel, S., "Cortical Folding Scales Universally with Surface Area and Thickness, Not Number of Neurons," *Science*, Vol. 349, No. 6243 (July 3, 2015), 77.
7 McQuaid, M. "Extreme Textiles: Designing for High Performance," Cooper-Hewitt, National Design Museum, April 8, 2005–October 30, 2005; Paola Antonelli, P., "SAFE: Design Takes On Risk," Museum of Modern Art, October 16, 2005–January 2, 2006.
8 Gould E., Tanapat, P., McEwen, B. S., Flugge, G., Gross, C. G. and Fuchs E., "Proliferation of Granule Cell Precursors in the Dentate Gyrus of Adult Monkeys Is Diminished by Stress," *Proc. Natl Acad Sci USA*, Vol. 95 (1998), 31689–3171.
9 Thompson, D'Arcy Wentworth, *On Growth and Form* (Cambridge, UK: Cambridge University Press, 1917), 672. Quoted in Reid, Robert G. B., *Evolutionary Theory: The Unfinished Thesis* (Ithaca, NY: Cornell University Press, 1985), 34.
10 Holmes, Christina, Siobhan M. Carlson, McDonald, Fiona, Jones, Mavis and Graham, Janice, "Exploring the Post-Genomic World: Differing Explanatory and Manipulatory Functions of Post-Genomic Sciences," *New Genetics and Society*, Vol. 35, No. 1 (2016), 49.
11 Epigenetics is the study of heritable changes in gene expression that do not involve changes to the underlying DNA sequence.

12 Medawar, P. B., "Postscript: D'Arcy Thompson and Growth and Form," in *D'Arcy Wentworth Thompson: The Scholar-Naturalist 1860–1948*, edited by Ruth D'Arcy Thompson (Oxford, UK: Oxford University Press, 1958), 232.
13 Esposito, Maurizio, "Problematic "Idiosyncrasies": Rediscovering the Historical Context of D'Arcy Wentworth Thompson's Science of Form," *Science in Context*, Vol. 27, No. 1 (March 2014), 81.
14 Gould, S. J., "D'Arcy Thompson and the Science of Form," *New Literary History*, Vol. 2, No. 2, Form and Its Alternatives (Winter 1971), 254; Esposito, M., "Problematic 'Idiosyncrasies': Rediscovering the Historical Context of D'Arcy Wentworth Thompson's Science of Form," *Science in Context*, Vol. 27, No. 1 (March 2014), 81.
15 Coen, D. R., "Living Precisely in Fin-de-Siècle Vienna," *Journal of the History of Biology*, Vol. 39, No. 3 (Autumn 2006), 495.
16 Coen, 496.
17 Keller, Evelyn Fox, *The Century of the Gene* (Cambridge, MA: Harvard University Press, 2000).

Acknowledgments

We thank the College Art Association (CAA) and Leonardo, The International Society for the Arts, Sciences and Technology (Leonardo/ISAST), for providing the initial conference platform to explore *On Growth and Form* on the occasion of its centenary in 2017. Many of those who participated in our two panels later became contributors to this anthology. We are grateful to the kind generosity of artists, their galleries, institutions, and museums named in the extensive List of Illustrations for the images that have brought the text to life. Many thanks to Janet Echelman for the powerful image she provided for the cover and to her Studio Manager, Melissa Henry, for assistance with selection. We thank Toby Way, Designer, Bloomsbury Academic and Professional Division, for the Series design on the cover. We note with sadness the passing of Carolee Schneemann (1939–2019), one of our contributors, during the preparation of this anthology. We thank the following individuals and their staffs at Bloomsbury Publishing and Ingenta for seeing the anthology through to completion: Commissioning Editor, April Peake, Production Editor, Barbara Cohen Bastos, and James Thompson, Publisher - Architecture, Art and Visual Culture, Bloomsbury Publishing. We are grateful for invaluable counsel from Yvonne Thouroude, Assistant Editor and Anita Iannacchione, Editorial Assistant, both of Art and Visual Culture, Bloomsbury Publishing. We are appreciative of the thoughtful guidance and attention to detail received from Karthiga Sithanandam, Project Manager Integra Software Services. We express gratitude to Margaret Michniewicz, former Visual arts Acquisitions Editor at Bloomsbury Academic, whose enthusiasm propelled our early involvement in the project.

Many individuals helped inform my ideas, including discussions with Stephen Jay Gould and Niles Eldredge. I thank Assimina Kaniari and Gemma Anderson for including me in the interdisciplinary workshops they organized to discuss *On Growth and Form*, held respectively at Oxford University (2010) and the Lorenz Center in Leiden (2017). I thank Matthew Jarron who invited me to undertake an art residency at Dundee, Scotland (2014) that provided access to D'Arcy Thompson's laboratory and specimens. I also thank my co-editor Charissa Terranova for her enthusiasm and expressive originality. Most essentially, I thank my husband, neurologist David E. Levy, with whom, in wonderfully generative conversations over decades, I developed my ideas.

Ellen K. Levy

This book would not have been possible without the participation and aid of many individuals. I would like to thank the speakers in the double panel Ellen K. Levy and I organized for the annual meeting of the College Art Association in February 2017, which became the kernel of this book: Hadas Steiner, Caroline O'Donnell, Matthew

Jarron, Justine Kupferman, Marjorie Senechal, Sarah Bonnemaison, Todd Siler, Ingeborg Reichle, and Roger Malina. Margaret Michniewicz, the former visual arts/art history acquisitions editor at Bloomsbury Press, brought enthusiastic, insightful, and supportive energies that are the foundation of this book. I would like to thank Matthew Jarron also for his special insights into the life of Thompson and his help as Curator of Museum Services at the D'Arcy Thompson Zoology Museum, University of Dundee, Scotland in my multiple visits there. Similar thanks go out to the librarians overseeing the Thompson archive at the University of St. Andrews Special Collection in St. Andrews, Scotland. I would like to thank my co-editor Ellen K. Levy for unfaltering focus throughout the making of this book. Finally, and most importantly, I thank my husband Trent J. Straughan for steadfast support of my life as his partner and scholar roving the world - for taking care of the house and our menagerie of pets on my multiple research trips overseas for this book.

Charissa N. Terranova

Timeline

SCIENCE

Theories and Approaches:
1801–: Inheritance of Acquired Characteristics
1813–: Catastrophism

1798: Malthus *Essay on the Principle of Population*

1855: Spencer *Principles of Psychology*

1858: Wallace sends Darwin *Ternate Essay*

1859: Darwin *On the Origin of Species*

1800 | 1820 | 1840 | 1860

1851: Paxton's Crystal Palace

Philosophy:
1700s– : Idealism
1800s–1900s: Positivism
1800s– : Rationalism

Art Movements and Styles:
1800–1850: Romanticism
1830s–: The Monstrous
1850s–: Modernism

Philosophy (pre-1800):
1720s–: Monism
1750s–: The Sublime

CULTURE

SCIENCE

Theories and Approaches:

1859–: Natural Selection

1870s–: Social Darwinism

1880s: Neo-Lamarckism

1883–: Germ Plasm (Weismann)

1896–: Baldwin Effect

1901–: Mutation Theory

1880s–1930s–: Eugenics Movement

1866: Haeckel *Generelle Morphologie*

1866: Mendel *Versuche über Pflanzen-Hypbirden*

1869: Galton *Hereditary Genius*

1871: Darwin *Descent of Man and Selection in Relation to Sex*

1910: Davenport *Eugenics*

1917: Thompson *On Growth and Form*

1860 | 1880 | 1900 | 1920

WW I

Philosophy:

1800s–1900s: Pragmatisim

1900s–: Phenomenology

1893: Riegl *Problems of Style*

1907: Bergson *Creative Evolution*

Art Movements and Styles:

1890–1910–: Art Nouveau

1910s–1930s: Art Deco

Modernism cont.

CULTURE

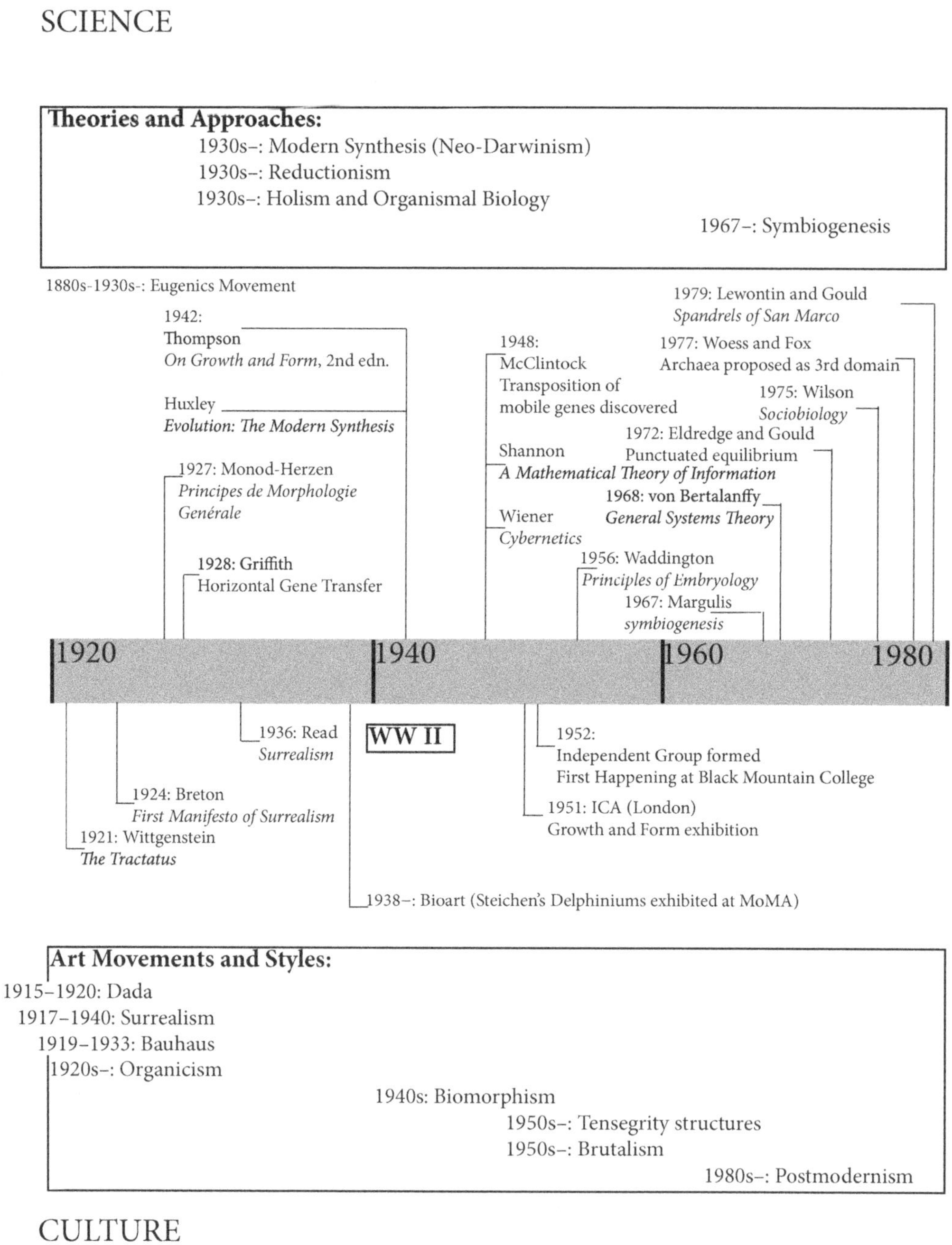
SCIENCE
Theories and Approaches:
1930s–: Modern Synthesis (Neo-Darwinism)
1930s–: Reductionism
1930s–: Holism and Organismal Biology
1967–: Symbiogenesis
1880s-1930s-: Eugenics Movement
1942:
Thompson
On Growth and Form, 2nd edn.
Huxley
Evolution: The Modern Synthesis
1927: Monod-Herzen
Principes de Morphologie Genérale
1928: Griffith
Horizontal Gene Transfer
1948:
McClintock
Transposition of mobile genes discovered
Shannon
A Mathematical Theory of Information
Wiener
Cybernetics
1956: Waddington
Principles of Embryology
1967: Margulis
symbiogenesis
1968: von Bertalanffy
General Systems Theory
1972: Eldredge and Gould
Punctuated equilibrium
1975: Wilson
Sociobiology
1977: Woess and Fox
Archaea proposed as 3rd domain
1979: Lewontin and Gould
Spandrels of San Marco
1920
1940
1960
1980
WW II
1921: Wittgenstein
The Tractatus
1924: Breton
First Manifesto of Surrealism
1936: Read
Surrealism
1938–: Bioart (Steichen's Delphiniums exhibited at MoMA)
1951: ICA (London)
Growth and Form exhibition
1952:
Independent Group formed
First Happening at Black Mountain College
Art Movements and Styles:
1915–1920: Dada
1917–1940: Surrealism
1919–1933: Bauhaus
1920s–: Organicism
1940s: Biomorphism
1950s–: Tensegrity structures
1950s–: Brutalism
1980s–: Postmodernism
CULTURE

SCIENCE

Theories and Approaches:
Modern Synthesis continued
1990s–: Epigenetics
2000–: Systems Biology
2003–: Evolutionary Developmental Biology (EvoDevo)
2010–: Extended Evolutionary Synthesis (EES)

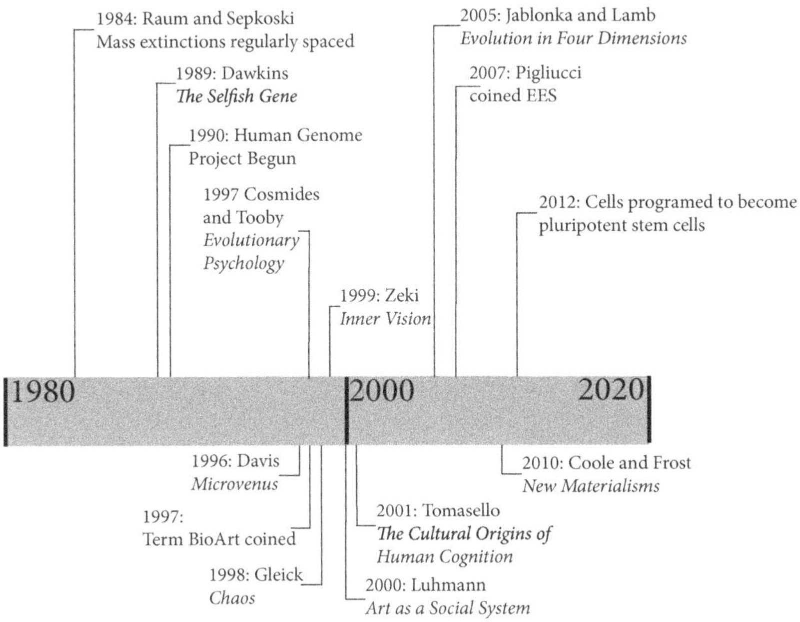

Approaches:
1980s–: Complex Systems
2000–: Dynamical Systems Theory
2010–: New Feminist Materialisms

Art Movements and Styles:
1990s–: Artificial Life Art
Installation Art
Bioart continued from 1938
1980s–: Bioarchitecture
1980s–: Biodesign
1990s–: New Media, including robotics, interactive art, virtual reality and augmented reality

CULTURE

Introduction

Ellen K. Levy and Charissa N. Terranova

The essays in *D'Arcy Wentworth Thompson's Generative Influences in Art, Design, and Architecture: From Forces to Forms* have been collected to show the longevity and flexibility of Thompson's methodology. They aim to emulate the organic, back-and-forth process that characterized his thinking. The anthology explores Thompson's forces-based view of morphogenesis and what it continues to mean to art, design, and architecture practices in the twentieth and twenty-first centuries. This compendium brings together fourteen chapters from authors working in the history, theory, and practice of art, architecture, science, and mathematics to create a dynamic portrait of how Thompson's pursuits helped expand significant cultural as well as scientific possibilities. His demonstration of how the mechanics of physical force are central to generating living forms presented an alternative to purely selectionist, Neo-Darwinist formulations of evolution that resonate today in many creative spheres.

Commemorating the centenary of Thompson's book, the compendium examines why *On Growth and Form* enjoys numerous ramifications in art, design, and architecture. It situates Thompson within both scientific and cultural domains, providing an opportunity to analyze the reciprocity in these fields and investigate several largely overlooked dimensions of evolutionary theory, including the roles of aesthetics, agency, contingency, and part to whole relationships. Our goal is to reveal Thompson's influences over the last century and in the present—and to more precisely situate his work within an expanded field of evolution. It follows, thus, that the texts included here are located within the cultural domain of evolutionary theories. For the sake of enriching transdisciplinary dialogues and knowledges, we seek like Thompson to develop the part of the world of evolutionary theory generated by effects *other than genetic inheritance*.

Language, Metaphors, and Visualizations

The anthology explores Thompson's rich deployment of language, metaphors, and visualizations. The classical erudition displayed by Thompson in *On Growth and Form* makes apparent his reverence for Aristotle, whose work *Historia Animalium* he translated in 1910. Like Aristotle, Thompson had an abiding concern with form and phrasing. In addition to constituting a logophile's aesthetic delight, words used by scientists can convey differing views of the processes surrounding evolution. Darwinian expressions such as natural selection, fortuitous variation, fitness, and adaptation diverge from typically Thompsonian expressions such as correlations, development, transformation, and growth. Changes in biological terminology impact arguments about evolution that rely on shared meanings between scientific fields.[1] For example, the expression "select" as in Darwinian "natural selection" suggests agency is involved, but it is unclear from what direction this agency comes, where it goes, and, more generally, what defines the process of selection. As Evelyn Fox Keller points out in the context of recent developments in evolutionary theory, genes lack any agency by themselves, stating, "What we think of as its causal powers are in fact provided by the cellular complex in which it finds itself."[2] She concludes, "Rather, it is more accurate to think of DNA as a standing resource on which a cell can draw for survival and reproduction, a resource it can deploy in many different ways, a resource so rich as to enable the cell to respond to its changing environment with immense subtlety and variety."[3] In a separate article Fox Keller states that macromolecular complexes have what amounts "to a kind of agency that comes directly out of molecular structure. Drawing energy from their interactions with their neighbors and the larger environment, these are molecular entities that act; indeed, they perform the work that is required for the survival of living systems."[4]

When terms are shared between the fields of art and evolutionary biology, the opportunity for ambiguity greatly increases. Consider the expression "form," which is basic to both disciplines. Questions of form, units of form, and the generation of form permeate both fields. As another example, the term "evolution" has varied greatly over the centuries. In the eighteenth century, it meant unfolding, referring to embryological development often conceived as an expansion of preexisting structures.[5] As late as the 1900s, evolution was conceived as species "transmutation" or transformism, a process by which one species became another. There has been a succession of evolutionary ideas through time, most without sufficient supporting evidence. For Thompson, evolution involved physical forces acting on form. As Stephen Jay Gould pointed out, Thompson argued that form achieved the force of a science through the application of analytic techniques of mathematics, engineering, and physics.[6] Gould further emphasized that "modern evolutionary theory has tended to deemphasize form."[7] In contradistinction, form is at the heart of Thompson's achievement. It is the main subject of Thompson's *On Growth and Form* and key to its pertinence in the cultural realm.

In art, the word "form" reverberates with the "forms" art historian Henri Focillon describes in *The Life of Forms in Art* (1942), an inquiry into how art forms change over time. Focillon, his well-known student, George Kubler, and Thompson all considered form to be moving, in process, and generative of other forms.[8] More recently, anthropologists of science Stefan Helmreich and Sophia Roosth point out that the expression *life form* has, since its earliest usage, "pointed to a space of possibility within which life might take shape, but the way that space is imagined and theorized in biology has substantially changed. Today, in the age of synthetic biology and astrobiology, it has come to signal conjectural and future possibilities."[9] This anthology shows how forms have assumed new actualities in the current landscape of complex systems, generative art, bioart, and bioarchitecture—developments all propelled in part by Thompson's investigations of matter. The materials used by artists, designers, and architects embody an inherent logic of forces coalescing as form. Form-making entails a process in which the ways works are made and what they are made of count as much as what they represent.

Thompson's take on the aesthetic realm involved structures, matter, and pictures, many of which he used in arguments of comparison. In general, this penchant effloresces in the visual. The fact that Thompson delighted in images is made apparent by the sheer number in his tome (over 550) and by his expansion of typical scientific parameters to include creative interpretation.[10] Thompson navigated between analysis of the inorganic and organic in both text and image. As one such example, his work on soap films helped to explicate the pattern of dividing cells of frog eggs. He drew on Joseph Plateau's law of behavior to resolve a problem known as the minimal surface problem, incorporating a diagram of Plateau's analysis of the lines of revolution generated by a soap film rolling across a wire. Thompson looked to art to study proportion and coordinates, including Albert Durer's studies of human proportion (1613), which show how parts of the face are modified relative to each other. One of the most remarked upon visual icons in Thompson's tome is the deformed Cartesian grid. It represents the transformation of biological forms between related species as products of grid coordinate distortions caused by differing rates of growth and actions of a variety of forces. While the hand-wrought deformed grids in Thompson's text caught the public's attention when published, today's ready availability of morphing and digital compositing computer software can achieve such transformations rapidly.

Many scientists distrust the visual, finding images unstable, mathematically uncertain, and too ambiguous to communicate scientific knowledge. Thompson, an exception to this attitude, often used images through applying stable rules, articulating a language of graphics. As a totality, the image comparisons in *On Growth and Form* tend to convey transformational, time-based results rather than vertical descent from a common lineage. The visual metaphors adopted by the public suggest how scientific paradigms communicate and change over time.[11] Consider the most prevalent Darwinian icon, the image of the branching tree. It is now being supplanted by a reticulated image made by biologist W. Ford Doolittle in 1999 in response to microbiologist Carl Woese's research, which confirmed the prevalence of lateral (horizontal) gene transfer in evolution (unlike vertical hereditary transfer from parent to child). This reticulated image has gradually caught the public's imagination, although, until recently, Woese was not known to the public. His work, like that of Thompson, speaks to an expanded field of evolutionary thought and challenges previously accepted tenets of evolutionary theory.

Cultural and Artistic Ramifications of *On Growth and Form*: From the Modernism of Weimar and the UK to Contemporary Global Bioart, Bioarchitecture, and Biodesign

In the context of Weimar Germany (1919–1933), Thompson's book echoed the influences and ideas that would coalesce in the form of biocentrism at the German

Bauhaus. The collective ideas about organic morphology and the workings of life developed by German naturalist Johann Wolfgang von Goethe, German biologist Hans Driesch, and Austro-Hungarian soil botanist Raoul Francé at the core of biocentrism complemented Thompson's take on morphogenesis and forces. While pejoratively branded "metaphysics" and "vitalism," their ideas were, rather, a matter of complexity within biology and evolutionary theory. They thus amount to a dialectic of resistance and demand: a rejection of natural selection and call for an approach to various biofunctionalisms that accounted for nonlinear activities within the cell, such as induction and regulation within genetics itself. From the perspective of the present, these ideas are a harbinger of things to come—early proto-complex systems thinking that would eventually lead to the "omics" within an expanded understanding of genetics. Such complexity within the natural sciences was borrowed and used with ease by architects and designers at the Bauhaus, such as Ludwig Mies van der Rohe, Walter Gropius, László Moholy-Nagy, and Hannes Meyer.[12] While in most instances, Thompson's influence materialized in architectural structures, his unity of ideas articulated in monism and holism held sway generally through biocentric thinking at the Bauhaus. For example, Paul Klee and Wassily Kandinsky examined pictorial form with regard to its regular features, governing principles, and genesis.

In the UK during the following decade of the interwar period, the sculptural work of Barbara Hepworth, Henry Moore, Naum Gabo, and Constantin Brancusi showed the influences of Thompson's book.[13] These artists paid tribute to *On Growth and Form* by way of their preoccupation with individual bones, whole skeletons, and biomorphism in sculpture, drawing, and painting. As the artists found inspiration for new form in the greater sciences, biology harnessed itself to an expansive but increasingly reductive sense of evolution called the Modern Synthesis (the merger of Mendelian genetics with Darwinian evolution that is sometimes referred to as Neo-Darwinism). While quietly present for most of the first half of the century, *On Growth and Form* became prominent in the postwar period in the United States, especially in American architecture and design schools. Yet for engineer-cum-architect Richard Buckminster Fuller, whose career spanned from the 1920s to the 1970s, Thompson's influences were always present. Thompson's thinking on forces within the structure of bridges and skeletons was foundational to Fuller's development of structures, such as the geodesic dome, embodying "tensegrity." Similarly fostered by the designer and space-frame sculptor Kenneth Snelson, tensegrity is a portmanteau of "tensional integrity," which describes a continuity of forces within an easily mountable and demountable structure. In such structures, "compression-resistant struts do not touch but instead are individually lifted, each embraced and interconnected by a system of continuously tensed cables."[14]

Thompson's influences were fixed and continual within the British context into the middle of the twentieth century. In the early 1950s, a younger generation of modern artists affiliated with the Institute of Contemporary Arts (ICA), including Eduardo Paolozzi, Richard Hamilton, Nigel Henderson, Magda Cordell, William Turnbull, and architects Alison and Peter Smithson, became especially fascinated with *On Growth and Form*. Known first as the Young Group and, then later, more famously as the Independent Group, these artists went on to pioneer Pop art before its appearance in the United States. The Smithsons became well known as forerunners of Brutalist architecture, a late-stage modern design vocabulary with the gritty textural surfaces of rough concrete and sometimes confusing volumetrics characterized by experimental geometries. In the United States, Thompson's influences on architecture are also felt through the Smithsons. While by later in the twentieth century, Thompson's *On Growth and Form* was widely read and standard fare in American architecture schools, at mid-century the book effloresced with unique energy, in particular at the Department of Architecture at the University of Pennsylvania. There, architects Louis Kahn, Anne Tyng, and Robert Le Ricolais used Thompson's work as a catalyst for further evolution of the space-frame within architecture.[15] If Kahn, Tyng, and Le Ricolais deployed Thompson's ideas through space-frame structures, using evermore inventive materials in their realization, Paul Rudolph reconsidered such thinking at the urban scale, giving shape to Thompson's forces-based morphological inquiry at a much larger scale of sheltering in the form of the endlessly repeatable modules of the megastructure.[16]

The biological legacies of Thompson's *On Growth and Form* within the fine arts are reflected in processes adapted from biology such as morphogenesis and in the construction of charts and taxonomies. Today the fine arts and design fields reflect the ongoing revolution in biology driven by complex systems and postgenomics. Exploiting an array of media and materials, including nanomaterials and biomaterials, artists may integrate aspects of biology, mathematics, computation, and design. Their works offer a wide range of structural, physical, and environmental situations that challenge many current approaches to the care of our natural resources.

Thompson's influence is present in artworks that exploit generative algorithms, including cellular automata, in which complex patterns can result from simple rules (e.g., Philippe Parreno). Stephen Wolfram described how cellular automata might conceivably yield results akin to evolution and mutation. As opposed to genetic algorithms that mimic some of the process of natural selection, generative algorithms tend to downplay the roles of adaptation in the shaping of an artwork. In general, cellular automata offer artists a bottom-up way to create models of biological phenomena and self-organization based on physical laws. More recent technology such as Clustered Regularly Interspaced Short Palindromic Repeats (CRISPR) has become increasingly accessible to artists who work with living organisms, enabling a merger of biology and technology. This merger echoes with Thompson's frequent juxtaposition of the living and nonliving. Artists such as Marta de Menezes deploy both genetic and nongenetic techniques, and this hybrid approach can increasingly be seen on an international scale.

The design and architecture fields brachiate in both "dry" and "wet" formations. Biodesign deploys living materials and generative growth principles in addition to its standard resources of synthetic polymers, natural materials (e.g., wood, leather, plants), glass, and ceramics. This has resulted in hybrid materials such as BioConcrete, a material embedded with microorganisms that generate limestone in order to initiate self-repair.[17] New fuels are created with synthetic biology that can spawn new approaches to urban pollution. As one example, designer Marin Sawa explores a synthesis of algal biotechnology and textiles, aiming to fuse an industrial technology with the natural world.[18] The medical field has profited by designers' use of smart materials, resulting in new kinds of internal and external prosthetic devices. Shape-memory polymers are created in suture threads for surgical procedures, and nanotechnology is being incorporated in biodesign practices.

Biodesign and bioarchitecture enlist "life" as a resource that comes from a variety of potential sources. It might be a matter of synthetic biology, self-healing concrete, phyto-tectonics (leafy green plant-based), myco-tectonics (mushroom-based), flesh-oriented, and/or other wet sources of building components. In addition to principles of engineering-based tensegrity, much contemporary architecture is biologically based. For example, Philippe Rahm uses physiological principles to create habitable spaces.[19] Bioarchitecture is defined as inhabitable space made from living components. This genre of Thompson-inflected architecture tends to draw young experimental designers to the field, such as Rachel Armstrong, Mitchell Joachim, Amber Frid-Jimenez, Joe Dahmen, and Dennis Dollens. Architect David Benjamin explicitly acknowledges Thompson's influence in harnessing processes found in plants and uses patterns in xylem cell growth to serve structural design problems.[20]

Within the sciences, Thompson's *On Growth and Form* has lived a life calibrated by battles that were at times ideological, and others pragmatic. His literary approach to arguing a thesis concerning physical forces and organic morphology made his work dubious to scientists in pursuit of hard and fast "results." Its literary style and unorthodox method situated the book in sometimes ambiguous positioning amid the drive for reductive scientific practices across the last century. In turn, it facilitated the dominance of other narratives within biology, such as the triumphalist stories or natural selection and the gene. However, within the creative world of art, architecture, and design, *On Growth and Form* flourished. The holism of his endeavor was expansive and inclusive, welcoming artist and scientist alike. In the decades after its publication, *On Growth and Form* has lived a many and varied life that is truly in keeping with Peter Medawar's description of the book as evidence of the organic nature of thinking itself. With each decade, its effects have morphed and trans-mutated according to emerging technological and cultural vectors combined, rejuvenating Thompson's voice in ways unforeseen.

Situating D'Arcy Wentworth Thompson within Evolutionary Theory and Culture

This anthology is a marked interrogation of morphology in terms of the physics of forces. We conceive Thompson's place within an expanded theory of evolution. Evolution needs to be understood in its historical context; it unfolds in a broad field of forces. Its history overlaps with the history of heredity and is replete with debates not only about the processes involved, but about what is conceived as being heritable, and what factors apart from natural selection might have a role in inheritance. We have provided a timeline that enables us to condense the long history of evolutionary theory and provides information about Thompson's position within the history of science. To this timeline we have added information that contextualizes Thompson's achievement relative to some of the problematic developments of his time such as the eugenics movement. Yet Thompson was largely apolitical and makes no mention of eugenics. In fact, his comments on the then-emerging field of genetics are rare. Based on this absence and, more importantly, his larger interests and expansive work on morphology and physical forces, it can be deduced that Thompson believed that genetics, and thus eugenics by connection, was a force of overweening and unyielding reductionist determinism. In the case of eugenics, such reductionism becomes dangerous and deleterious to humankind. Thompson's position, albeit more scientific and scholarly than political and doctrinaire, was in opposition to the reductionism of genetics and eugenics.

The timeline begins with Jean-Baptiste Lamarck and culminates in the present integrating an Expanded Synthesis of Darwinian pluralism and epigenetic findings with cultural practices. To be clear, the contemporary meaning of epigenetics is not the same as epigenesis, which Aristotle originated in the fourth century BC. The early theory held that an individual animal or plant develops by the gradual elaboration of a fertilized egg cell. Epigenetics today refers to the study of factors that regulate patterns of gene expression. Epigenetics explores heritable phenotypic changes that do not involve alterations in the DNA sequence.[21] Many of these factors appear to act in accordance with physico-chemical forces akin to those identified by Thompson, but at cellular and molecular levels. The cell's capacity for action allies Thompson with current ideas about epigenesis.

The genetic structure was not worked out until 1953. Unsurprisingly, genes play an almost nonexistent role in *On Growth and Form*. Nevertheless, Thompson's theory holds implications for current evolutionary ideas that are traced in the timeline. Thompson demonstrated the actions of organisms throughout *On Growth and Form*, and his descriptions of the ways cells and organisms are generative set him apart from both Darwin and the Neo-Darwinists. Our timeline singles out several pertinent developments. In his theory of germ plasm (1892), August Weismann reasoned that the germ line was unaffected by changes undergone by somatic cells during the organism's lifetime.[22] For Weismann, germ cells were opposed to soma cells, which serve as hosts to receive the information from germ cells that dictate the form of an organism's body.[23] While a watershed moment in the history of biology, it is a now in many ways an obsolete conception of biology rooted in the fixity and reliability of genes across generations of deep time. His concept did not account for the effects within evolution of Thompsonian physical forces, environmentally driven transformations within and between generations, or perturbations caused by the ongoing development of an organism. August Weismann's germ-plasm theory provided support for the constancy of populations for given characteristics as opposed to indicating actions of the environment. In this sense it bolstered eugenic views.[24] Note also our inclusion of the development of the Baldwin effect. The theory was that learned behaviors could be converted to genetic adaptations or that genetic adaptations could support learned behaviors. The theory involves organic selection and references both Darwin and Lamarck, but it is neither. In contrast to germinal selection, organic selection was considered "a process by which acquired individual characters are sometimes considered to protect heritable variations while these are still insufficiently developed to be perpetuated by natural selection."[25] Organic selection did not claim that a genotype is being altered by a phenotype. The idea was that learned behaviors acting as vectors of evolutionary forces can at times influence natural selection through opening a channel for further development.

Defining Biology: From Simplistic to Complex Reductionisms

In producing an anthology about the zoologist-mathematician D'Arcy Wentworth Thompson, it is without question evident that we are part of the continued effort to hew art, architecture, and design to the field of biology. Yet, if the greater field of biology is the epistemological backdrop of this compendium of essays, then it must be clarified we are working in the plural. We ask that readers think of biology in the manifold as "biologies." This means that the field of biology, like science itself, is always embodied in rhetorical positions and shot-through with language, which in turn means that there are ideological biases within the science of biology, its evidences, discoveries, and constructions. We celebrate this fact. Like biology itself, the union of science and rhetoric creates a fathomless fount of protean possibilities for knowledge production.

Elizabeth A. Wilson writes in *Gut Feminism* (2015) about the opportunities that go missed in the persistently separated and siloed state of these fields—the humanities and sciences, art and biology—within the academy. It is a situation which has placed feminist theory and biology immemorially at loggerheads. From Michel Foucault's biopolitics to Judith Butler's "sexual traffic," biology within the humanities, and feminist theory in particular, has been reified, objectified into a monolithic oneness of dominance, imposition, and wealth extraction because of the genitalia- and gender-based reductionism of sociobiology and evolutionary psychology. While created in the name of critical thinking, the anti-biologism and anti-scientism of this now older strain of critical theory operate against the core claims of criticality. We seek to cleave our text to a bumpy, craggy, and more topological biology in order to make sundry other interpretations of the data. In keeping with this, Wilson argues that "embedding motive or deliberation in biological substance is one way of broadening questions of causality beyond narrowly mechanistic definitions of influence."[26] Set in this light the critical theorists of yore fall prey to a very similar strain of simplistic reductionism as that of the gene-centric evolutionary theorists against which this volume of essays is situated. In *Gut Feminism*, which revamps critical theory by way of science studies, Wilson moves directly into the heart of the sciences, generating a new strain of textual political activism by querying the properties and functions of the brain–gut axis, bringing this information to bear upon feminist theory. *Gut Feminism* investigates the dialogical ups and downs, hills and valleys, of the gastrointestinal and central nervous systems, and the connective energies therein of the vagus nerve and the dis-connective ones of the pharmaceutical industry. The biology with which we work here is informed by a similarly interactive science, one that is in connection and in extension across fields. We, like Wilson, "reconsider the nature of thinking in the usual sense," and harness the chapters in this book to a biology with an underbelly, or what she calls after Sándor Ferenczi, the "biological unconscious"—the idea that "a nascent kind of psychic action (motive, deliberation) is nonetheless native to biological substance."[27]

Based on the "plastic nature of all organic substrate," it is a biology in which mind extends across the body and through matter: it is one in which the "gut thinks."[28]

Let us circumscribe the pluralism of biology within a spectrum delimited by two extremes: simplistic reductionism on one end and complex reductionism on the other. In sum, we identify our argument, including Thompson's *On Growth and Form* and the essays included in this volume, in terms of complex reductionism *within the field of evolutionary theory*. Before going further, an explanation of our use of "reductionism" is in order. The word exists, similar to biology, in the plural. Reductionism has meaning at varying registers, largely in the field of science and more specifically in biology. As for art, reductionism there exists in large part as an absence. Very generally, in everyday parlance, it means reducing something of great complexity to its rudiments. *Merriam-Webster's Dictionary* defines it as an "explanation of complex life-science processes and phenomena in terms of the laws of physics and chemistry."[29] Since the mapping of the human genome in the early aughts, this form of reductionism has been declining in influence and relevance. Professor of Biotechnology at the University of Strasbourg Marc H. V. Van Regenmortel cogently argues, "however, many biologists now realize that this approach has reached its limit."[30] In keeping with certain aspects of our argument, he further explains that "biological systems are extremely complex and have emergent properties that cannot be explained, or even predicted, by studying their individual parts."[31] He argues that reductionism in much research is past its usefulness as during the early days of molecular biology.[32] We distance ourselves from the propensity to reduce the complexity of behavioral mores and cultural practices to false causes given scientific legitimacy by mere association with genes. In fact, reductionism within any of these frameworks is usually anathema to art, and historically has held little influence or sway within the art world.

An advocate of distancing art from deterministic interpretations, the American biologist E.O. Wilson said, "the love of complexity without reductionism makes art; the love of complexity with reductionism makes science."[33] We find reductionism useful and recognize the shared importance in art and science of reductionism defined very generally according to willful and objective goals. At the same time, we do not rule out the possibility that there might be a gene–culture feedback loop: that culture and genes reciprocally influence each other to varying degrees. We are not anti-reductionism but rather interested in a better, more accurate understanding of evolution and inheritance that accounts for genetic as well as epigenetic, behavioral, and symbolic forces—and by proxy, the role of art and architecture within this multidimensional evolutionary framework.

Within the united field of art-and-science we seek to approach situations, seek results, and concretize form through recognizing and moving *within states of complexity*. Thus, we argue in contradistinction to Wilson that artists and scientists share a love of complexity, which, in the processes and methodologies of both, is punctuated by moments of reductionism followed by flow, the concretizing of objects and discoveries followed by research and experimentation. This rhythm of form and flow, form and flow, is aesthetic for artist and scientist alike. A work of art or architecture always culminates in a finite thing—a willful arrangement, process, performance, or concept—even while drawn out over ecologies and deep evolutionary time.

Our template for complex reductionism is modeled after Thompson's work itself, in particular the way in which within *On Growth and Form* he allowed and recognized the important role of heredity, even while not focusing on it. While Thompson mentions "heredity" in *On Growth and Form* sporadically, it is nonetheless present in several passages. Thompson explains late in the book, in the penultimate chapter "On Form and Mechanical Efficiency":

> although I have tried throughout this book to lay emphasis on the direct action of causes other than heredity, in short to circumscribe the employment of the latter as a working hypothesis in morphology, there can still be no question whatsoever but that heredity is a vastly important as well as a mysterious thing; it is *one* of the great factors in biology, however we may attempt to figure to ourselves, or howsoever we may fail even to imagine, its underlying physical explanation.[34]

Thompson's ideas in *On Growth and Form* foreshadow Susan Oyama's *The Ontogeny of Information Developmental Systems Theory* (1985). Together, the two works are part of a greater field of organismic developmental understanding bearing possible openings to granting power and influence to forces within evolution other than natural selection—including the impact of species on their own development and environment and the role of culture and aesthetics therein. Questions beg. Can there be a convincing science of culture? Can there be a convincing science of aesthetics? By "convincing," we mean scientific approaches to culture and aesthetics that do not bring wholesale misinterpretation, violence, or destruction to either field in bearing its methodologies. So, we conclude this introduction with an aperitif of what such a convincing science of culture and aesthetics might look like. In doing so, we have gone willingly to the birds!

In *The Evolution of Beauty: How Darwin's Forgotten Theory of Mate Choice Shapes the Animal World—and Us* (2017), ornithologist Richard O. Prum proffers a persuasive science of aesthetics that has explicit effects when brought to the human cultural context. At the same time, Prum's contemporary thinking resonates well with Thompson's now century-old book. Prum's ideas are based on decades of birding, careful studies in the science of ornithology, and Darwin's *The Descent of Man, and Selection in Relation to Sex* (1871). Not unlike Thompson's *On Growth and Form*, Darwin's *Descent* went ignored by the scientific community, but for slightly different reasons. While Darwin's ideas about sexual selection and the evolution of aesthetics therein went unappreciated by the scientific community because they veered from natural selection, they were considered downright risqué by Victorian England because of the word "sexual." In agreement with Darwin's lost tome, Prum describes an autonomous science of beauty based on the fact that sexual selection and mate choice among birds are explicitly aesthetic, and not a matter of natural selection. "The evolutionary origin of beauty in nature" is a "consequence of the fact that animals ha[ve] evolved to be beautiful to *themselves*."[35] Prum uses years of bird watching and scientific study of birds to prove this point. The rhythmic pulsions, swoops, and dives of the male manakin of Suriname, the complex structures built by the male Great Bowerbird of Queensland, and the fanciful smiley face that unfolds on the wingspread of the male Superb Bird of Paradise in Papua New Guinea are not intended to communicate genetic fitness, but rather the delight of convincing form: movements, shapes, and morphologies that function as communications intended for aesthetic pleasure, and not natural selection. For Prum, the flight, construction, and feathers will preserve the autonomy and safety of the female of the species. This complex set of forces forgoes the species-destructive tendencies of certain male birds to bring violence, even death, through sexual force on the female of the species. Aesthetics in evolution helps to block the destructive effects of rape and violence against females of a given species. "Female sexual autonomy and same-sex behavior have both evolved," Prum explains, "to be disruptive to male hierarchical power and control."[36] Within this system, aesthetics plays the important role of cultivating sexual autonomy for females: the female chooses the male who has developed the most keen and creative set of aesthetic tools, be they, as with the birds listed above, acrobatic flying, architecture building, or colorful plumage.

How does this final tangent into the realm of aesthetic evolution affect our understanding of D'Arcy Wentworth Thompson? First, we argue that Prum's thinking sits within the greater expanse of pluralist Darwinism as does Thompson's. Prum's insistence on the importance of sexual selection and the important role of aesthetics, forces as powerful but distinct from natural selection, is of the same stripe as Thompson's argument about forces within morphology. A century ago, not unlike Prum in 2017, Thompson insisted scientists look beyond the monologic of natural selection to the bigger world of physical forces within morphogenesis. Like Darwin and Prum's sexual selection, Thompson's forces bear aesthetic ramifications within the history and theory of evolution. Thompson's book tells of a world in which aesthetics are necessary, but not reducible in simple fashion. Aesthetics within evolution bear a function that is complex in its reductionism to reason and function. In *On Growth and Form*, he described this composite necessity as the "harmonious complexity of the world," meaning it is an open totality of forces with multiple guy ropes pulling it in all directions and, thus, an approach far distinct from the "teleology without telos" that is natural selection.[37] We believe, like Thompson:

> that it is no less an exaggeration if we tend to neglect these direct physical and mechanical modes of causation altogether, and to see in the characters of a bone merely the results of variation and of heredity, and to trust, in consequence, to those characters as a sure and certain and unquestioned guide to affinity and phylogeny.[38]

We traverse, make connections, and find meaning within a space that includes physical and ontogenetic forces as well as genetic and phylogenetic forces. It is an inclusive and integrative domain. It is one that is holistic.

There is a mixture of voices present within this volume, with writers bringing to the table information and insights about Thompson and his work from across disciplines. Mathematician Stephen Wolfram explores biographical information about Thompson in Chapter 1 and elaborates on Thompson's education, training, and professional activities. In addition, he analyzes the text and the reception of *On Growth and Form* over time. Wolfram shows how he developed computation and simple programs that could simulate forms found in nature. Wolfram then expands on his own work in the field of cellular automata. Wolfram's program has produced a range of forms that one sees in biological organisms inspired by Thompson's initiative.

Geneticist Evelyn Fox Keller explores how Thompson's physical science is relevant to our current understanding of biology in Chapter 2. She examines why genetics generated a false hope that an explanation of macro phenomena could ultimately be "reduced" to an understanding of the properties and behavior of the lowest-level entities. By positioning Thompson as an important predecessor for a current scientific model of evolutionary development, Fox Keller implicitly validates the works of practitioners now motivated by Thompson.

Architectural historian and theorist Hadas A. Steiner writes lyrically in Chapter 3 of the holism expressed by Thompson and elaborates on the critical connections between parts and wholes. Beginning with the discussion of the beauty of the vulture's metacarpal that functioned like a three-dimensional Warren's truss, Steiner investigates the implications of D'Arcy Wentworth Thompson's conclusion that the whole animal will always be more than a sum of its parts.

Curator and art historian Matthew Jarron describes in Chapter 4 that, while Thompson was completing the research for *On Growth and Form*, he was connecting to artists in Dundee. Jarron revisits Thompson's commissioning in 1900 of the decorative artist George Dutch Davidson to create a series of murals for his study at University College Dundee. The friendship of Davidson and Thompson continued, placing Thompson close to a significant group of Celtic Revival and symbolist artists in Dundee inspired by nature, pattern, and the ancient Celtic culture that linked Scotland to mainland Europe.

Mathematician Marjorie Senechal examines Thompson's mentoring in Chapter 5 of the young British mathematician and member of the Theoretical Biology Club, Dorothy Wrinch. For Senechal, Thompson is necessarily identified as "D'Arcy." Through exploring his long correspondence with Wrinch, we see D'Arcy as an early feminist. Senechal enables us to get a close view of his keen thought processes when considering the mathematical and physical forces behind such structures as gelatin cubes and birds' wings. She shows how Wrinch probed the nature of inorganic crystals while receiving council from Thompson.

Art historian Brandon Taylor looks explicitly to painting in Chapter 6. He revisits modern art 1920 to 1950, focusing on the connections between Thompson and Surrealism in painting and art theory. Taylor compares Thompson's accommodation of nature's discrepancies (so-called monsters and other exceptional cases) to the genealogy of forms within the work of Picasso, Miró, and Dalí. He queries the role of anamorphosis, the geometry of projection systems discovered in the Renaissance and taken up again in Surrealism, in Thompson's method of coordinates.

Art historian and theorist Assimina Kaniari delves in Chapter 7 into Thompson's illustrations in the context of display and art practice of the Independent Group in the 1950s at the ICA and in Nigel Henderson's photographic practice. She traces the photographic imagery in Thompson's 1917 book, in particular Arthur Worthington's splash imagery, and, for a second edition in 1942, she compares Harold E. Edgerton's splash, captured by time-lapse photography to images of jewels formed directly by laborious craft processes but also composed of natural materials. Her discussion elaborates upon the insights of Claude Lévi-Strauss, whose ethnographic analogies compare splash imagery to artifacts that reference a range of social orders across different cultural and historical contexts.

Art historian and theorist Charissa N. Terranova analyzes in Chapter 8 the exhibition Growth and Form at the ICA in London that was based on Thompson's book and curated by artist Richard Hamilton in 1951. She differentiates organic art from biomorphism by viewing it as a series of actions in lived time. She particularly explores the exhibition in terms of evolutionary theory and concepts of holism. Focusing on embryologist Albert M. Dalcq and curator and critic Herbert Read, Terranova reframes Hamilton's ICA exhibition as an "extended organism" bestowing viewers with a sense of evolutionary agency.

Architect Caroline O'Donnell analyzes in Chapter 9 Thompson's diagrams of deformed grids and the resultant organisms generated through topologically stretching and compressing the coordinates. She queries what invisible forces are active in these transformations. In addition, she compares the way variations within species focus can be compared to the generation of architectural types, looking to the architecture of Greg Lynn. O'Donnell explores how Lynn has used evolutionary transformations to argue for a dynamic formalism which is a radical change from the typological thinking that preceded it. This chapter proposes a "drawing in" of those elements that motivate the forces acting upon the organisms, in order to address questions of context.

Architect Philip Beesley builds on the wet side of Thompson's thinking in Chapter 10, using his famous maxim "form is a diagram of forces" in order to explore interactive, quasi-living architecture. Beesley positions Thompson's visions of dynamic systems at the center of his investigations for collaborations within the Living Architecture Systems Group. The group is guided by concepts stemming from Ilya Prigogine's dissipative forms, Gavin Crooks's renewed definition of dissipative adaptation, and Jeremy England's writing on evolutionary dynamics. The aim is that a new understanding of entropy will contribute to new architecture.

Artist and theorist Ellen K. Levy analyzes in Chapter 11 the "pluripotent morphology" in Thompson's *On Growth and Form* and its influence on contemporary art and theory. She examines works of nine contemporary artists using a wide range of media who address morphogenesis, taxonomy, and structure. She explores their works by way of Thompson and more recent research by microbiologists Lynn Margulis and Carl Woese who expanded the field of evolutionary biology as they uncovered the extent of the occurrence of Lateral (Horizontal) Gene Transfer across organisms.

Artist Todd Siler investigates how materials work at the nano-scale in Chapter 12. He describes his long-standing collaboration with nanochemist Geoffrey Ozin and explores how nanoscience continues explorations initiated by Thompson. Ozin states that Thompson's basic paradigm for the growth and form of silica and calcareous microskeletons remains significant. Siler compares Thompson's work in *On Growth and Form* to the current work of Ozin and his colleagues' descriptions of six nano-concepts (size, shape, surface, self-assembly, degree of imperfection, and utility). Siler explains how he translates these concepts into potent art works.

Architect and architectural historian Sarah Bonnemaison explores in Chapter 13 how the structure of soap bubbles and their analogies to spherical dwellings caught the attention of Thompson, architect Frei Otto, and engineer Robert Le Ricolais. She explores conceptions of the organic via August Schlegel, Peter Sloterdijk, and Buckminster Fuller. From the 1960s to the 1980s, architecture schools explored nature via Thompsonian principles. *On Growth and Form* was required reading, and Bonnemaison recalls her own experiences working with organic paradigms from soap bubbles to spider webs.

The final chapter is an assemblage of short vignettes about Thompson and contemporary practice written by seven practitioners and theorists in the arts and sciences. Each elaborates upon how *On Growth and Form* has significantly informed or currently informs their professional practice. Philip Ball, Justine Kupferman, Carolee Schneemann, Bart Kahr, Ellen K. Levy, Meredith Tromble, and Henry Petroski offer living testimony to the question of why Thompson's work continues to matter.

Notes

1 Fox Keller, Evelyn and Lord, Elisabeth A. (eds.), *Keywords in Evolutionary Biology* (Cambridge, MA: Harvard University Press, 1992), 4.
2 Fox Keller, Evelyn, *The Mirage of a Space between Nature and Nurture* (Durham & London: Duke University Press, 2010), 6.
3 Fox Keller, Evelyn, 51.
4 Fox Keller, Evelyn, "Self-Organization, Self-Assembly, and the Inherent Activity of Matter," The Hans Rausing Lecture, Uppsala University, SALVIA SMÅSKRIFTER, No. 12 (Swepografiska, Stockholm, 2009), 22.
5 Bowler, P. J., "The Changing Meaning of 'Evolution,'" *Journal of the History of Ideas*, Vol. 36, No. 1 (January–March 1975), 95–114, 98.
6 Gould, Stephen Jay, "D'Arcy Thompson and the Science of Form," *New Literary History*, Vol. 2, No. 2, Form and Its Alternatives (Winter 1971) (Baltimore, MD: The Johns Hopkins University), 236.
7 Gould, 232.
8 Helmreich, Stefan and Roosth, Sophia, "Lifeforms in Representations," *Representations*, Vol. 112, No. 1 (Fall 2010), 27.
9 Helmreich and Roosth, 27.
10 Whyte, Lancelot Law, "Towards a Science of Form," *Hudson Rev*, Vol. 23, No. 4 (Winter, 1970–1971), 613–32.
11 Archibald, J. David., *Aristotle's Ladder, Darwin's Tree: The Evolution of Visual Metaphors for Biological Order* (New York: Columbia University Press, 2014), 114–32.
12 See Botar, Oliver A. I., "The Biocentric Bauhaus," in *Routledge Companion to Biology in Art and Architecture*, edited by Charissa N. Terranova and Meredith Tromble (London: Routledge, 2016), 17–51; Mertins, Detlef, *Mies* (London: Phaidon, 2014), 324–5; and Terranova, Charissa N., "Bauhaus Biology: The Beginnings of Biofunctionalism," in *Art as Organism: Biology and the Evolution of the Digital Image*, edited by Charissa N. Terranova (London: I. B. Tauris, 2016), 19–66.
13 Juler, Edward, *Grown but Not Made: British Modernist Sculpture and the New Biology* (Manchester, UK: Manchester University Press, 2015).
14 http://www.scholarpedia.org/article/Tensegrity (accessed May 23, 2017).

15 See Leslie, Thomas, *Louis Kahn: Building Art, Building Science* (New York: George Braziller, 2005); Mimram, Marc, *Structures et formes: Études appliquée à l'oeuvre de Robert Le Ricolais* (Paris: Dunod, 1983); and Del Rey, Luis et al., *Robert Le Ricolais: Visions and Paradox* (Madrid: Cultural Foundation FOAM, 1997).

16 Banham, Reyner, *Megastructure: Urban Futures of the Recent Past* (New York: Harper & Row, 1976) and Rohan, Timothy, *The Architecture of Paul Rudolph* (New Haven, CT: Yale University Press, 2014), 141–69.

17 Myers, W., "Beyond Mimickry," in *Bio Design*, edited by William Myers (New York: Museum of Modern Art, 2012), 3–31.

18 Interviews in *Bio Design* (New York: Museum of Modern Art, 2012), 266–7.

19 Rahm, Philippe, "Polarized House," *The [Gen] Home Project*, Philippe Rahm, MAK Center for Art and Architecture, 46–51, exh. cat., October 29, 2006–February 25, 2007.

20 "Bio-Processing," in *Bio Design*, 2012, New York, Museum of Modern Art., 83.

21 A definition of epigenetics from the U.S. National Library of Medicine; https://ghr.nlm.nih.gov/primer/howgeneswork/epigenome (accessed December 2018).

22 Conklin, Edwin G., "Weismann on Germinal Selection," *Science 12*, Vol. 3, No. 76 (June 1896), 853–7.

23 Stanford, P. K., "August Weismann's Theory of the Germ-Plasm and the Problem of Unconceived Alternatives," *Hist. Philos Life Sci*, Vol. 27, No. 2 (2005), 163–99; https://www.ncbi.nlm.nih.gov/pubmed/16602485 (accessed January 2019).

24 Kevles, Daniel, *In the Name of Eugenics: Genetics and the Uses of Human Heredity* (Berkeley and LA: University of California Press, 1985), 70–1.

25 Depew, David J., "Baldwin Boosters, Baldwin Skeptics," in *Evolution and Learning The Baldwin Effect Reconsidered*, edited by David J. Depew and Bruce H. Weber (Cambridge, MA: MIT Press, 2003), 3–31.

26 Wilson, Elizabeth A., *Gut Feminism* (Durham, NC: Duke University Press, 2015), 56.

27 Wilson, 55–8.

28 Wilson, 59; 170–4.

29 https://www.merriam-webster.com/dictionary/reductionism (accessed October 30, 2020).

30 Van Regenmortel, Marc H. V., "Reductionism and Complexity in Molecular Biology," *EMBO Reports*, Vol. 5, No. 11 (December 2004), 1016. See also Darden, Lindley and Craver, Carl F., "Reductionism in Biology," in *Encyclopedia of Life Sciences* (Chichester, UK: John Wiley & Sons, Ltd, 2009).

31 Van Regenmortel, 1016.

32 Van Regenmortel, 1016.

33 Wilson, E. O., *Consilience: The Unity of Knowledge* (New York: Vintage Books, 1998), 59.

34 Thompson, D'Arcy Wentworth, *On Growth and Form: The Complete Revised Edition* (New York: Dover Editions, 1992 [1942]), 1023.

35 Prum, Richard O., *The Evolution of Beauty: How Darwin's Forgotten Theory of Mate Choice Shapes the Animal World—and Us* (New York: Doubleday Press, 2017), 23.

36 Prum, 332.

37 Thompson, 5–6.

38 Thompson, 1023.

1

Are All Fish the Same Shape If You Stretch Them? The Victorian Tale of *On Growth and Form*

Stephen Wolfram

Is there a global theory for the shapes of fishes? It's the kind of thing I might feel encouraged to ask by my explorations of simple programs and the forms they produce.[1] But for most of the history of biology, it's not the kind of thing anyone would ever have asked. With one notable exception: D'Arcy Wentworth Thompson. And it's now 100 years since D'Arcy Thompson published the first edition of his magnum opus *On Growth and Form*[2]—and tried to use ideas from mathematics and physics to discuss global questions of biological growth and form. Probably the most famous pages of his book are the ones about fish shapes.[3] Stretch one kind of fish, and it looks like another. Yes, without constraints on how you stretch, it's not quite clear what this is telling one, and I don't think it's much. But just to ask the question is interesting, and *On Growth and Form* is full of interesting questions—together with all manner of curious and interesting answers.

D'Arcy Thompson was in many ways a quintessential British Victorian academic, steeped in the classics, and writing books with titles like *A Glossary of Greek Fishes*[4] (i.e., how were fish described in classical Greek texts). But he was also a diligent natural scientist, and he became a serious enthusiast of mathematics and physics. And where Aristotle (whom D'Arcy Thompson had translated[5]) used plain language, with perhaps a dash of logic, to try to describe the natural world, D'Arcy Thompson tried to use the language of mathematics and physics. At Christmas time, according to his daughter, he used to entertain children by drawing pictures of dogs on rubber sheets and stretching them from poodles to dachshunds. But it was not until the age of fifty-seven that he turned such pursuits into the piece of scholarship that is *On Growth and Form*.

The first edition of the book was published in 1917. In many ways it's like a catalog of biological forms—a kind of geometrical analog of Aristotle's books on natural history. It's particularly big on aquatic life—from plankton to fish. Land animals do make a showing, though mostly as skeletons. And ordinary plants make only specific appearances. But throughout the book the emphasis is on "why does such-and-such a thing have the form or shape it does?" And over and over again the answer that's given is: "because it's following such-and-such a physical phenomenon, or mathematical structure."

An earlier version of this text was published in the December 2018, 40 (4) issue, 39–61 of the *Mathematical Intelligencer*.

Much of the story of the book is told in its pictures. There are growth curves—of haddock, trees, regenerated tadpole tails, etc. There's a long discussion of the shapes of cells—and especially their connection with phenomena (like splashes, bubbles, and foams) where surface tension is important. There are spirals—described mathematically and appearing in shells and horns and leaf arrangements. And finally there's a long discussion of the "theory of transformations—about how different forms (like the shape of fishes or primate skulls) might be related by various (mathematically rather undefined) "transformations."

In D'Arcy Thompson's time—as still to a large extent today—the dominant form of explanation in biology is Darwinism: essentially the idea that things are the way they are because they've somehow evolved to be that way, in order to maximize some kind of fitness. D'Arcy Thompson didn't think that was the whole story, or even necessarily the most important part of the story. He thought instead that many natural forms are the way they are because it's an inevitable feature of the physics of biological tissue, or the mathematics of geometrical forms.

Sometimes D'Arcy Wentworth Thompson's explanations fall a little flat. Leaves aren't really shaped much like polar plots of trigonometric functions. Jellyfish aren't convincingly shaped like drops of ink in water. But what he says often rings true. Hexagonal arrangements of cells[6] are like closest geometrical packings of disks. Sheep horns and nautilus shells[7] form logarithmic (equiangular) spirals.

D'Arcy Wentworth Thompson uses basic geometry and algebra quite a bit—and even sometimes a little combinatorics or topology. But he never goes as far as calculus (and, as it happens, he never learned it), and he never considers ideas like recursive rules or nested structures. But for me—as for quite a few others over the years—D'Arcy Thompson's book is an important inspiration for the concept that even though biological forms may at first look complicated, there can still be theories and explanations for them.

In modern times, though, there's a crucial new idea, that D'Arcy Thompson did not have: the idea of using not traditional mathematics and physics, but instead computation and simple programs as a way to describe the rules by which things grow. And—as I discovered in writing my book *A New Kind of Science*[8]—it's remarkable to what extent that idea lets us understand the mechanisms by which complex biological forms are produced, and lets us finish the bold initiative that D'Arcy Thompson began a century ago in *On Growth and Form*.

Who Was D'Arcy Thompson?

D'Arcy Wentworth Thompson was born in Edinburgh on May 5, 1860. His father, who was also named D'Arcy Wentworth Thompson, had been born in 1829, aboard a ship captained by his father, that was transporting convicts to Tasmania. D'Arcy senior was soon sent to boarding school in England and eventually studied classics at Cambridge. Though academically distinguished, he was apparently passed over for a fellowship because of perceived eccentricity—and wound up as a (modernizing, if opinionated) schoolteacher in Edinburgh. Once there, he soon met the lively young Fanny Gamgee, daughter of Joseph Gamgee, an early and distinguished veterinary surgeon—and in 1859 they were married.

D'Arcy (junior) was born the next year—but unfortunately his mother contracted an infection during childbirth and died within the week. The result was that D'Arcy (junior) ended up living with his mother's parents, taken care of by one of his mother's sisters. When D'Arcy (junior) was three years old, his father then got a university professorship (of ancient Greek) in Ireland and moved there. Still, D'Arcy (junior) stayed in close touch with his father through letters and, later, visits. And indeed, his father seems to have doted on him, for example, publishing two children's books dedicated to him.

In a foreshadowing of his later interests, D'Arcy (junior) learned Latin from his father almost as soon as he was able to speak and was continually exposed to animals of all sorts in the Gamgee household. There was also a certain math/physics theme. D'Arcy Thompson senior's best friend in Edinburgh was Peter Guthrie Tait—a distinguished mathematical physicist (mechanics, thermodynamics, knot theory, etc.) and friend of Maxwell, Hamilton, and Kelvin—and D'Arcy (junior) often hung out at his house. Joseph Gamgee was also engaged in various scientific pursuits, for example, publishing the book *On Horseshoeing and Lameness* based in part on a statistical study he'd done with the then ten-year-old D'Arcy (junior). Meanwhile, D'Arcy Thompson (senior) began to travel, as D'Arcy (junior) would later do, for example, visiting Harvard in 1867 to give the Lowell Lectures—which D'Arcy (junior) would also give, in 1936, sixty-nine years later.

At the age of eleven, D'Arcy went to the school where his father had previously taught. He did well in academic studies, but also organized a natural history ("Eureka") club, where he and his friends collected all sorts of specimens. And by the end of his time at school, he published his first paper: the eleven-page (with photographs) "Note on Ulendron and Halonia," describing the regular pattern of growth scars on two kinds of fossil plants.

At eighteen, D'Arcy started at Edinburgh University as a medical student. His grandfather—while distinguished—was not wealthy, with the result that D'Arcy had to support himself by tutoring Greek and writing articles for the Edinburgh-published *Encyclopedia Britannica* (the ninth edition, from 1889, contains an extensive article by D'Arcy on John Ray, a British naturalist of the 1600s). But D'Arcy's real passion at the time was the then-hot field of paleontology, and after two years he abandoned his medical studies—and left to instead study natural science at the place his father had been years earlier: Trinity College, Cambridge.

D'Arcy did well at Cambridge, had an interesting circle of friends (including the future co-author of *Principia Mathematica*, Alfred North Whitehead), and quickly became something of a fixture in the local natural history scene. This led Macmillan & Co. to commission D'Arcy (still an undergraduate) to produce his first book: a translation from German of Hermann Muller's *The Fertilization of Flowers*. The publisher thought that the book—which was a fairly traditional work of descriptive natural history, based in part on observing about 14,000 visits of insects to flowers—

would be of popular interest, and (in one of his last published appearances) got no less than Charles Darwin to write a preface for it.

At Cambridge, D'Arcy hung out a lot at the new Museum of Zoology, and was particularly influenced by a young professor named Frank Balfour who studied comparative embryology, and for whom a new Department of Animal Morphology was being created—but who died trying to climb Mont Blanc right when D'Arcy was finishing Cambridge.

D'Arcy began to pursue all sorts of projects, giving lectures on topics such as "Aristotle on Cephalopods," and making detailed studies of "hydroid zoophyte" specimens (aquatic animals like sea anemones that look like plants) brought back from expeditions to the Arctic and Antarctic. He applied for a fellowship in Cambridge, but—like his father before him—didn't get it.

In 1884, though, the newly created and new-thinking (non-religious, co-ed, young professors, etc.) University College in Dundee, Scotland, advertised for a professor of biology (yes, combining zoology and botany!). D'Arcy applied, and got the job—with the result that at age twenty-four he became a professor, a role in which he would remain for nearly sixty-four years.

D'Arcy the Professor

D'Arcy was immediately popular as a teacher, and continued to do a certain amount of rather dry academic work (in 1885 he published *A Bibliography of Protozoa, Sponges, Coelenterata, and Worms*, which was, as advertised, a list of about 6,000 publications on those subjects between 1861 and 1883). But his real passion was the creation of his own Museum of Zoology, and the accumulation of specimens for it. He was soon excitedly writing that "within the last week, I have had a porpoise, two mongooses, a small shark, an eel 8ft long … a young ostrich and two bagfuls of monkeys: all dead of course." His archive (among its 30,000 items) contains extensive evidence of all sorts of trading of specimens from around the world.[9] But in Dundee he found a particularly good local source of specimens. Dundee had long been a center of international textile trade and had also developed a small whaling industry. And when it was discovered that by mixing jute with whale oil it could be turned into fabric, whaling in Dundee grew dramatically. Some of the hunting they did was local. But whaling ships from Dundee went as far as Canada and Greenland (and once even to Antarctica). And befriending their captains, D'Arcy persuaded them to bring him back specimens (as skeletons, in jars, etc.) from their expeditions—with the result, for example, that his museum rapidly accumulated the best arctic collection around.

The museum always operated on a shoestring budget, and it was typical in 1886 when D'Arcy wrote that he'd personally been "working all day on a baby *Ornithorhynchus*" (platypus). In his early years as a professor, D'Arcy published only a few papers, mostly on very detailed matters—like the strangely shaped stomach of a type of possum, or the structure of the porpoise larynx,[10] or the correct taxonomic placement of a duck-like dinosaur. And he always followed the prevailing Darwinian paradigm of trying to explain things either by their evolutionary connections, or by their fitness for a particular function.

The Matter of the Alaskan Seals

In Dundee, D'Arcy joined various local clubs, like the Dundee Naturalists' Society, the Dundee Working Men's Field Club, the Homeric Club, and, later, also the Freemasons. He became quite active in university and community affairs, notably campaigning for a medical school (and giving all sorts of statistical evidence for its utility), as well as for education for the local poor. But mostly D'Arcy lived the life of an academic, centered around his teaching and his museum. Still, as a responsible member of the community, he was called on in various ways, and in 1892, he joined his first government commission—formed to investigate a plague of voles in Scotland[11] (conclusions included: "don't shoot hawks and owls that eat voles," and "it's probably not a good idea to set loose a 'virus' to infect the voles"). Then in 1896—at the age of thirty-six—D'Arcy was tapped for a piece of international scientific diplomacy.

It all had to do with seals, and the fur trade based on them. When Russia sold Alaska to the United States in 1867 it also sold the rights to the seals which bred on some of the islands in the Bering Sea. But by the 1890s Canadian ships (under British protection) were claiming the right to catch seals in the open ocean—and too many seals were being killed for the population to be maintained. In 1893 a treaty was made to clarify the situation. But in 1896 there was a need to analyze more carefully what was going on (and, yes, to claim what ended up being $10M in damages for Canadian/British sealers).

Lord Salisbury, the British Prime Minister at the time, who happened to be an amateur botanist, knew of D'Arcy and asked him to travel to the Bering Sea to investigate. D'Arcy had by that point traveled a bit around Europe, but this was a complex trip. At first he went to Washington, DC, dropping in at the White House. Then across Canada, and then by Coast Guard ship (and dog sled) to the seals. D'Arcy did well at making friends with his American counterparts (who included the president of the then-a-decade-old Stanford University), and found that at least on the American-controlled islands (the Russian-controlled ones were a different story) seals were being herded a bit like sheep in Scotland, and that though there was "abundant need for care and prudent measures of conservation," things were basically OK. In Washington, DC, D'Arcy gave a long speech, and helped broker a kind of "seal peace treaty"—that the British government was pleased enough with to give D'Arcy a (medieval-inspired) "Companion of the Bath" honor.

Statesman of Science

Being a professor in Dundee wasn't a particularly high position in the pecking order of the time. And after his Bering Sea experience, D'Arcy started investigating moving up. He applied for various jobs (for example at the Natural History Museum in London), but perhaps in part because he didn't have fancier academic credentials (like a PhD)—and also had spent so much of his time organizing things rather than doing research—he never got any of them. He was nevertheless increasingly sought after as a kind of statesman of science. And in 1898 he was appointed to the Fishery Board for Scotland (a role in which he continued for forty-three years), and the next year he was the British delegate to the first International Conference on Oceanography. D'Arcy was a serious collector of data. He maintained a team of people at the fish market, keeping track of the catches brought in from boats.

And then he took this data and created graphics and statistical analyses. And over the years he became well known as a negotiator of fishing rights, both locally and internationally. He was also a collector of oceanographic data. He saw to it that there were detailed tide measurements made. And had the data analyzed and decomposed into harmonic components—much as it is today. The Scottish government even provided for him a research ship (a steam trawler named the SS *Goldseeker*) in which he and his students would go around the Scottish coast, measuring ocean properties and collecting specimens.

D'Arcy the Classical Scholar

D'Arcy always had many interests. First and foremost was natural history. But after that came classics. And indeed, back in his undergraduate days, D'Arcy had already started working with his classicist father on translating Aristotle's works on natural history into English. One of the complexities of that task, however, was to know what species Aristotle meant by words he used in Greek. And this led D'Arcy into what became a lifelong project—the first output of which was his 1894 book *Glossary of Greek Birds*.[12] It is an interesting exercise—trying to fit together clues to deduce just what modern bird some passage in classical Greek literature was talking about. Often D'Arcy succeeds. Sometimes by using natural history; sometimes by thinking about mythology or about configurations of things like constellations named for birds. But sometimes D'Arcy just has to describe something as "a remarkable bird, of three varieties, of which one croaks like a frog, one bleats like a goat, and the third barks like a dog"—and he doesn't know the modern equivalent.

Over the years, D'Arcy continued his efforts to translate Aristotle, and finally in 1910 (eight years after his father's death) he was able to publish what remains to this day the standard translation of Aristotle's main work on zoology, his *History of Animals*.[13] This project established D'Arcy as a classical scholar—and in 1912 he even got an honorary PhD (DLitt) at Cambridge on the basis of it. He also began a long association with what's known as Liddell & Scott,[14] the still-standard dictionary of ancient Greek. (Liddell had been notable for being the father of Alice, of Wonderland fame.) But D'Arcy's interests in Greek science extended beyond natural history, and into astronomy and mathematics. D'Arcy explored such things as ancient methods for computing square roots—and also studied Greek geometry. So in 1889, when D'Arcy was investigating Foraminifera (protozoa that live in sediment or in the ocean and often form spiral shells) he was able to bring his knowledge of Greek mathematics to bear, declaring that "I have taken to Mathematics ... and discovered some unsuspected wonders in regard to the Spirals of the Foraminifera!"

Toward Something Bigger

When he was forty-one, in 1901, D'Arcy married his stepmother's niece, the then twenty-nine-year-old Ada Maureen Drury (yes, small world that it is, she was named after "Byron's" Ada, because an ancestor had reputedly been a romantic interest of Byron's). They bought a small house somewhat outside of town—and between 1902 and 1910 they had three children, all daughters. By 1910, D'Arcy was fifty years old, and an elder statesman of science. He kept himself busy teaching, managing his museum, doing administrative and government work, and giving public lectures. A typical lecture—given at Oxford in 1913—was entitled "On Aristotle as a Biologist." It was charming, eloquent, ponderous, and Victorian.

In many ways, D'Arcy was first and foremost a collector. He collected natural history specimens. He collected Greek words. He collected academic references—and antiquarian books. And he collected facts and statements—many of which he typed onto index cards, now to be found in his archive. Still, in his role as elder statesman, D'Arcy was called upon to make broad pronouncements. And in many ways the great achievement of the later part of his life was to connect the disparate things he collected and identify common themes that could connect them.

In 1908, he had published (in *Nature*) a two-page paper entitled "On the Shapes of Eggs and the Causes Which Determine Them."[15] In a sense the paper was about the physics of egg formation. And what was significant was that instead of accounting for different egg shapes in terms of their evolutionary fitness, it talked about the physical mechanisms that could produce them. Three years later D'Arcy gave a speech entitled "Magnalia Naturae: or the Greater Problems of Biology"[16] in which he took this much further, and started discussing "the possibility of ... supporting the observed facts of organic form on mathematical principles [so as to make] morphology ... a true natural science ... justified by its relation to mathematics." In 1910, Cambridge University Press had asked D'Arcy if he'd like to write a book about whales. He said that instead perhaps he should write a "little book" about "The Forms of Organisms" or "Growth and Form"—and he began the process of assembling what would become *On Growth and Form*. The book had elements that drew on D'Arcy's whole range of interests. His archives contain some of what went into the assembly of the book, like the original drawings of fish-shape transformations (D'Arcy wasn't a great sketch artist). There were also other, more impressionistic images—like the one illustrating transformations between zebra-related animals (e.g., quagga) or one showing tortoise shell structure. D'Arcy did not contact his publisher again for several years, but in 1915—in the middle of the First World War—he wrote them again, saying that he had finally finished the book "on a larger scale," and soon signed a publishing contract (that's shockingly similar to the way modern ones look). It took a couple more years, between D'Arcy's last-minute changes and paper shortages associated with the war—but finally in 1917 the book (which had by then swelled to 800 pages) was published.

The Book

On Growth and Form opens with a classic D'Arcy "Prefatory Note": "This book of mine has little need of preface, for indeed it is 'all preface' from beginning to end." He goes on to apologize for his lack of mathematical skill—and then launches, beginning with a discussion of the relation of the philosophies of Kant and Aristotle on the nature of science. The reviews were positive, and surprisingly sensible, with the *Times Literary Supplement*, for example, writing (Figure 1.1):

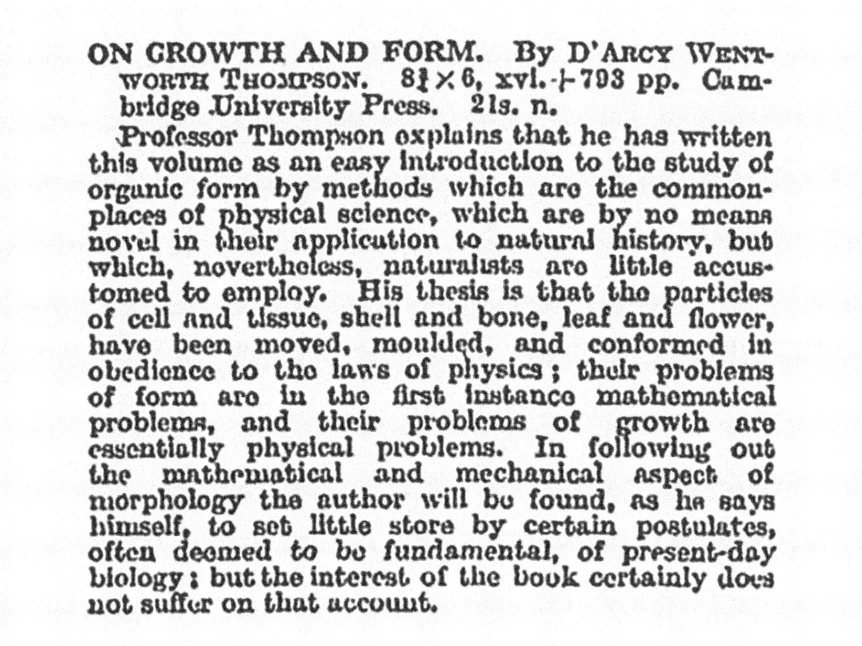

ON GROWTH AND FORM. By D'ARCY WENTWORTH THOMPSON. $8\frac{3}{4} \times 6$, xvi.+793 pp. Cambridge University Press. 21s. n.

Professor Thompson explains that he has written this volume as an easy introduction to the study of organic form by methods which are the commonplaces of physical science, which are by no means novel in their application to natural history, but which, nevertheless, naturalists are little accustomed to employ. His thesis is that the particles of cell and tissue, shell and bone, leaf and flower, have been moved, moulded, and conformed in obedience to the laws of physics; their problems of form are in the first instance mathematical problems, and their problems of growth are essentially physical problems. In following out the mathematical and mechanical aspect of morphology the author will be found, as he says himself, to set little store by certain postulates, often deemed to be fundamental, of present-day biology; but the interest of the book certainly does not suffer on that account.

Figure 1.1 *Times Literary Supplement* review for *On Growth and Form. Source:* Public domain.

D'Arcy was fifty-seven years old by the time *On Growth and Form* was published—and he could have used it as a closing act in his career. But instead it seemed to make him more energetic—and seemed to encourage him to take mathematical methods as a kind of personal theme.

In his study of the shapes of biological cells, D'Arcy had gotten very interested in polyhedra and packings, and particularly in Archimedean solids (such as the tetrakaidecahedron). His archives contain all sorts of investigations of possible packings and their properties, together with actual cardboard polyhedra, still ready to assemble. D'Arcy extended his interest in number theory, collecting properties of numbers a little like he'd collected so many other things. He dipped into chemistry, thinking about it in terms of graphs, like those derived from polyhedra. And even when he worked on history, D'Arcy used mathematical thinking, studying the distribution of when famous people lived, in connection with writing about the Golden Ages. As an administrator he brought in math as well, analyzing what in today's world would be called a grading curve—and comparing exam results between different years. He worked extensively on tides and tide computations. He collected data from harbors and came up with theories about the various components of tides, some of which turned out to be correct.

The mathematics he used was always a bit circumscribed—and, for example, he never learned calculus, even to the point of apparently getting confused about growth rates versus finite differences in plots in *On Growth and Form*. (There seems to be just a single sheet of calculus-like work by him in his archives, and it's simply an exercise copied without solution from the famous Whittaker & Watson[17] textbook.) But what about systems based on pure computational rules—of the kind that, for example, I have spent so much time studying?[18] Well, in the archive there are things like a version of a space-filling curve. And back from 1897 there's a curious cardboard object that D'Arcy described as a "reasoning machine." It is not completely clear what this was (though its wheel still turns nicely!). It seemed to involve a diagrammatic way of determining the truth value of a logical expression, perhaps following the work of Jevons from a couple of decades earlier. But so far as I can tell it was D'Arcy's sole excursion into the world of logic and rule-based processes—and he never connected anything like this to biology.

The Later D'Arcy

Before *On Growth and Form*, D'Arcy had published only quite sporadically. But after it, as he entered his sixties, he began to write prodigiously, publishing all over the place on a wide range of topics. He gave lectures, in person and on the radio. And he also began to receive all sorts of honors (he became Sir D'Arcy in 1937)—and was invited to events all over the world (he did a grand tour of the United States in the 1930s, and was also received as a celebrity in places like the Soviet Union). *On Growth and Form* was considered a commercial success. Its original print run was 500 copies (of which at least 113 are now in academic libraries around the world), and by 1923 it had sold out. The publisher (Cambridge University Press) wanted to reprint it. But D'Arcy insisted that it needed to be revised—and in the end it took until 1942 before he got the revisions done. The second edition added 300 pages to the book—including photographs of splashes (obtained directly from Harold Edgerton at MIT), analysis of teeth, and patterns on animal coats. But the main elements of the book remained exactly the same.

D'Arcy had published a second edition of his *Glossary of Greek Birds* in 1936 (more birds, more interpretations), and in 1947, based on notes he started collecting in 1879, he released a kind of sequel: his *Glossary of Greek Fishes*. (Oxford University Press, in the flap copy for the book, says charmingly that "it is highly improbable that there is any other scholar who has studied Greek fishes over so long a period as

Sir D'Arcy Thompson.") Even into his eighties, D'Arcy continued to travel all over the place—with his archives containing some typical travel documents of the time. His travel was interrupted by the Second World War (which is perhaps why the second edition of *On Growth and Form* finally got finished in 1942). But in 1947, with the war over, at the age of 87, D'Arcy went to India for several months, notably lecturing on the skeletal structure of birds while holding a somewhat impatient live hen in a box. But in India D'Arcy's health began to fail, and after returning to Scotland, he died in June 1948—to the last corresponding about specimens for his museum.

Aftermath

D'Arcy's wife (who seemed in frail health through much of her forty-seven-year marriage to D'Arcy) lived on for only seven months after his death. None of D'Arcy's daughters ever married. His oldest daughter Ruth became a music teacher and administrator at a girl's boarding school, and in 1958 (when she was fifty-six) published a biography of D'Arcy. His middle daughter Molly moved to South Africa, wrote children's and travel books, and lived to the age of 101, dying in 2010—while his youngest daughter Barbara wrote a book on healing and herbalism and died in a freak river accident in 1990. *On Growth and Form* was D'Arcy's most notable output, and it has been reprinted many times over the course of a hundred years. The museum D'Arcy created in Dundee was largely dismantled in the 1950s, but has now been to some extent reconstituted, complete with some of the very specimens D'Arcy collected, with labels duly signed "DWT" (yup, that's me next to the same orangutan as in the old picture of the museum) (Figure 1.2):

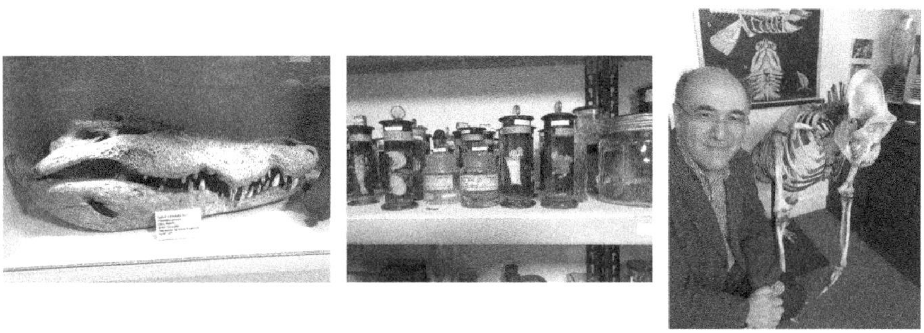

Figure 1.2 Stephen Wolfram with specimens at the D'Arcy Thompson Zoology Museum. Courtesy of Stephen Wolfram.

In 1917 D'Arcy moved from Dundee to the nearby but more distinguished and ancient university in St Andrews, where he took over another museum. It too fell upon hard times, but still exists in a reduced form. And now some of the D'Arcy's specimens are being 3D-scanned (yes, that's the same crocodile) (Figure 1.3):

```
In[1]:= ResourceData["D'Arcy Thompson Zoology Museum 3D Anatomical Models",
        "SaltwaterCrocodileSkull"]
```

Out[1]=

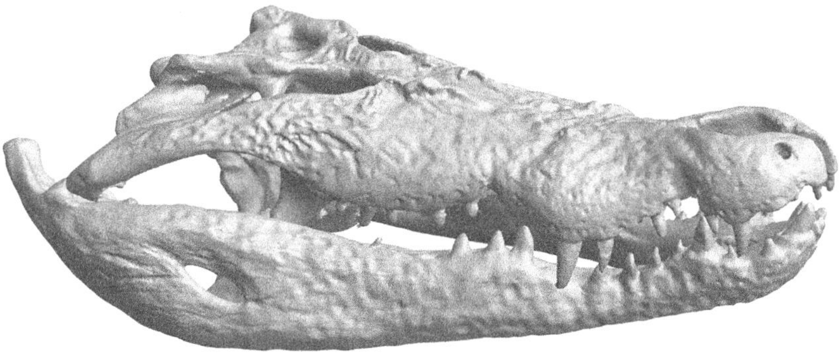

Figure 1.3 D'Arcy Thompson's 3D-scanned specimen. Courtesy of Stephen Wolfram.

What Was D'Arcy Like?

D'Arcy had an imposing physical presence. He stood 6'3" and had a large head, on which he often wore a black fedora. He had piercing blue eyes, and in his youth, he had red hair—which he grew into a large beard when he was a young professor. He often wore a long coat, which could sometimes seem moth-eaten. Later in his life, he would sometimes walk around town with a parrot on his shoulder.[19] He was renowned as an engaging speaker and lecturer—known both for his colorful and eloquent content (he could regale the audience with the tale of a walrus he had known, or equally well discuss Aristotle's year by the seaside), and for the various physical (and biological) demonstrations he would use. Many stories are told of his eccentricities, especially by his former students. It is said, for example, that he once came to give a lecture to his students which began with him pulling a dead frog out of one pocket of his coat—and then a live one out of the other pocket. Despite having spent most of his life in Scotland, he didn't have a Scottish accent.

He was charming and jolly, and even in his eighties he was given to dancing when he could. He was tactful and diplomatic, if not particularly good at sensing other people's opinions. He presented himself with a certain modesty (for example always expressing his weakness in mathematics), and—perhaps to his detriment—did little to advocate for himself.

He led a fairly simple life, centered around his work and family. He worked hard, typically until midnight each day. He always liked to learn. He enjoyed children and the young, and would happily play with them. When he walked around town, he was universally recognized (the shoulder parrot helped!). He was happy to chat with anyone, and in later years, he carried candy in his pocket, which he gave out to children he encountered.

D'Arcy was a product of his age, but also of an unusual combination of influences. Like many of the members of his adopted family, D'Arcy aspired to be a scientist. But like his father, he aspired to be a classical scholar. He did diligent and detailed academic work for many years, in natural history, in classics, and in ancient science. But he also enjoyed presentation and lecturing. And it was in large part through his efforts to explain his academic work that he came to make the connections that would lead to *On Growth and Form*.

What Happened After

If you search the scientific literature today, you'll find about 4,000 publications citing *On Growth and Form*. Their number relative to the total scientific literature has remained remarkably fairly even over the years (with a peak around the publication of the second edition in 1942, and perhaps a dip in the 1960s when genetics began to dominate biology) (Figure 1.4):

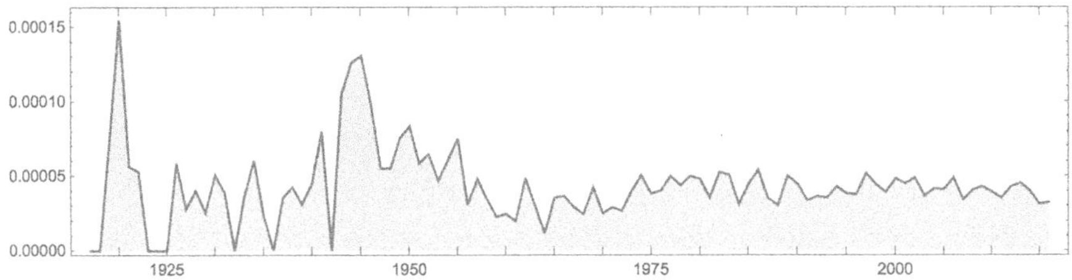

Figure 1.4 Graph of publications citing *On Growth and Form* in 2017. Courtesy of Stephen Wolfram.

There's quite a diversity in the topics, as this random sample of titles indicates (Figure 1.5):

Autecology of Silurian Gastropods and Monoplacophorans

Mechanics of Posture and Gait of Some Large Dinosaurs

Modeling and Inferring Cleavage Patterns in Proliferating Epithelia

Adaptive Versatility and Skeleton Construction

A Framework for Understanding Generative Art

An Experimental—Study of Cracking Induced by Desiccation

Geometry of Human Cerebellar Vermis

Aging of Cities, Corporations, and Civilizations

Folded Tubes as Compared to Kikko (Tortoise-Shell) Bamboo

Evolution of Functional Design in Rudists (Hippuritacea) and Its Taxonomic Implications

Multivariate Allometry as a Model of Morphometric Harmony of 4 Skinfolds in the Newborn

Modulatory Complexity of the Feeding Repertoire in Scincid Lizards

Patterns of Individual Growth of Gray Garden Slug Deroceras Reticulatum

Inverting Dedevelopment: Geometric Singularity Theory in Embryology

Multiagent Control of Self-Reconfigurable Robots

Physicochemical Factors in the Organization of Multicellular Aggregates and Plants

Design and Creativity in Synthesis of Multivalent Neoglycoconjugates

Figure 1.5 Random sample of titles citing *On Growth and Form*. Courtesy of Stephen Wolfram.

Most concern specific biological systems; some are more general. Making word clouds from titles by decade, one sees that "growth" is the dominant theme—though centered in the 1990s there are signs of the discussion that was going on about the "philosophy of evolution," and the interplay between natural selection and "developmental constraints" (Figure 1.6):

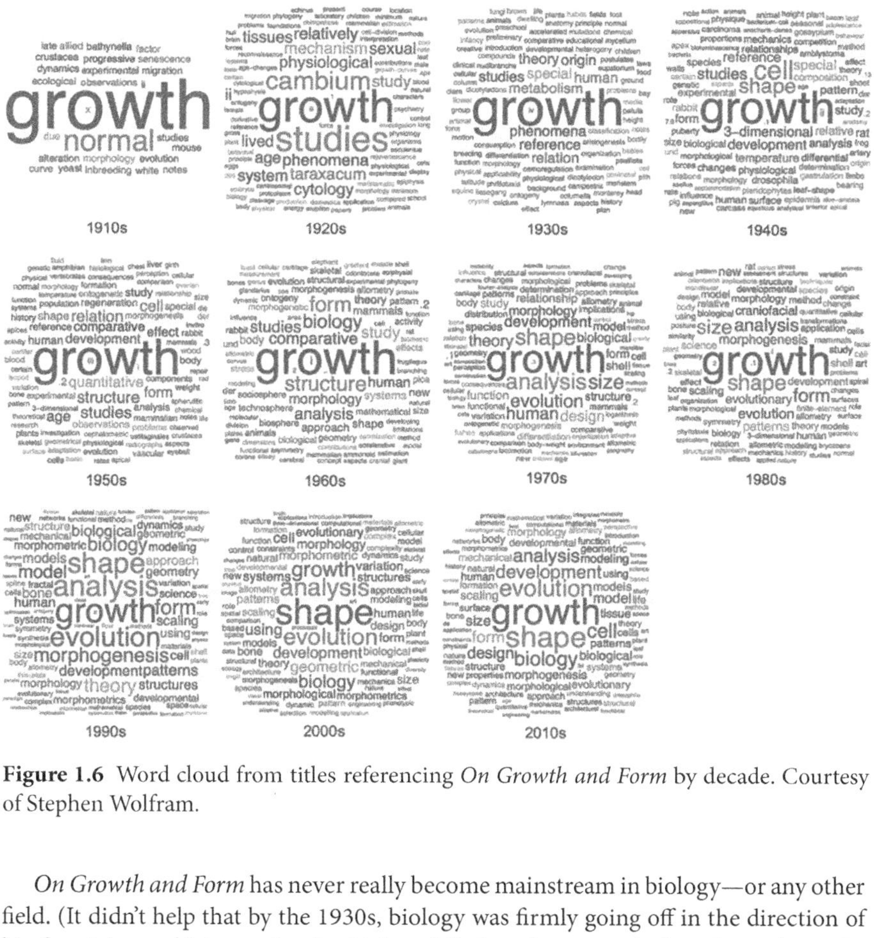

Figure 1.6 Word cloud from titles referencing *On Growth and Form* by decade. Courtesy of Stephen Wolfram.

On Growth and Form has never really become mainstream in biology—or any other field. (It didn't help that by the 1930s, biology was firmly going off in the direction of biochemistry and later molecular biology.) So how have people found out about *On Growth and Form*? Indeed, as I write this, I'm wondering: how did I myself find out about *On Growth and Form*? I can tell I knew about it by 1983, because I referenced it (somewhat casually) in my first long paper about cellular automata[20] and the patterns they generate. I also know that in 1982 I bought a copy of the (heavily abridged) version of *On Growth and Form* that was available then. (I was thrilled in 1992 when I chanced upon a complete second edition of *On Growth and Form* in a used bookstore; I'd never seen the whole book before.) But how did I first become aware of D'Arcy, and *On Growth and Form*? My first hypothesis today was that it was in 1977, from the historical notes of Benoit Mandelbrot's *Fractals* book[21] (yes, D'Arcy had actually used the term "self-similar," though only in connection with spirals). Then I thought perhaps it might have been around 1980, from the references to Alan Turing's 1952 paper on the chemical basis of morphogenesis.[22] I wondered if perhaps it was from hearing about catastrophe theory, and the work of René Thom, in the mid-1970s. But my best guess as of now is that it was actually around 1978, from a little book titled *Patterns in Nature*,[23] by a certain Peter S. Stevens, that heavily references *On Growth and Form*, and that I happened across in a bookstore.

I have almost never seen mentions of *Patterns in Nature*, but in some ways it's a simplified and modernized *On Growth and Form*, full of photographs comparing biological and non-biological systems, together with diagrams about how various structures can be built. But what was the path from D'Arcy to *Patterns in Nature*? It's a typical kind of history question that comes up. The first thing I noticed is that Peter Stevens (born 1936) was trained as an architect and spent most of his career around Harvard. In his book, he thanks his father, Stanley Stevens (1906–1973), who was a psychoacoustics expert, who was at Harvard from 1932 on, and who organized a "Science of Science" interdisciplinary discussion group there. But recall that D'Arcy visited Harvard to give the Lowell Lectures in 1936. So that's no doubt how Stevens, Sr. knew about him. But in any case, from his Harvard connections came, I believe, the references to D'Arcy by evolutionary biologist Stephen J. Gould, and by John Tyler Bonner, who was the person who created the abridged version of *On Growth and Form* (sadly, omitting, for example, the chapter on phyllotaxis). I suspect D'Arcy's influence on Buckminster Fuller also came through Harvard connections. And maybe Benoit Mandelbrot heard about D'Arcy there too. (One would think that with *On Growth and Form* being out there as a published book, there wouldn't be need for word-of-mouth communication, but particularly outside of mainstream areas of science, word of mouth remains surprisingly important.)

But what about Turing? How did he know about D'Arcy? Well, I have at least a guess here. D'Arcy had been good friends in high school with a certain John Scott Haldane, who would go on to be a well-known physiology researcher, and who had a son named J. B. S. Haldane, who became a major figure in evolutionary biology and in the presentation of science to the public. Haldane often referenced D'Arcy, and notably introduced him to Peter Medawar (who would win a Nobel Prize for immunology), of whom D'Arcy (in 1944) would say, "I do believe that more than any man you have understood what I have tried to say!"[24] Both Medawar and evolutionary biologist (and originator of the term "transhumanism") Julian Huxley encouraged D'Arcy to think about continuity and gradients in connection with his shape transformations (e.g., of fish). I don't know the whole story, but I suspect these two connected with C. H. Waddington, a developmental biologist (and inventor of the term "epigenetics") who interacted with Turing in Cambridge. (Small world that it is, Waddington's daughter is married to a distinguished mathematician named John Milnor, with whom I discussed D'Arcy in the early 1980s.) And when Turing came to write about morphogenesis in 1952, he referenced D'Arcy (and Waddington), then proceeded to base his theory on (morphogen) gradients.

In another direction, D'Arcy interacted with early mathematical biologists like Alfred Lotka and Vito Volterra and Nicolas Rashevsky. And though their work was heavily based on differential equations (which D'Arcy didn't really believe in), he took pains to support them when he could. *On Growth and Form* also seems to have been popular in the art and architecture community, with people as diverse as the architects Mies van der Rohe, Le Corbusier, the painter Jackson Pollock, and the sculptor Henry Moore also mentioning its influence.

Modern Times

So now that it has been 100 years since *On Growth and Form* was published, do we finally understand how biological organisms grow? Lots of work has certainly been done at the genetic and molecular scale, and great progress has been made. But when it comes to macroscopic growth, much less has been done. And a large part of the reason, I suspect, is that it's needed a new paradigm in order to make progress. D'Arcy's work was, more than anything, concerned with analogy and (essentially Aristotelian-style) mechanism. He didn't really pursue traditional "theory" in the sense of the exact sciences. In his day, though, such theory would normally have meant writing down mathematical equations to represent growth, and then solving them to see what would happen. And the problem is that when one looks at biological forms, they often seem far too complex to be the results of traditional mathematical equations. But starting in the 1950s a new possibility emerged: perhaps one could model biological growth as following not mathematical equations but instead rules like a program for a computer. And when I started my systematic investigation of the computational universe of possible programs in the early 1980s, I was immediately struck by how "biological" a lot of the forms created, say, by simple cellular automata seemed (Figure 1.7):

```
In[2]:= CellularAutomaton[{1635, {3, 1}}, RandomInteger[2, 1400], 400] // ArrayPlot
```

Out[2]=

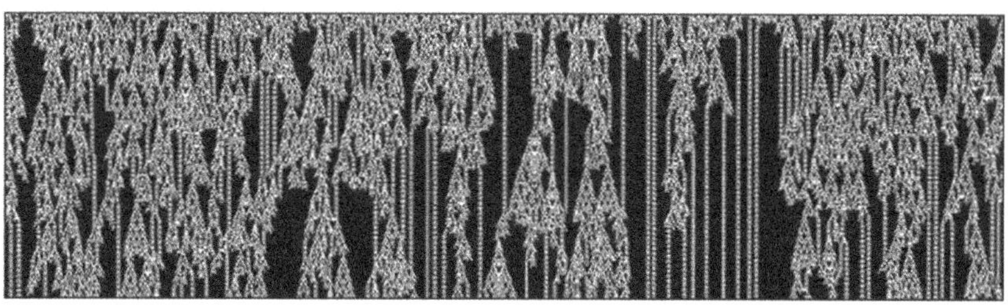

Figure 1.7 Cellular automata. Courtesy of Stephen Wolfram.

This is how I came to study *On Growth and Form*. I viewed it almost as a catalog of biological forms—that I wondered if one could explain with computational rules. I even started collecting specimens—in a very pale shadow of D'Arcy's efforts (and with no animal skeletons!). Occasionally I would find one that just seemed to cry out as being from something like a program. But more than that, I kept on exploring spaces of possible programs—and discovering that the range of forms they produced seem to align remarkably well with the actual range of forms one sees across biological organisms. (I looked particularly at shell shapes and patterns, as well as other pigmentation patterns, and various forms of plants.) And in a sense what I found strongly supports a core idea of D'Arcy's: that the forms of organisms are not so much determined by evolution, as by what it's possible for processes to produce. D'Arcy thought about physical processes and mathematical forms; 60+ years later I was in a position to explore the more general space of computational processes. And it so happened that, like D'Arcy, I ended up presenting my main results in a (big) book, that I called *A New Kind of Science*. My main purpose in the book was to describe what I'd learned from exploring the computational universe. And I devoted two sections (out of 114) respectively to "Growth of Plants and Animals"[25] and "Biological Pigmentation Patterns"[26]—producing something that looks a bit similar to *On Growth and Form* (Figure 1.8):

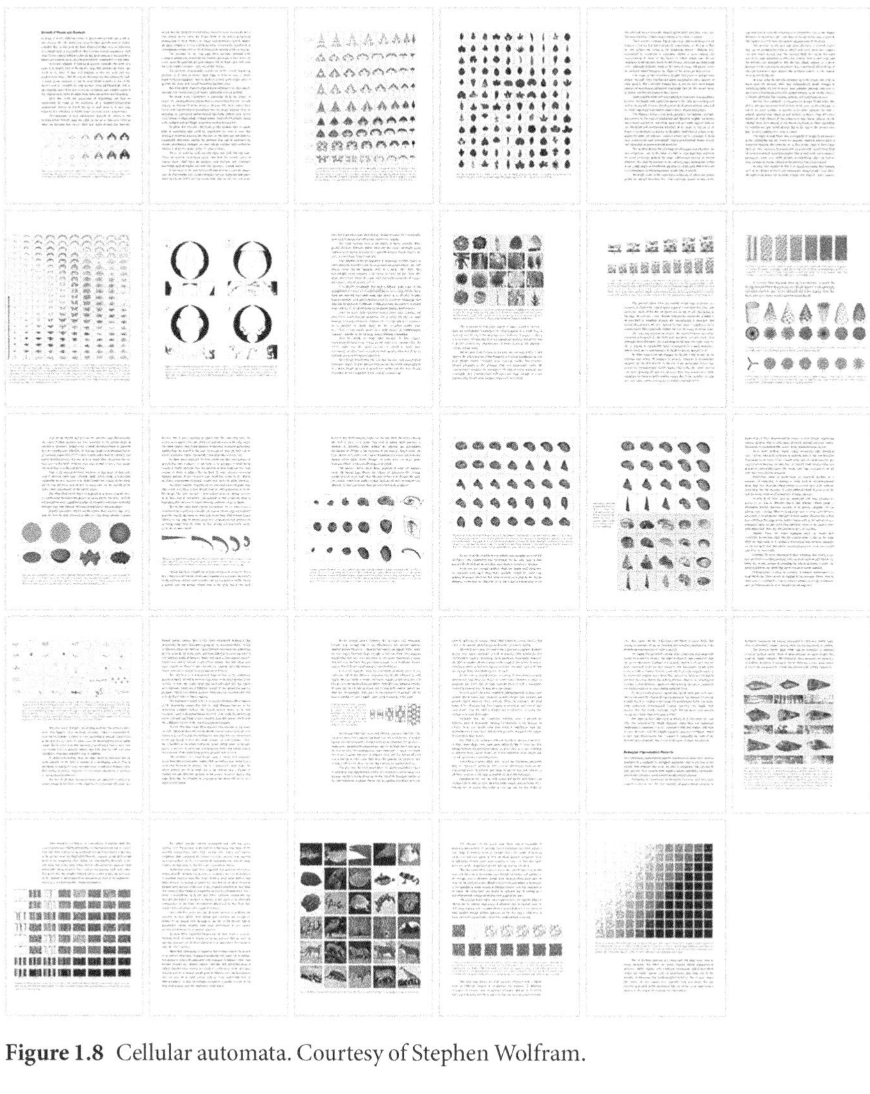

Figure 1.8 Cellular automata. Courtesy of Stephen Wolfram.

So, in the end, what about the fish? Well, I think I have managed to understand something about the "morphospace" of possible mollusk shells. And I've made a start on leaves—though I am hoping one of these years to be able to get a lot more data. I have also looked at animal skeletons a bit. But, yes, I at least still don't know about the space of possible fish shapes. Though maybe somewhere inside our image identification neural net (which saw plenty of fish in its training) it already knows. And maybe it agrees with what D'Arcy thought—a hundred years ago.

(*For help with facts and materials I'd like to thank Matthew Jarron, Maia Sheridan, Isabella Scott, Special Collections at the University of St Andrews Library, and the On Growth and Form 100 conference in Dundee/St Andrews.*)

Notes

1 Wolfram, Stephen, "Chapter 8: Implications for Everyday Systems," in *A New Kind of Science* (Champaign, IL: Wolfram Media, 2002), 400. https://www.wolframscience.com/nks/.
2 Thompson, D'Arcy Wentworth, *On Growth and Form* (Cambridge: Cambridge University Press, 1917).
3 *On Growth and Form*, August 2016. https://www.ongrowthandform.org/files/2016/08/OGF-fig-375-376.jpg (accessed November 20, 2018).
4 Thompson, D'Arcy Wentworth, *A Glossary of Greek Fishes* (London: Oxford University Press, 1947).
5 Aristotle. n.d., *The History of Animals*, trans. D'Arcy Wentworth Thompson. *The Internet Classics Archive*. http://classics.mit.edu/Aristotle/history_anim.html (accessed November 1, 2018).
6 Wolfram, Stephen, "Chapter 8: Implications for Everyday Systems, Section 6: Growth of Plants and Animals, Notes: Shapes of [biological] Cells," in *A New Kind of Science* (Champaign, IL: Wolfram Media, 2002). https://www.wolframscience.com/nks/.
7 Wolfram, Stephen, "Chapter 8: Implications for Everyday Systems," in *A New Kind of Science* (Champaign, IL: Wolfram Media, 2002), 413–14. https://www.wolframscience.com/nks/.
8 Wolfram, Stephen, *A New Kind of Science* (Champaign, IL: Wolfram Media, 2002). https://www.wolframscience.com/nks/.
9 University of St Andrews Special Collections. University of St Andrews. https://www.st-andrews.ac.uk/media/special-collections/documents/Darcy Wentworth Thompson Index.pdf (accessed November 13, 2018).
10 Thompson, D'Arcy Wentworth, "On the Cetacean Larynx," Essay, in *Studies from the Museum of Zoology in University College, Dundee*, Vol. 1. (Berlin: R. Friedlander & Son, 1890).
11 *The Spectator*, "The Plague of Voles in Scotland," June 12, 1892. http://archive.spectator.co.uk/article/4th-june-1892/12/the-plague-of-voles-in-scotland.
12 Thompson, D'Arcy Wentworth, *A Glossary of Greek Birds* (London: Milford, 1936).
13 Aristotle, n.d, *The History of Animals*.
14 Liddell, H. G. and Scott, R., *Greek-English Lexicon* (Oxford: Clarendon Press, n.d.).

15 Thompson, D'Arcy Wentworth, "On the Shapes of Eggs and the Causes Which Determine Them." *Nature*, November 2014. https://www.nature.com/articles/078111b0.pdf.
16 Thompson, D'Arcy Wentworth, 1911, "Magnalia Naturae: or the Greater Problems of Biology." *Report of the British Association for the Advancement of Science, 80th. Meeting*. Speech presented at the Report of the British Association for the Advancement of Science, 80th Meeting.
17 Whittaker, E. T., *A Course of Modern Analysis* (Cambridge: University Press, 1920).
18 Wolfram, Stephen, "Chapter 2: The Crucial Experiment," in *A New Kind of Science* (Champaign, IL: Wolfram Media, 2002). https://www.wolframscience.com/nks/.
19 *On Growth and Form*, August 2016. https://www.ongrowthandform.org/files/2016/08/ms50127_4.jpg (accessed November 20, 2018).
20 Wolfram, Stephen, "Statistical Mechanics of Cellular Automata," *Reviews of Modern Physics*, Vol. 55, No. 3 (1983), 601–44. https://www.stephenwolfram.com/publications/academic/statistical-mechanics-cellular-automata.pdf.
21 Mandelbrot, Benoit B., *Fractals: Form, Chance, and Dimension* (San Francisco, CA: W.H. Freeman, 1977).
22 Turing, Alan M., "The Chemical Basis of Morphogenesis," *Philosophical Transactions of the Royal Society of London, B*, Vol. 237, No. 641 (1952), 37–72. https://www.jstor.org/stable/92463.
23 Stevens, Peter S., *Patterns in Nature* (Harmondsworth: Penguin Books, 1977).
24 "Letters from D'Arcy Thompson." n.d. Letters from D'Arcy Thompson. Wellcome Library. https://wellcomelibrary.org/item/b17083114#?c=0&m=0&s=0&cv=0&z=-0.6905,0.4286,2.6536,1.6669 (accessed November 20, 2018).
25 Wolfram, Stephen, "Chapter 8: Implications for Everyday Systems, Section 5: Fundamental Issues in Biology," in *A New Kind of Science* (Champaign, IL: Wolfram Media, 2002). https://www.wolframscience.com/nks/.
26 Wolfram, Stephen, "Chapter 8: Implications for Everyday Systems, Section 7: Biological Pigmentation Patterns," in *A New Kind of Science* (Champaign, IL: Wolfram Media, 2002). https://www.wolframscience.com/nks/.

2

Physics in Biology—Has D'Arcy Thompson Been Vindicated?

Evelyn Fox Keller

One hundred years ago, D'Arcy Wentworth Thompson began the first edition of *On Growth and Form* with a claim that, to us at least, seems impossible to refute: "In general no organic forms exist save such as are in conformity with physical and mathematical laws."[1] But he also claimed that "of the construction and growth and working of the body, as of all that is of the earth earthy, physical science is, in my humble opinion, our only teacher and guide."[2] And on that little word *only*, he was insistent, even while being abundantly aware that he was bucking a well-established tradition in biology. Indeed, even his contemporary, and enthusiastic reviewer, J. Arthur Thomson, demurred, suggesting that "it will be difficult to justify the word 'only'."[3]

But the word was important to D'Arcy, for it was precisely the tradition of insisting there was "something more" to biology—something that neither physics nor mathematics can explain—that was his target. D'Arcy Thompson was no vitalist. And if we want to understand why readers have been tempted to say (as his reviewer suggested they would), "Magnificent, but not biology"—we need to ask, what exactly do we mean by the term "biology"?

Indeed, the problem arises with the very introduction of a new term, a term coined to designate the study of life as a distinctive endeavor. J-B Lamarck, writing at the beginning of the nineteenth century, was one of the first, and certainly most influential, to use the term. He defined biology as a subdivision of "terrestrial physics," one that included "all which pertains to living bodies." Yet, despite explicitly placing the new subject under the rubric of terrestrial physics, and despite devoting the bulk of his *Philosophie Zoologique* (1809) to "an enquiry into the physical causes which give rise to the phenomena of life," he was also committed to the need for a separation between the subdivisions. "Between crude or inorganic bodies and living bodies," he wrote, "there exists an immense difference, a great hiatus, in short, a radical distinction such that no inorganic body whatever can even be approached by the simplest of living bodies."[4] Lamarck, it seems, was of two minds: on the one hand he vehemently rejected

An earlier version of this text was published in the December 2018, 40 (4) issue, 33–8 of the *Mathematical Intelligencer*.

the evocation of extra-natural causes: "Nature," he wrote, "has no need for special laws, those which generally control all bodies are perfectly sufficient for the purpose." Yet, he continued, "if we wish to arrive at a real knowledge of … what are the causes and laws which control so wonderful a natural phenomenon, and how life itself can originate those numerous and astonishing phenomena exhibited by living bodies, we must above all pay very close attention to the differences existing between inorganic and living bodies."[5]

Thompson, while sharing many of Lamarck's concerns, sought understanding of biological form by focusing not on the differences but on the similarities between inorganic and living bodies; and in doing so, he surely incurred the wrath of those invested in the inviolability of the gap between the two domains. One might read Lamarck as offering something of a comfort zone in which vitalists could hide (as many surely did). Thompson, however, did not allow his readers such a luxury. Faced with the choice between vitalism and mechanism, his answer was unambiguous. Mechanism could be his only guide.

But in any case, by the time Thompson embarked on his magnum opus, vitalism was already on the rout. Other debates (e.g., debates over the place of final vs. efficient causation; preformation vs. epigenesist)—closely related debates that were of even longer standing—surely continued to divide biological scientists and on these divides. Thompson also held strong views. But I suspect that none of them carried quite the same sense of mutual exclusion that attitudes often expressed in the twentieth century concerning the relation between biological and physical science. And this development, I want to argue, was a byproduct of the emergence in the twentieth century of a new science of genetics, a science that was to radically reconfigure the entire landscape of Biology.

The term "genetics" was coined by William Bateson in 1905 to refer to the study of inheritance and the science of variation, but it does not appear in the 1917 edition of *On Growth and Form*. Thompson does, however, refer to certain precursors of genetics, especially to Darwin's theory of pangenesis, and in rather scathing terms. For example, he writes,

> [I]n all such hypotheses as that of "pangenesis," in all the theories which attribute specific properties to micellae, idioplasts, ids, or other constituent particles of protoplasm or the cell, we are apt to fall into the error of attributing to matter what is due to energy and is manifested in force.[6]

Indeed, he is utterly dismissive of all efforts to locate the roots of morphogenesis in the structures of the germ cell: "In an earlier age," he wrote, "men sought for the visible embryo, even for the homunculus, within the reproductive cells; and to this day, we scrutinize these cells for visible structure, unable to free ourselves from that old doctrine of "pre-formation."

This dismissal seemed to include all those (and perhaps August Weismann especially) who

> "speak of a "hereditary substance," a substance which is split off from the parent-body, and which hands on to the new generation the characteristics of the old; we

> can only justify our mode of speech by the assumption that that particular portion of matter is the essential vehicle of a particular charge or distribution of energy, in which is involved the capability of producing motion, or of doing "work."
>
> But as Newton said, to tell us that a thing "is endowed with an occult specific quality, by which it acts and produces manifest effects, is to tell us nothing."[7]

Clearly, Thompson had little interest in such arguments. Indeed, in a debate with J. S. Haldane before the Aristotelian Society, he went so far as to acknowledge that "I for my part look forward, in faith and hope, to the ultimate reduction of the phenomena of heredity to much simpler categories, to explanations based on mechanical lines … that the special science which deals with it has at least found, in Mendel, its Kepler, and only waits for its Newton."[8]

Twenty-five years later, however, the successes of genetics could no longer be ignored. And in his second edition, Thompson acknowledges that "The efforts to explain 'heredity' by the help of 'genes' and chromosomes, which have grown up in the hands of Morgan and others since this book was first written, stand by themselves in a category which is all their own and constitute a science which is justified of itself."[9] But he himself is still not much interested:

> To weigh or criticize these explanations would lie outside my purpose, even were I fitted to attempt the task I leave this great subject on one side not because I doubt for a moment the facts nor dispute the hypotheses nor decry the importance of one or other; but because we are so much in the dark as to the mysterious field of force in which the chromosomes lie, far from the visible horizon of physical science, that the matter lies (for the present) beyond the range of problems which this book professes to discuss, and the trend of reasoning which it endeavors to maintain.[10]

So what has changed between the two editions? Clearly, Thompson's focus on the physical forces involved in morphogenesis has not changed. However great the successes of genetics, and however useful its focus on genes and chromosomes may have been in explaining heredity, he has little confidence that it will prove useful for understanding the problems of morphology. Indeed, he warns the reader that "To look on the hereditary or evolutionary factor as the guiding principle in morphology is to give to that science a one-sided and fallacious simplicity." The path he chooses is thus "to leave this great subject to one side" (1942: 340), "to rule 'heredity' or any such concept out of our present account" (1942: 284). Perhaps the most conspicuous illustration of this decision is to be seen in his discussion of the differences between smooth and wrinkled peas. Here, he saw no need to invoke Mendelian factors; instead, he wrote, "The difference between a smooth and a wrinkled pea, familiar to Mendelians, merely depends, somehow, on amount and rate of shrinkage" (1942: 534).

But if Thompson was not impressed by the achievements of genetics, neither were geneticists impressed by his own contributions. As the cell biologist J. W. Wilson remarked in his review (1944) of the 1942 edition,

> The ideas of *On Growth and Form* have played little part in the spectacular advances of biology since the book was first written. For this I think there is a very good reason: the point of view which the book represents is out of fashion and is indeed the antithesis of the one now in vogue, to which these advances have been due.

As Wilson goes on to explain:

> When the book was first published in 1917, experimentation in genetics was just beginning to produce its brilliant results ... A rich harvest has been reaped in these fields with very slight if any influence of Thompson's book ... [On] the other hand, almost every reviewer has complained that the new edition has not been influenced by this research, which is for the most part hardly mentioned ... The failure to take the results of modern biological research into consideration ... is related to the failure to make an important impact upon this research. They are both due to the antithesis of the fundamental ideas. On growth and form harks back to an older habit of thought ... it is not an essential part of contemporary biological advance.[11]

If the rise of genetics did not compel a change either in Thompson's focus or in his argument, and if Thompson's arguments had no impact on the development of that field, what IS its relevance to this history? What relation DID exist between the rise of genetics and the fate of Thompson's work? What I want to suggest is that, for the problems with which Thompson was concerned, the primary change induced by the rise of genetics was in its reconfiguration of the conceptual landscape of Biology. Where earlier biological thought had surely been shaped by the divides between epigenesis and preformation, and later, between vitalism and mechanism, the advent of genetics in the twentieth century brought with it a new kind of antithesis, one that inherited much of the oppositionality of those earlier debates, but added to that oppositionality a new sense of mutual exclusion—the need to choose between genetic and physicalist explanations. Wilson claimed that:

> The nature of this organization [i.e., of the germ] may be considered from two points of view: biologically, from the point of view of heredity, as with the trend begun by Weismann and leading to modern genetics, or mathematically and physically as in Thompson's book.[12]

Thompson's interest, he writes, "is not in the biological analysis of the organization of the germ, ... He approaches the problem from the opposite direction."[13] My question is this: what set Thompson's approach in opposition to genetics, what made it "antithetical" to contemporary developments? The separation of the biological from the physical had already been implied by the very coining of the term *biology*, to name a distinct discipline. As you all know, J. B. Lamarck was one of the first to use this term. But Lamarck had repeatedly insisted that separation need not (or at least not to him) imply opposition; one kind of analysis need not exclude the other. "Nature has no need for special laws," he wrote, "those which generally control all bodies are perfectly sufficient for the purpose."

In a similar vein, it should also be acknowledged that the separation of the animate from the inanimate was also of long standing: Thompson's complaints about the traditional reluctance of zoologists to compare the living with the dead, the reluctance to abandon the expectation of "something more," were well grounded, and in evidence long before the advent of genetics. But I want to suggest that genetics, itself often represented as a reductionist science, in fact provided support for such reluctance; I would even argue that it offered a kind of fulfillment of the expectation of a "something more." To be sure, geneticists had no use for a vital force, but they had a seductive alternative, and that alternative was the gene. I submit that the concept of the gene provided the "something more"—the crucial element that set biological organisms apart, and that seemed inevitably (perhaps necessarily?) to be missing from physicalist accounts. The gene was immortal, and it seemed all but totally resistant to the effects of physical force. Most importantly, it seemed to have what Newton referred to as an "occult specific quality, by which it acts and produces manifest effects." What was the basis of life? Not osmosis, shrinkage, torsion, or tension, but the gene. What was the source of biological characters? The answer generally proffered by classical genetics was "Gene Action." The genes to which particular characters were associated—particles that one might even imagine seeing under the microscope—that were assumed, somehow, to produce their effects by their action. Indeed, these entities ushered in not only a new science, but a new kind of reductionist explanation. Where the standard form of reductionism that had long prevailed in the physical sciences assumed that the smallest (microscopic or submicroscopic) components of the system lay at the bottom of the causal chain, genetics introduced a new kind of causal chain, one which placed genes at the bottom of the causal chain. From genes one gets proteins and RNA, from which are formed cells, then tissues, organs, and finally organisms.

Two different causal chains, both linear and unidirectional, but standing apart, each from the other. Indeed, by introducing its own causal chain, genetics introduced the appearance of an incommensurability between the explanatory goals of the two sciences. In both cases, the assumption of a linear causal chain promised explanatory reductionism, with the hope that an explanation of macro phenomena could ultimately be "reduced" to an understanding of the properties and behavior of the lowest level entities—atoms (or lower) in the case of physics, and genes, in the case of biology. But until (or unless) the properties and behavior of genes could be fully characterized in physical terms, how could these two different models of explanation possibly be integrated?

As genetics continued to triumph, and in mid-century, began to give way to molecular biology, confidence in the explanatory power of genes (or of DNA) grew ever stronger. The dramatic discovery of a code for translating DNA sequences to proteins lent particular strength to the belief in the causal power of DNA, and with that triumph, so too did the tendency to see biology and genetics as coterminous grow stronger. Indeed, the solidification of that equation was, I think, essential to the growing opposition between biology and physics.

How could this be, you might ask? After all, didn't physicists play a central role in the creation of molecular biology? Well, yes they did, but not necessarily qua physicists. Molecular genetics certainly welcomed the participation of physicists. (Especially, it welcomed their imprimatur, their prestige). But the contributions of most of these physicists (crystallographers aside) turned out to rely precious little on the application of either the tools or the principles of physics to the problems of biology. One might even ask of Schroedinger's widely celebrated contribution to the emergence of molecular biology, namely, of his offer of the notion of a "hereditary codescript" embedded in an "aperiodic crystal" as an answer to the question of "What is Life," where is the physics in this answer? Surely, the discovery of a protein code was not in itself sufficient to substantiate the notion of codescript capable of determining "the entire pattern of the individual's future development and of its functioning in the mature state." Schroedinger's attention to the subject has been credited by many for encouraging the movement of physicists into molecular biology, but how much of his influence was due to the legitimation provided by his own prestige, and how much to his demonstration of the usefulness of the concepts of physics to biology? Over the next fifty years, during the first half-century of molecular biology, I can find little evidence of the application of physics to the study of biological development and morphogenesis (especially when construing physics, as D'Arcy seems to have done, as an inquiry into the role of physical forces). Indeed, this was a period in which people interested in such subjects had an especially hard time finding support of any kind. I recall an anecdote from the late 1960s or early 1970s (a time when I myself was working in mathematical biology) told me by my colleague Mimi Koehl. Mimi, now a renowned leader in the field of biomechanics, had applied to the NSF for funding, but she'd been turned down. As she continued her account, she had just received a letter of explanation. "This applicant," wrote the reviewer, "seems to be under the impression that physics has something to do with biology." I might note here the tacit equation—already explicit in Wilson's 1942 remarks but now only grown stronger—between biological and genetic.

In mid-twentieth century, I think we can all agree that THAT equation was new. If we couple that equation with the representation of biology as alternative to (even opposed to) physics, what room is left for a project such as Thompson's—a project that chooses to leave genetics aside, and focus instead, and exclusively, on the physical and mathematical laws with which organisms must conform? To be sure, *On Growth and Form* is a work to which a great deal of homage has been paid, but for most of its readers, it remained, as Stephen Jay Gould put it, "an unusable masterpiece."

Even toward the end of the twentieth century, when work in both mathematical biology and biomechanics first began to rebound (see, e.g., Odell et al. (1981); Koehl, M. A. R. (1990)), such efforts continued to meet with puzzlement: where was the genetics? They might have asked (though I do not actually recall anyone doing so in the very early days), Is there a way we can integrate this work with what we have learned about genetics? Only in the 1990s did we begin to see the beginnings of sustained effort to bridge this gap. (I'm thinking, e.g., of the work of people like Maria Leptin [1991], Donald Ingber, and Viola Vogel; also, of the emergence of a few papers on the role of mechanical forces in the regulation of gene expression [see, e.g., the work of Emmanuel Farge].) But these efforts proved to be just the tip of the iceberg.

For the role of mechanics in gene expression, what in my view really opened up the flood gates was the turn, within molecular biology, from the search for "genes for" particular effects (in other words, the search for genetic causation), to a focus on the regulation of gene expression, and, more specifically, to the roles of chromatin, epigenetic markers, and ncRNA in this regulation.

Today we are in a new century, one in which the landscape of biological thought is once again being reconfigured. With the maturation of molecular biology, the genome has been transformed from the executive director of biological development, or, as Schroedinger put it, from "architect's plan and builder's craft" to the immensely complex physical-chemical structure that seems to need no new laws of physics, only a Herculean effort of sifting through the jungle of physical and chemical interactions responsible for what it can and cannot do. This transformation is itself largely the result of the turn of research attention from questions about nucleotide sequences coding for amino acids to the much harder questions about how particular proteins are produced in the quantities, at the times, and in the places in which they are needed for the normal functioning of the cell. We may claim to account for the difference between wrinkled and smooth peas by Mendelian factors, but where is the account of how these factors actually give rise to their effects? If genes "act," how do they act? If sequences of DNA inform, how and what do they inform?

With this shift—a shift that focuses attention on the complex relation between genotype and phenotype, we can finally return to Thompson. DNA, it has become clear, is not a mystical entity, no more so than is any other part of the organism. It is a very particular kind of physical/chemical structure, and to the extent that it can be said to "act" it can do so solely through its internal interactions (i.e., interaction among its parts) and the interaction of those parts with other physical/chemical components. What Thompson once said about the structure and conformation of organic forms can also be said about the structure and conformation of all the molecular components of organisms, including the organic forms of both DNA and what we have been calling genes: "no organic forms," he wrote, "exist save such as are in conformity with physical and mathematical laws."

How DNA is "read" for protein synthesis depends on its conformation as a physical structure. Its morphology differs from that of the forms on which Thompson focused in at least two ways: first, it is highly dynamic, and second, the processes involved operate on a molecular rather than a macro scale. Notwithstanding, it has recently become a tremendously fertile field for physicists looking to extend their horizons.

One of the most fruitful domains of investigation has been in the physical dynamics of the processes by which epigenetic markers regulate gene expression. This they do by way of regulating genomic conformation: exposed sequences of DNA can be transcribed and sequences that are hidden by the topology of the chromosome cannot. Of the many epigenetic markers we know, perhaps the most extensively studied is that of methylation. Methylation means the attachment of a methyl group to a nucleotide (cytosine may be the frequently discussed example).

This alteration in the chemical composition of a nucleotide can change the mechanical properties of the DNA (it can change, e.g., its rigidity or flexibility), and hence it can change the conformation of the genome. In turn, and, accordingly, such changes regulate the availability or non-availability of particular sequences for transcription (gene silencing). Methylation also has many other effects: e.g., it alters resistance to strand separation (necessary for both replication and transcription); it also helps regulate the positioning of nucleosomes, the basic unit of DNA packaging. Now all these changes are problems in physics, albeit on the molecular scale, where the physics involved seems to be that of polymer folding. But through their role in regulating genomic structure, they have profound—and trackable—effects on large-scale biological properties of the entire system.

There are other kinds of epigenetic markers as well—e.g., DNA supercoiling is sometimes regarded as example of a purely physical epigenetic marker, as is the condensation of chromatin into highly compacted forms known as heterochromatin. (This process too is implicated in gene silencing.) In a recent review of "The Physics of Epigenetics," the authors conclude:

> By putting the rich and diverse biological literature under the new light of a physical approach, the emerging picture is that a limited set of general physical rules play a key role in the epigenetic regulation of gene expression … Mainly, epigenetics display an intrication of physical mechanisms and specific biological entities, devised in the course of evolution to achieve an exquisitely coordinated and adaptable regulation of transcriptional activity. Our review demonstrates the need to take into account both aspects, within a dialogue between physics and biology, theory and living-cell experiments.[14]

Even more recently, yet another branch of physics has emerged as a source of tools that can be put to use to understand other aspects of genomic and cellular organization—aspects that are crucial to the dynamics of the regulation of gene expression. I think of this work as belonging to condensed matter physics. To find the literature that has begun to attract so much attention, just try Googling "phase separation in biology."

Basically, the problem is this: cells contain tens of thousands of different molecular components, the activities of which need to be coordinated in time and space to sustain cell function. This coordination is achieved by compartmentalization of molecular components into diverse membrane-bound and non-membrane-bound cell organelles that perform distinct functions. One question that has particularly confounded cell biologists is that of how such organelles are assembled and maintained. And it is here that studies of liquid–liquid phase separation have proven useful.

One of the first such efforts to attract attention comes from the Max Planck Institute of Molecular Cell Biology and Genetics (MPI-CBG), where researches have demonstrated the role of phase transitions in droplet formation in the specification of germ cell lineages. More recently yet (indeed, in the last couple of months), several papers have appeared that discuss the role of condensation and phase separation in the architecture of heterochromatin (where genes are mostly silent) and its separation from euchromatin (where sequences can be transcribed). As Mark Buchanan puts it [10], "we're learning something about how biology uses geometry [and, of course, physics] for control."

Because liquid–liquid phase separation is so sensitive to changes in environmental parameters, this work may be of particular interest for illuminating a mechanism for gene regulation by environmental cues. In doing so, it provides one of the few handles we have available for understanding the role of environmental variation in heredity, in development, and in evolution.

But will the tools that physicists have developed in other contexts prove equal to the task of providing a physical analysis of systems as complex as we find in biological morphogenesis? Almost surely not. Biology introduces us to states of matter, to material processes, that are not often seen in non-biological systems, and there is no reason to expect that available methods will prove adequate to the challenges posed by this unfamiliar terrain: new tools will almost certainly need to be developed.

Now, finally, we are brought back to the original question: Do such efforts, preliminary as they may so far have been, help to heal the divide in which Thompson was caught? Do they point to a way out of the need to choose between, e.g., a Mendelian and a biophysical account of the difference between smooth and wrinkled peas?

I think they do. Indeed, I think they do so by offering a fundamental challenge to the simplistic models of explanatory reductionism, be they genetic or atomic—models that have traditionally prevailed in both disciplines. Earlier I suggested that these models, both of them linear causal chains, may be incommensurable. But now it has begun to appear out that, in general, and even without such incommensurability, complex multi-causal systems cannot be adequately described by linear causal chains of any kind, in whatever discipline they are found. The multi-causal systems we encounter in biology seem to be especially resistant to unidirectional causal accounts because of the kinds of feedback, built into by evolution, that introduce crucial links between the different levels.

Annick Lesne is a mathematical physicist who has authored numerous efforts to model biological processes in physical terms; she is the director of a research group in theoretical biology at the CNRS. In a recent review (2013), she writes:

> Biological systems are a special instance ... of complex systems; [they are] assemblies of interacting elements where emergent features directly or indirectly modify the elements. Such a ... scheme is often termed circular causality, [a term that refers to] the coupling of bottom-up and top-down relationships, leading to self-organized and possibly adaptive behaviors. Typically, elements collectively modify their surroundings, in a way sufficient to influence back the elementary interactions, which in turn may change the collective behavior of the assembly. Only a few purely physical systems display such features.[15]

By contrast, biological systems do, and they do so typically. Indeed, it is just these features that render them what Lesne describes as "intrinsically and irreducibly multiscale processes"—in other words, that lead us to conclude that no explanatory effort that focuses on single scale can suffice. "Regulation of biological function," she writes:

> has to bridge the state of the cells and ... surrounding features ... in an adaptive way. [The] cell itself has to perform a multiscale integration. For instance, transcription in eukaryotes relies at the same time on information about DNA sequence and bound proteins, histone chemical status, ... chromatin conformation ..., nuclear localization (e.g. near a nuclear pore), cell state and its surroundings, [mechanical constraints, and so on ... ; it involves influences from each of these various levels, either directly [or indirectly] ..." "Our analysis and modeling," she concludes," should follow the same line.[16]

Similar conclusions are reached by philosophers of science Sara Green and Robert Batterman in their recent review of the rapprochement of physics and biology. Their review focuses on the multilevel causal dynamics of early morphogenesis, and they, too conclude, "There is no single approach that can account for all relevant aspects of multi-scale systems." But what does all this imply about the prospects of integrating the different approaches to morphogenesis? Here is what Green and Batterman write:

> Connecting the models formulated at different levels « does not involve a process of reduction of one model to a more fundamental one. [Rather] The different models ... are explanatorily independent but epistemologically interdependent ... The combination of models forms a pluralistic mosaic of different strategies rather than a reductive explanation." The different strategies are "explanatorily independent in the sense that they describe different processes at characteristic scales while drawing on different (and often conflicting) theoretical frameworks. At the same time, the models are interdependent in the sense that the success of one model depends on sources of information that are not explicitly represented but covered via other models or sources of information.[17]

In these formulations, reduction from one causal level to another goes out the window, effectively prohibited by the feedback between levels. But so too does the hope (or, for some, the fear) of the reduction of genetic explanations to the kinds of causality conventional in physics—that hope (or fear) is also radically undermined. Though I remain convinced that our accounts of biological effects in terms of "genes for" needs updating, and so, even I am not ready to dispense with genetic explanations. These may not belong in the linear hierarchy of explanatory reductionism traditional in physics, but they almost certainly do belong in the "pluralistic mosaic of different strategies." And this sort of mosaic might, after all, be the best we can aim for. It may not dispense with the need for genetic explanations, but it might, finally, dispense with that ancient dichotomy lurking behind that between physics and genetics, namely, the dichotomy between the animate and inanimate.

Notes

1 Thompson, D. W., *On Growth and Form*. 1st edn. (Cambridge, UK: Cambridge University Press, 1917), 10.

2 Thompson, 9.
3 Thomson, J. A., Review D'Arcy W. Thompson, *On Growth and Form*. *Nature* 100.
4 Lamarck, J. B., *Zoological Philosophy: An Exposition with Regard to the Natural History of Animals*. translated, with an introduction by Hugh Elliot Lamarck (New York: Hafner, 1809 [1963]), 194.
5 Lamarck, 202.
6 Thompson, 110.
7 Thompson, 158.
8 Thompson, D. W., *Proceedings of the Aristotelian Society*, n. s., 18 (1918), 419–61. https://archive.org/stream/proceedings18arisuoft/proceedings18arisuoft_djvu.txt (accessed April 11, 2017).
9 Thompson, D. W., *On Growth and Form*. 2nd edn. (Cambridge, UK: Cambridge University Press, 1942), 340.
10 Thompson, 340–1.
11 Wilson, J. Walter, "Review: D'Arcy W. Thompson," *On Growth and Form*. *Bull. Amer. Math. Soc*, Vol. 50, No. 3 (1944), 164.
12 Wilson, 165.
13 Wilson, 166.
14 Cortini, R., Barbi, M., Caré, B. R., Lavelle, C., Lesne, A., Mozziconacci, J. and Victor, J.- M., "The Physics of Epigenetics," *Rev. Mod. Phys*, Vol. 88 (2016), 27.
15 Lesne, A., "Multiscale Analysis of Biological Systems," *Acta Biotheoretica*, Vol. 61, No. 1 (2013), 9.
16 Lesne, 4.
17 Green, S. and Batterman, R., "Biology Meets Physics: Reductionism and Multi-scale Modeling of Morphogenesis," *Studies in History and Philosophy of Science*, Part C, Vol. 61 (2017), 33.

References

Cortini, R., Barbi, M., Caré, B. R., Lavelle, C., Lesne, A., Mozziconacci, J. and Victor, J. M. (2016). The Physics of Epigenetics, *Rev. Mod. Phys*, 88, 1–27.

Green, S. and Batterman, R. (2017). Biology Meets Physics: Reductionism and Multi-scale Modeling of Morphogenesis. *Studies in History and Philosophy of Science*. Part C, 61, 20–34.

Koehl, M. A. R. (1990). Biomechanical Approaches to Morphogenesis, *Sem. Dev. Biol*, 1, 367–78.

Lamarck, J. B. (1809 [1963]). *Zoological Philosophy: An Exposition with Regard to the Natural History of Animals*. translated with an introduction by Hugh Elliot Lamarck, New York: Hafner.

Lesne, A. (2013). Multiscale Analysis of Biological Systems. *Acta Biotheoretica*, 61(1), 3–19.

Odell, G., Oster, G., Alberch, P. and Burnside, B. (1981). The Mechanical basis of morphogenesis. *Dev Biol*, 85, 446–62.

Thompson, D. W. (1917; 1942 [2nd edn.]). *On Growth and Form*, Cambridge, England: Cambridge University Press.

Thompson, D. W. (1918). *Proceedings of the Aristotelian Society*, n. s., 18, 419–61. https://archive.org/stream/proceedings18arisuoft/proceedings18arisuoft_djvu.txt (accessed April 11, 2017).

Thomson, J. A. (1917). Review: D'Arcy W. Thompson. *On Growth and Form*. *Nature*, 100, 21–2.

Wilson, J. Walter. (1944). Review: D'Arcy W. Thompson, *On Growth and Form*. *Bull. Amer. Math. Soc.* 50(3), 163–8.

3

The Beauty of the Metacarpal

Hadas A. Steiner

Ineluctable modality of the visible ... Open your eyes now. I will. One moment. Has all vanished since?
... See now. There all the time without you: and ever shall be, world without end.
James Joyce, *Ulysses*

From the Part

The first industrialized war, the First World War, muddled the reception of D'Arcy Wentworth Thompson's now-classic work, *On Growth and Form* (1917).[1] Particularly rendered suspect was his approach to the living body through the abstract language of calculation. Ambivalence toward his methods came as no surprise to Thompson, who addressed the reaction that his effort would receive upfront. "To treat the living body as a mechanism was repugnant," he wrote in the introduction, "even ludicrous."[2] Thompson nonetheless made the case that such analytic treatment of the organism was vital to the expanded field of biology in the modern era. In a segment on the varied effects of force from the human scale to that of the bacillus, he observed: "We have come to the edge of a world of which we have no experience, and where all our preconceptions must be recast."[3] With this turn of phrase, Thompson acknowledged the sense of epistemological crisis that accompanied the unpredictable, even malignant, outcomes of mechanization. Moreover, its experimental approach to organic life suited the anxieties of the zeitgeist. Here was a mechanically driven theory of evolution that was not couched in the rhetoric of improvement, but predicated on the arbitrary circumstances of concurrence. *On Growth and Form* stood the ideology of progress on its head: it offered a teleology without the succor of a *telos*.[4] Yet a salve was embedded in the vivid nature of the eloquent language used to express the perfection of formal adaptation to the forces at hand, no matter how ordinary the example. Thompson's delightful descriptions of what he assessed as formal beauty could be addressed even to the most functional parts of what might seem like the most vulgar of species.

Despite the elusiveness of a definitive classification for what constituted the beautiful that ranged across scholarly domains, the correspondence between beauty and the suitability of form to its purpose was not novel to *On Growth and Form*. Darwin had explained the role of visual preference in terms of sexual attraction; Thompson, his critique of the author notwithstanding, was born shortly after the publication of *Origin of the Species* (1859).[5] Moreover, morphological pursuits had merged biological inquiry and aesthetic judgments for over a century, in science and architecture.[6] In the words of its instigator Johann von Goethe: "Beauty is the manifestation of secret laws of nature which, were it not for their being revealed through beauty, would have remained unknown forever."[7] Thompson's familiar premise was that beauty resulted from the effective adaptation of form to a myriad of forces that acted upon a body from within, as well as without. Beauty, however, was promoted above any ancillary status as a passive by-product of use in Thompson's analysis. The morphological system of language was masterfully deployed to support this valuation, resulting in a prose quite unlike that generally deployed in scientific literature. The redolence of language was often at its most pronounced when the subject turned to birds.[8] A flock of flamingos, for example, "wearing on rosy breast and crimson wings a garment of invisibility, fades away into the sky at dawn or sunset like a cloud incarnadine."[9] Thompson often drew on avian species to illustrate his points regarding formal optimization because of their "exquisite adaptation … to the navigation of the air."[10] From wing to plumage, birds charmed as they went about their routines. And where avian behavior had been scientifically misunderstood, as was long the case with regard to the phenomenon of migration, for example, deliberation could lead to evocative "speculations regarding lost continents, sunken islands, or bridges across ancient seas."[11]

That a reflection on birds might in the end be one on bridges was more than a literary flourish, as this essay will unfold. As birds had been honed for flight in response to the forces acting upon a moving body, so had the structure of bridges evolved to express the loads that they bore. Indeed, the bridge would return as a central focus of a chapter that would most inspire architects because it drew directly on the principles of built form. Aptly, the groundwork for that chapter, entitled "On Form and Mechanical Efficiency," was laid by avian example. Specialization for flight was described right down to the minutiae of the ligaments that run from feather to feather and the entanglements of quill to quill. Thompson appraised the "automatic mechanism" of the wing as having nothing "prettier in all anatomy."[12] As the wing was broken down to its constituent parts, Thompson's language pressed further along the aesthetic spectrum and beyond the conventional associations conjured up by the classification of "pretty." "[N]othing can be more beautiful," Thompson concluded, "than the construction of a vulture's metacarpal bone."[13] Beauty was not the attribute ascribed to flocks of birds at sunset or even the collocation of parts that together made flight possible, but one that was isolated to a wing bone of a rather homely bird. It lay concealed in the structural anatomy of a bird that scavenged its livelihood from the dead.

The particulars of this example aside, the extrapolation of a theory of the whole from the logic of a single bone—including the reverberation of that hypothesis in the political milieu—had a long-standing history in both science and architecture. Bones, whether human, mastodon, or bird, filled many books; skeletons of vertebrates charged the halls of natural history museums. The study of bones was an elaborative art in which the form of an entire animal was deduced from a solitary component, or even a fragment thereof. The potential of the extrapolation from the individuated part to the structure of the body as a whole was not lost on architects who were seeking a basis on which to form a contemporary material aesthetic. "Scientific observation led Cuvier to reconstruct a whole animal out of jaw or limb," noted the architect Eugène Emmanuel Viollet-le-Duc, just as "the artist who seeks the beautiful, or harmony, feels a greater interest in knowing how nature proceeds, and what *logical consequences are deduced from the construction of a part*."[14] With this celebration of the whole as deduced from its part, Viollet-le-Duc deviated from the traditional permutations of beauty as based in the universal laws of proportion and symmetry handed down the centuries from Vitruvius and Alberti. The novelty of iron construction demanded, according to Viollet-le-Duc, an equally new architecture in which every constituent part was embedded with the rationale that determined the overall result. Skeletons were thus related to the logic of the iron anatomies that they inspired and that were at times designed to contain them.

There is more than meets the eye to the appraisal by Viollet-le-Duc—and, of course, Thompson—of the role played by the individual element in determining the shape of the collective. Deductive theories of structure often led to deliberations over social reform. The famous generational quarrel that erupted in 1830 between the architects Antoine-Chrysostome Quatrèmere de Quincy and Henri Labrouste, for example, was over the contention that form could act as a vehicle for communal expression. The architectural debate had a more contentious and high-profile counterpart at the *Académie des sciences* where the Georges Cuvier and Étienne Geoffroy Saint-Hilaire disagreed over the conclusions of a paper regarding the possibility of a common precedent for the anatomy of mollusks and vertebrates.[15] The backdrop of the July Revolution cast a political shadow over this contentious argument regarding the shared characteristics of the two phyla and rendered the nature of environmental adaptation acutely relevant for the human condition. The prolonged duration required for shell growth in a mollusk, for example, was likened in the course of the debate to the extended course of human history. Again, the whole was extrapolated from the individual part. The growth lines of any particular shell encapsulated the stages of individual formation, but as part of an aggregate it was a record of geological time. By extension, it was argued, architectural "shells" recorded human history. At stake in this debate between competing models of biology was the potential of revolution to bring about social transformation. To this end in the mollusk the rigidity of the collective register was countered by the malleability of the body within.

When Thompson—who had published a bibliography of almost 300 pages of the scholarship on invertebrates in 1885—weighed in on the subject in *On Growth and Form*, he prioritized the compliance of tissue to form, adaptability over revolutionary change. "In short," he wrote, "it is the shell which curves the snail and not the snail which curves the shell."[16] Having made its home, the snail is then shaped by it, thus becoming the product of its own material environment. That physical force leaves its imprint on the process of structural adaptation was indeed the main point of *On Growth and Form*, a text that demonstrated the role of physical laws and mechanics on the development of form at a variety of scales—forces that had been underplayed in Thompson's view in both the theory of evolution and the comparative morphology that theory replaced. Form was *the* central concern of biology, or as Stephen J. Gould put it in an essay on the contemporary relevance of Thompson: "Form pervades the material world. Aristotle was willing to consider form without matter for prime movers and demiurges, but not matter without form."[17] For Thompson, form was a topic of "deep *theoretical* significance."[18]

Thompson's prioritization of the adaptability of form, however, differed in a meaningful way from those naturalists that preceded him who were content to draw conclusions about living creatures from the dead casings and bones that were catalogued in the museum. The extrapolation from the part to the whole was incomplete in Thompson's view without the supple, connective tissue between parts. In language that could describe a cybernetic system, Thompson wrote: "ligament and membrane, muscle and tendon, run between bone and bone; and the beauty and strength of the mechanical construction lie not in one part or in another, but in the harmonic concatenation which all parts, soft and hard, rigid and flexible, tension bearing and pressure-bearing, make up together."[19] Structural anatomy did not sufficiently acknowledge the adaptive capacities of living bodies in which the hard and the soft composed an integrated system. The implications of Thompson's work for computational models to come were recognized early on. Alan Turing, for example, would cite Thompson as one of only six sources for his classic paper on morphogenesis.[20] As the dialectic of hard and soft continued to evolve, the relation between the rigid qualities of form—now called hardware—and the soft ones of usage—now called program—would be especially of interest among artists and architects that found support for their cause on the pages of *On Growth and Form* after the Second World War. The consequences of the debate of the 1830s still lay in the balance for the digital age.

To the Whole

As part of the "harmonic concatenation" of which the vulture was composed, the distinction of its metacarpal bone stemmed from a similarity to the Warren truss, patented 1848. The Warren truss was a sequence of inverted equilateral triangles in which every element was in either tension or compression, with no detracting bending or torsional forces. While the long, thin metacarpal bone of a vulture carried little weight, its hollow construction had to evolve in such a way so as to provide the stiffness sufficient to support the wing, as well as the flexibility needed to avoid any snapping or buckling from the wind pressure produced when a bird takes flight. The engineered form of this truss, too, combined an economy of materials with the strength that made it ideal for the prefabricated construction of bridges. Thompson thus inverted the usual flow of analogy and used an engineered element to explain a biological phenomenon.

Thompson extrapolated from manmade structures—mostly bridges of different structural kinds—to how the forces of tension and compression might act upon the configuration of a bone. Bridges demonstrated a range of succinct responses to both tensile and downward force of the kind that different parts of a skeleton might endure. The ropes and wires of a suspension bridge, for instance, are in tension, while the piers at either end carry the overall weight in compression. The section thus begins with the need, as Thompson explained, "for strength of two kinds, strength to resist compression or crushing, and strength to resist tension or pulling asunder."[21] The minute latticework of which bone is composed, for example, was interpreted as a stress diagram. I-beams, H-beams and tubular struts were further galvanized to understand the morphology of the mammalian backbone, which was described as a truss that functions as a beam. Human engineering illustrated how natural processes had arrived at particular solutions.

Thompson mostly spoke hypothetically of bridge construction techniques, but a specific bridge from which Thompson felt that anatomists should learn a lesson was the cantilevered Forth Bridge at Queensferry, Scotland. The Forth Bridge, designed by Benjamin Baker with Sir John Fowler and completed in 1890, was notable for several reasons. Firstly, it was the longest single cantilever bridge in the world until it was superseded by the Quebec Bridge in 1919. It was also the first major structure to be constructed of mild steel in Britain. In addition to the technical challenge of providing for an 8000-foot span that rested on a silty foundation, the engineers were asked to represent the expansion of the east coast route in such a way that could be lucidly reproduced on posters and bank notes at a time when British engineering was losing ground to its French and German counterparts. Finally, a recent and dramatic failure of a nearby suspension bridge had rendered that prototype suspect.

Thompson saw the highly legible and easily calculable stress pattern of this cantilever bridge as a demonstration of the consequences of force on a monumental scale. As Michael Baxandall has articulately summarized:

> In the cantilever piers such as the three at Queensferry the top lateral members are in tension, the bottom lateral members are in compression, as are the central vertical members. Baker decided that the strongest form for the girders in tension would be the lattice girder composed of L-section beams and the strongest for girders in compression would be the circular tube. This decision is clearly a basic and determining element in the rationale of the middle forms of the structure. Each cantilever tells its story of tension and compression in this vocabulary.[22]

The bridge, in short, concisely corresponded to Thompson's model of fitness of a body to perform the functions that are advantageous to its existence. In his elaboration of the structural role of each element, Thompson applied the analytical techniques of engineering to organisms, thus relating the forms of hollow girders and tubular struts to those of stems, and, repeatedly, of the bones of birds. By using an inorganic example to illuminate the function of organic structure, Thompson emphasized the fact that for him the purpose of understanding form was not in order to trace ancestry, as it had been for the unified theories of nature of Goethe or Saint-Hilaire. Rather, Thompson was looking at form for the purpose of isolating "designs for modern existence."[23]

While he was not advocating a methodology that captured the ur-forms of the past in the present, Thompson was also wary of offering up form by way of objective. "In Aristotle's parable, the house is there that men may live in it; but it is also there because the builders have laid one stone upon another," Thompson wrote. "All the while, like warp and woof, mechanism and teleology are interwoven together, and we must not cleave to the one nor despise the other; for their union is rooted in the very nature of totality."[24] The nature of this totality writ large has served to shift the relevance of *On Growth and Form* from the field of biology, where for the most part the concept of good design had been abandoned in favor of the microscopic and genetics, to that of the digital, where the unity of the organic and the inorganic across physical scales resonates especially now that the technology has caught up with Thompson's insights.

For all the celebration that the Forth Bridge encountered upon its completion, the structure was judged by William Morris to be the "supremist specimen of all ugliness"—in other words, the antithesis of what, for the very same reason, Thompson had judged as the highest beauty in the metacarpal.[25] And, indeed, the designers had a bone to pick with Morris of the form-follows-function variety, arguing that "Mr. Morris would judge the beauty of a design from the same standpoint, whether it was for a bridge a mile long, or for a chimney ornament."[26] Beauty, Baker continued, was only ascertainable from knowing the function of an object, without which a complex body could be judged by the same terms as a decorative fragment. Force, then, like beauty, had "no independent objective existence;"[27] it is an inherent in form, which is itself dependent on matter.

Thompson drew the conclusion that fragmentation exaggerates the properties of its parts. To "divide the body into organs, the skeleton into bones," or even judgment from the mind, *deforms* the outcome:

> we may go so far as to say that even bones themselves are only in a limited and even in a deceptive sense, separate and individual things. The skeleton begins as a *continuum*, and a *continuum* it remains all life long ... A bridge was once upon a time a heap of pillars and rods and rivets of steel. But the identity of these is lost, just as if they were fused into a solid mass, when once the bridge is built; their separate functions are only to be recognized and analyzed in so far as we can analyze the stresses, the tensions and the pressures.[28]

A bridge disassembled is no longer a bridge; a bone is not a body; a truss is only significant as part of a "diagram of forces." This was the "dynamical aspect" of morphology, which Thompson also called "the operations of Energy," and is the quality by which form could transcend the theoretical limitations of the purely material. "The biologist, as well as the philosopher, learns to recognize that the whole is not merely a sum of its parts," Thompson concluded. "It is this, and much more than this."[29] Morphology's whole is much, much more than a bundle of individual histories. Organization itself becomes—and transcends—metaphysical conception.

Notes

1 And the expanded second edition came out in 1942, right in the midst of the next world war in which the mechanization of death was pushed to an extreme.

2 Thompson, D'Arcy Wentworth, *On Growth and Form*. A new edition. (Cambridge at the University Press New York: The Macmillan Company, 1945), 2.

3 *On Growth and Form*, 1945, 77.
4 "But in this last, and very important case, we have reached a teleology without a *telos*, as men like Butler and Janet have been prompt to shew, an 'adaptation' without 'design,' a teleology in which the final cause becomes little more, if anything, than the mere expression or resultant of a sifting out of the good from the bad, or the better from the worse, in short of a process of mechanism." OGF, 6.
5 Darwin wrote the introduction to Thompson's 1883 translation of Hermann Mueller's, *The Fertilization of Flowers by Insects*. For more on the subject of beauty, see Prum, Richard O., *The Evolution of Beauty* (New York: Doubleday, 2017).
6 Thompson acknowledged his debt to Johann von Goethe's theories of rational morphology that extended beyond external contours to internal structure and the structural relationship they brought to bear on each other.
7 quoted in Nisbet, H. B., *Goethe and the Scientific Tradition* (London: Institute of Germanic Studies, University of London, 1972), 35.
8 Note Thompson's glossary of Greek Birds here.
9 OGF, 959–60.
10 OGF, 966.
11 OGF, 3.
12 OGF, 963. As J. D. Watson wrote of the double helix, that it was "too pretty not to be true." (*The Double Helix* (New York: Simon & Schuster, 1968), 38).
13 OGF, 981.
14 Viollet-le-Duc, E.-E., *L'histoire d'un dessinateur comment on apprend a dessiner* (Paris: J. Hetzel, 1879), trans. by Benjamin Bucknall as *Learning to Draw* (London: Sampson Low, Marston, Searle & Rivington, 1880), 226.
15 See Paula Young Lee, "The Meaning of Molluscs: Leonce Reynaud and the Cuvier-Geoffroy Debate of 1830, Paris," *The Journal of Architecture*, Vol. 3, No. 3 (1998), 211–40, DOI: 10.1080/136023698374189. See also Appel, Toby A., *The Cuvier-Geoffroy Debate: French Biology in the Decades before Darwin* (Oxford: Oxford University Press, 1987).
16 OGF, 187.
17 Gould, Stephen J., "D'Arcy Thompson and the Science of Form," *New Literary History*, Vol. 2, No. 2 (Winter, 1971), 231.
18 OGF, 958.
19 OGF, 969.
20 Turing, Alan M., "The Chemical Basis of Morphogenesis," *Philosophical Transactions of the Royal Society: B*, Vol. 237 (1952), 37–72. For an account of Thompson's inspiration, See Boden, Margaret A., "D'Arcy Thompson: A Grandfather of A-Life," *The Mechanical Mind in History*, edited by P. Husbands, O. Holland and M. Wheeler (Cambridge, MA: MIT Press, 2008).
21 OGF, 967.
22 Baxandall, Michael, *Patterns of Intention: On the Historical Explanation of Pictures* (New Haven, CT: Yale University Press, 1985), 20.
23 Gould, 233.
24 OGF, 7.
25 Baxandall, 24.
26 Baxandall, 24.
27 OGF, 15.
28 OGF, 1018.
29 OGF, 1019.

4

"Drawn from Structures Living and Dead"—Art Collections and Connections, Growing and Forming

Matthew Jarron

In January 1901, the twenty-one-year-old artist George Dutch Davidson wrote to his mentor, John Duncan: "Professor D'Arcy Thomson [*sic*] came up and introduced himself to me, and asked me to decorate a space above his fire-place in the College. There will be three fair-sized panels: an Orpheus, a Neptune and a Juno."[1] As it turned out, this would be the last letter Davidson wrote before his untimely death, and the commission was never completed, but it provides clear evidence of D'Arcy Thompson's engagement with contemporary artists, and his part in a fascinating nexus of art and science that had developed in Dundee at the turn of the century. A key starting point for this was Thompson's close friend and colleague at University College, Dundee (UCD), Patrick Geddes (1854–1932).

Geddes came to Dundee in 1888 to assume the Chair of Botany at UCD that had been specially created for him and endowed by an important patron of the arts and sciences in Dundee, James Martin White. Born in 1857, White had started his career in the family textile business, J.F. White & Co, before inheriting the estate of Balruddery on the death of his father in 1884. A Fellow of the Royal Physical Society, White's scientific endeavors included lighting his house with electricity.[2] He became actively involved in educational causes, and was a keen supporter of UCD, which had been founded in 1881. White's wife, Mary Macrae, was a painter and the couple were also art collectors, owning works by G.F. Watts and Gaston la Touche, among others, as well as a notable collection of oriental art.[3] He would later become Honorary President of Dundee Art Society, his wife already being a professional member.

White was also an active member of the Dundee Naturalists' Society and would have known Geddes earlier from his lectures to the Society, the first of which was in January 1881. One of Geddes's biographers describes them as boyhood friends.[4] Certainly as a youth Geddes attended science classes at the YMCA in Dundee and may have met White then. In 1884 Geddes had applied for the Chair of Biology at UCD but lost out to D'Arcy Thompson, though this did not prevent them from being good friends later.[5] White was evidently determined to bring Geddes to Dundee, and this part-time chair seemed the ideal way. Geddes was required to be present only for the summer term, leaving him free to pursue his many other interests during the rest of the year.

Although Geddes had trained as a biologist, he was already developing much wider political and sociological ideas by the time he took up the Dundee chair. His published papers by that time included "An Analysis of the Principles of Economics," "Conditions of the Progress of the Capitalist and of the Labourer," and "Co-operation vs Socialism."[6] He had also developed unusual abilities as a visual thinker, using sheets of folded paper to classify ideas into tables that he called "thinking machines"—a legacy from a period of temporary blindness that he suffered while in Mexico in 1879–1880.

It is not surprising, therefore, that Geddes took a swift interest in Dundee's increasingly vibrant art culture. In 1890 the city was described by one leading Scottish painter as "perhaps the most vital centre of art appreciation in Scotland."[7] A museum and art gallery opened in the Albert Institute in 1873, and became home to major art exhibitions that were the biggest of their kind outside London, attracting artists across the country. At the same time, Dundee was home to one of the country's largest newspaper industries, and in 1880 the *Dundee Advertiser* became the first daily paper in Britain to employ a regular staff artist. Art education was also expanding at that time thanks to the Dundee Technical Institute founded in 1888, out of which would grow Dundee College of Art (now Duncan of Jordanstone College of Art & Design).[8]

Dundee's artists were also coming together to promote their work, in particular through the Graphic Arts Association (GAA) founded in 1890. Geddes took a quick interest in this society, becoming an Associate Member in 1891. In May that year he approached the society with a request to create a series of oil and watercolor paintings of plant life that he could show to his students. One of the members who approved the scheme was the painter John Duncan (1866–1945), so if he had not already met Geddes, he must surely have done so at the GAA meeting of June 1, 1891, which Geddes hosted at UCD.

Under Geddes's influence, Duncan would go on to become the leading artist of the Celtic Revival in Scotland. Geddes had become convinced that the study of biological systems could help in understanding the evolution and complexities of city life. When not teaching at UCD, he was spending much of his time on social and urban renewal projects in Edinburgh's Old Town. Geddes believed in the principle that "cities flourished or declined according to the people who lived in them."[9] As a first step toward practical regeneration of the area he left his comfortable New Town flat and moved into a dilapidated tenement in the Old Town in 1886, which he and his wife proceeded to renovate. In 1887 he started hosting annual Summer Meetings, which brought together leading intellectuals from various countries "interested in the reconciliation of specialisms with synthesis of knowledge."[10] The programs featured lectures, seminars, and excursions, and embraced both science and the arts. Duncan is known to have attended from 1891 and would soon be asked by Geddes to take charge of the art content of the meetings.

Initially, Geddes channeled many of his ideas for renewal through the Edinburgh Social Union, which championed the idea of bringing art and decoration into the lives of working people. A key starting point for this was the Arts & Crafts movement and the ideas of William Morris, but Geddes believed that Morris had failed in his intentions to give art to the people, recognizing that his products were too expensive for the average working man to afford.[11] By 1892, Geddes had distanced himself from the Arts & Crafts movement, believing it to be "essentially dominated by capitalistic consumption" and had also left the Edinburgh Social Union.[12] He would now pursue his own cultural agenda, the "Celtic Renascence," with John Duncan as his principal ally.

Geddes had observed the social and cultural developments that had taken place in Ireland and Wales.[13] Although there had been some development in Scotland (particularly the popularity of the Kailyard school of literature), it had largely continued the trend begun during the Enlightenment and then made fashionable by Walter Scott of re-imagining Scotland as "North Britain." Geddes, who was generally opposed to Enlightenment thinking as symptomatic of the New Town rather than the Old, preferred to stress Scotland's pre-Union Celtic culture, which linked it internationally to Ireland and Continental Europe. He hoped that through remembering and celebrating the past, people's outlook for the future could be strengthened – "in some young soul here and there the spirit of the hero and the poet may awaken" – and it was this that he referred to as "our Scottish, our Celtic Renascence."[14]

Geddes found an ideal collaborator in Duncan. Born and educated in Dundee, he had quickly developed an interest in scenes from mythology and fairy tales, and during further study on the Continent he became fascinated by the symbolist art that was then much in fashion. Geddes encouraged Duncan to draw on traditional Celtic stories and songs, linking them visually with the unique and distinctive style of decoration that Duncan would already have known from the many Pictish stones in the countryside north of Dundee. In 1892, Duncan moved to Edinburgh, where he was soon hard at work creating an elaborate series of murals for the halls of residence which Geddes founded, as well as other Old Town buildings. He also acted as art editor for the centerpiece of Geddes's Celtic Revival movement, the seasonal publication *The Evergreen*. Distinctively Scottish, it was also truly international, featuring Irish, English, American, French, and Dutch contributors, and from a wide range of disciplines, including biology and the social sciences. Geddes also appointed Duncan as director of his Old Edinburgh School of Art, founded to teach Celtic design and ornament.

By 1897, most of Geddes's Edinburgh projects had come to an end due to lack of money, and Duncan returned to Dundee, where he encouraged his fellow artists and designers to embrace the Celtic Revival in their work. Many of Dundee's leading artists, including Stewart Carmichael, Alec Grieve, and George Dutch Davidson, had also studied on the Continent and shared Duncan's love of symbolism. A distinctive style of Celtic-infused symbolist art developed in Dundee which was shown off (usually to a fairly hostile critical reception) at GAA exhibitions.

Davidson in particular worked closely with Duncan, sharing his studio in Albert Square and enthusiastically embracing Celtic design. Fellow Dundee artist David Foggie recalled: "Under John Duncan's teaching he became enthusiastic in the study of Celtic art, a style fascinating to him from its essential decorative character and its weird beauty; he liked to feel a personal relationship with it, and often associated his own imaginative gifts with the thought that his ancestors in some far back time were Highlandmen."[15]

Born in 1879, Davidson had originally intended to follow his father in becoming an engineer, but a severe bout of flu in 1896 left him with a serious heart condition which prohibited him from working. He turned to watercolor painting and created some astonishingly original pieces including an intense psychological self-portrait and some purely abstract paintings that now seem way ahead of their time (all now in The McManus: Dundee's Art Gallery & Museum).

The commission from D'Arcy Thompson would have been the most ambitious undertaking of Davidson's career: three decorative panels for his study at UCD featuring figures from Classical mythology surrounded by representative animals. Davidson's sudden death at the age of just twenty-one meant these were never completed, and all that survives today is a pencil study for the Orpheus panel (now in The McManus: Dundee's Art Gallery & Museum) (Figure 4.1). In its use of the four elements, the twelve animals of the zodiac, and possible references to the Golden Ratio, it shows that Davidson clearly understood Thompson's interests in natural history, mathematics, and Classics, and is the first example of an artist drawing inspiration from his work—though many more would follow.

Thompson never sought a replacement for this project, but he clearly had an interest in supporting local artists. On taking up his post at UCD in 1885, he employed

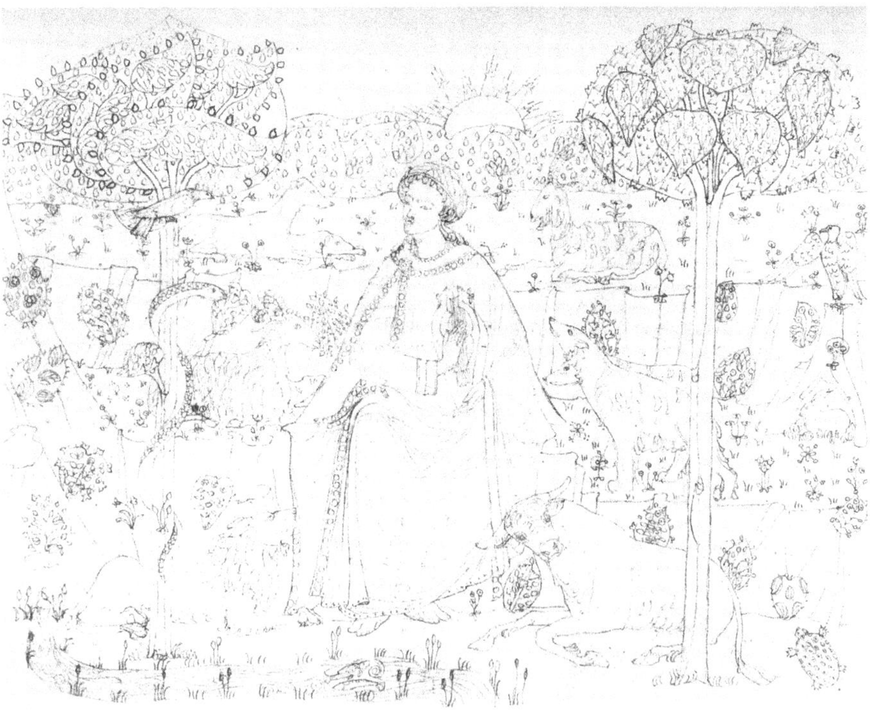

Figure 4.1 George Dutch Davidson, *Orpheus*, pencil study for decorative panel, 1900, reproduced in *George Dutch Davidson 1879-1901—A Memorial Volume* (1902), Dundee: Graphic Arts Association.

the designer and illustrator James Eadie Reid to create drawings and diagrams for him.[16] This seems to have been a full-time post since Reid gave up his job as a newspaper artist with the *Dundee Advertiser* in order to take it, but it evidently did not last long as Thompson would later use various other illustrators for his publications, mostly amateurs. Madge Valentine, for example, drew diagrams for his *Glossary of Greek Birds*.[17] Thompson also commissioned teaching models from local sculptors J. Gonnella & Co as well as companies further afield such as Fric of Prague.

Thompson soon got to know the art community in Dundee. In 1886 he chaired the annual prize-giving festival of Dundee School of Art, which must have been a somewhat awkward affair since the prizes themselves had not arrived yet![18] In his speech, however, Thompson extolled the virtues of what we would now call an interdisciplinary education in art and science. In 1917, he gave the introductory speech at the opening of a Dundee Art Society exhibition, referring to "my artist friends," but saying that "I have neither technical nor historic knowledge of art to impart to you, and nothing in the world but my share of that love of art and of natural beauty which brings us all here together."[19]

In one area, however, Thompson did have a detailed knowledge of art, one which combined his knowledge of zoology and Classics. In 1898, he presented a paper to the Royal Society of Edinburgh on "The Emblems of the Crab in Relation to the Sign Cancer," one of a number of talks given about the symbolic representation of animals in classical and ancient art. Two years earlier Geddes had written to him saying, "I am glad to see that you are holding forth on Symbolism, and write to remind you of my suggestion that you should contribute a short paper to the Summer [issue of the] *Evergreen* on this subject."[20] Sadly Thompson never took Geddes up on this invitation.

In 1910, Thompson became the prime mover to have the British Association for the Advancement of Science return to the city for the first time since 1867, convincing the Lord Provost to hold a public meeting to discuss the matter.[21] Two years later his efforts were rewarded, and Thompson not only acted as the local organizing secretary for the meeting and vice president of the zoology section, but also served on the Fine Art Committee that organized the accompanying exhibition at the Victoria Galleries. Four engravings from his own collection were included in the show—a reproduction of Raphael's *Madonna and Child and the Infant St John* by Francesco Bartolozzi (1727–1815); *A River in France* by Gaston Coindre (1844–1914) and two etchings (*London River* and *Dundee from Tayport*) by Frank Laing (1862–1907), a local artist with an international reputation.[22] Exactly what else Thompson's collection consisted of is unclear, but his very traditional tastes can clearly be seen from surviving photographs of Gowrie Cottage (Figure 4.2), a small house in Barnhill, just outside Dundee, where Thompson moved after his marriage in 1901. Although the walls are lined with pictures, a closer inspection suggests that almost all of them are reproductions. Perhaps he started to acquire originals only later in life, or perhaps such purchases were just rare exceptions.

In 1933, by which time he had moved to St Andrews, Thompson wrote to Stanley Cursiter, director of the National Gallery of Scotland, to help identify a painting he had acquired; Cursiter replied suggesting it was a Flemish work of the 1680s.[23] An invoice from the art dealers Doig, Wilson & Wheatley in 1935 refers to a picture acquired from

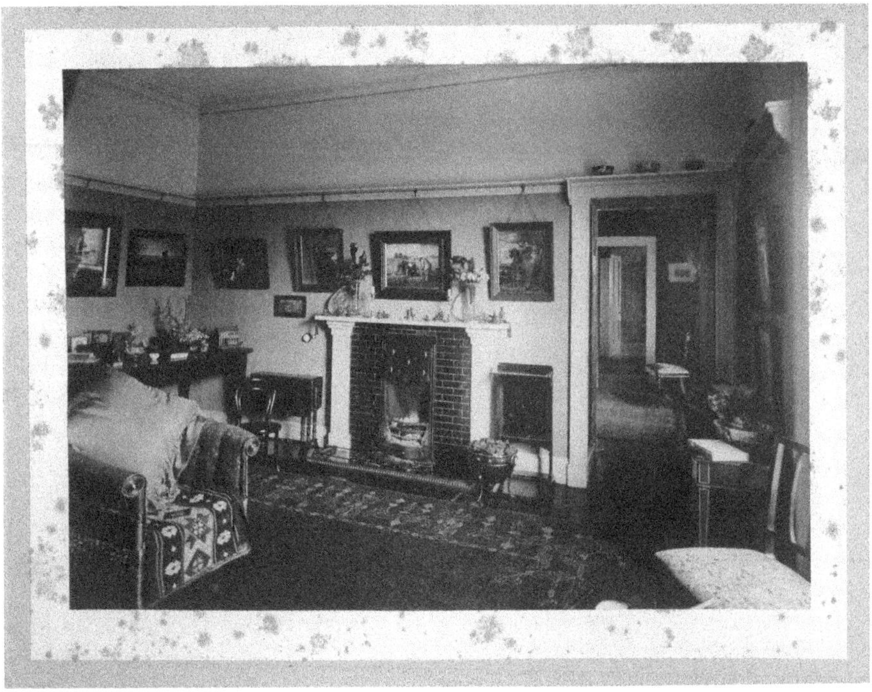

Figure 4.2 Interior of D'Arcy Thompson's home at Gowrie Cottage, *c.* 1901. Courtesy of University of St Andrews Library Special Collections.

the Society for Scottish Artists exhibition as well as the varnishing of a "small Sienese painting," again suggesting an interest in the Old Masters.[24]

Regarding contemporary artists, in 1913 Thompson employed one of Dundee's leading figurists, David Foggie, to do a portrait of his children,[25] and in 1936 would sit for a portrait himself by Foggie, which, according to one reviewer, "captures something of the overflowing kindliness of that genial personality."[26] As his reputation grew in the 1930s and 1940s he was often asked to be the subject of portraits, including sculptural pieces by Gilbert Coleridge (1859–1953) and Alfred Forrest (1909–1995) and a drawing by Hubert Freeth (1913–1986). Most notable, however, were the portraits by Glasgow painter David S. Ewart (1901–1965), for whom Thompson sat (or rather stood) in 1938. Ewart would ultimately produce three versions of the painting: one for the Royal Society of Edinburgh (of which Thompson was elected President in 1934), one for the University of St Andrews, and one for UCD (Color Plate 1). "I had a magnificent subject" Ewart wrote, "and no artist would have felt anything but inspired."[27]

Throughout his life, Thompson befriended and corresponded with artists. This geniality dates back to an exceptional childhood, when several of Scotland's leading artists were close friends of his family. These included the landscape painter Sam Bough (1822–1878), the portrait and genre painter Robert Herdman (1829–1888), the painter and photographic pioneer David Octavius Hill (1802–1870), the historical and mythological painter Joseph Noel Paton (1821–1901), whose son Diarmid was at school with Thompson and would later become a notable physiologist, and the painter and sculptor George Clark Stanton (1832–1894), who was married to Thompson's aunt (Figure 4.3). Later correspondence from the engraver Henry Macbeth-Raeburn (1860–1947) suggests that he too was a childhood friend.[28]

In later years, he knew well the English realist painter Dame Laura Knight (1877–1970). In 1931, Knight wrote to thank him for a walk he had taken her on featuring lots of ducks![29] He also corresponded with, among others, the English painter William Rothenstein, the Irish-based painter Elizabeth Rivers, and the Scottish painters John

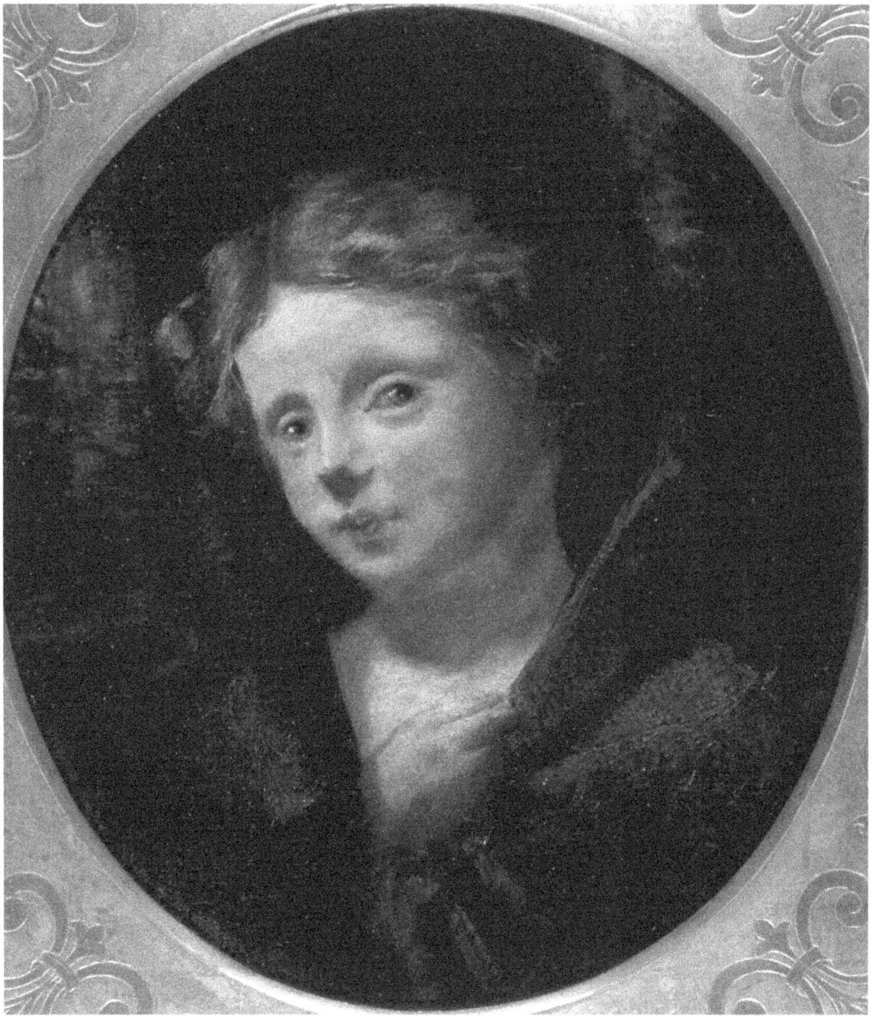

Figure 4.3 George Clark Stanton, Portrait of D'Arcy Wentworth Thompson aged five, 1865. Courtesy of University of St. Andrews Museum Collections.

Bowie and Keith Henderson. Other notable figures from the art world that he was in contact with included Daniel V. Thompson of the Courtauld Institute, Paul Lambotte of the Musée des Beaux-Arts in Paris, and Beaumont Newhall of the Museum of Modern Art in New York.[30]

There is no doubt that Thompson's love of art and his affinity with artists contributed to the interdisciplinary nature of his work. As a new and small institution, UCD was peculiarly suited to encouraging crossovers between the disciplines. Many of Thompson's colleagues shared his artistic interests. The anatomy professor A. M. Paterson, for example, taught a course on Artistic Anatomy and was strongly outspoken on the current state of art education and appreciation.[31] The mathematics professor J. E. A. Steggall was a keen amateur photographer and served on the council of Dundee Art Society. Most passionate, however, in uniting art and science was Patrick Geddes, whose teaching "was somewhat of a shock to the student who came expecting to get notes which he could learn by heart and recite at his forthcoming examination." According to one former student, a typical lecture "might quite as likely have been Ancient History or Fine Art or Political Economy, as a note about the structure or habits of one of the prescribed plants."[32]

To modern readers, Geddes and Thompson may seem visionary in their insistence on interdisciplinary or transdisciplinary approaches to education. For them, such holistic thinking was connected to a long tradition going back to antiquity. Increasingly, however, they were swimming against the tide. As Thompson himself recognized in 1903, "there is a cry in the land for specialisation," which soon affected the courses he was required to teach. "I grow more and more impatient of the customary syllabus and the conventional examination," he claimed. "The great principles of Biology are pleasant to teach and good to learn, but petty details are so much easier to examine upon." He demanded that each of his students should be "reinforced by a breadth of knowledge beyond his own specialty."[33]

Like Geddes, Thompson also loved to bring art and the Classics into his teaching, but it is clear that the two disagreed on how far this could be taken. "I thoroughly recognise the contrast you urged between the two methods of Biological teaching of Art and Science respectively," Geddes reassured Thompson in 1894. Geddes continued, however, "In Dundee I have been obliged (I thought by force of circumstances) to specialise too much on the scientific side."[34] Two years later Thompson complained that "the time has gone by for the popular or artistic work which you have hitherto encouraged," but Geddes insisted that "here I must continue my ways. Popular and artistic teaching are specially and peculiarly for a chair of botany, beyond all other chairs of science ... Furthermore in our town and college there is no point at which any of us can be more obviously, practically, & industrially useful, than at this one of mine—the botanist as designer."[35]

Geddes later wrote a series of articles in the *Dundee Advertiser* titled "Recent Art Movements in Dundee," the last of which included a plea for scientists to learn from artists:

> where can we look for real light, for practical leading, than to the arts? But to these not merely ... in their humbler and more mechanical forms, but their supreme

> ones, the arts of co-ordination and of expression, the arts of skill, the art of seeing. And who ... has ever expressed an ideal in most manifest and enduring yet subtlest form, in fullest mastery of matter, like the sculptor? Who has ever seen things real, and shown them as they truly are, like the painter, and who else can so worthily display them as they may be? ... [As] Science herself rises from her analyses of nature considered as dead, to comprehend everything from actual evolution in nature and man and, in his cities, all discerned as living, she will be first to claim for art a renewed and higher place.[36]

Although Geddes is not directly referenced in *On Growth and Form*, his influence pervades it. Throughout the book, Thompson writes in aesthetic terms about the organisms and mathematical patterns he describes, and frequently uses artistic analogies, comparing organic forms to the work of the potter or the glassblower. In contrast to most of the scientific textbooks that preceded it, *On Growth and Form* is richly illustrated. Indeed its illustrations have arguably become more influential than its text. It is important to note here that despite his love of art, Thompson himself was no artist, and relied on others to create illustrations for him. For *On Growth and Form* these were principally the work of his assistant at UCD, Doris Mackinnon (1883–1956), and one of his former students, Helen Ogilvie (1880–1960).

It is no surprise that many of the others whose work helped to shape *On Growth and Form* also brought together art and nature. These included Ernst Haeckel (1834–1919), the German biologist whose extraordinary illustrations of radiolaria in *Art Forms in Nature* (1899–1904). Thompson drew on liberally; indeed, he owned a set of plaster models of radiolaria based on Haeckel's artworks.[37] Another important precursor was the art critic Theodore Cook (1867–1928), whose books *Spirals in Nature and Art* (1903) and *The Curves of Life* (1914) contained an admirably wide range of examples, although Thompson dismissed the "mystical conceptions" of those like Cook who saw in the logarithmic spiral "a manifestation of life itself,"[38] and bluntly dismissed the notion that the sequence of numbers dictating it converged on the Golden Mean as "a mathematical coincidence devoid of biological significance."[39]

Also relevant here is Thompson's friendship with the Danish artist and paleontologist Gerhard Heilmann (1859–1946). Having abandoned his medical studies to become a professional artist, Heilmann's series of papers on the origin of birds (1913–1916) were dismissed by the biological establishment in Denmark but found an enthusiastic welcome from Thompson, who described them as "beautiful and original" and began a correspondence which led to Heilmann contributing several of the celebrated transformation diagrams in *On Growth and Form*.[40]

These iconic diagrams also relied on Thompson's knowledge of art, taking as his starting point Albrecht Dürer's work on geometry and proportion. In his *Four Books on Human Proportion* (1512–1528), according to Thompson, "the manner in which the human figure, features, and facial expression are all transformed and modified by slight variations in the relative magnitude of the parts is admirably and copiously illustrated."[41] As well as dealing with transformations of the human form, Dürer's work on proportion also encompassed his theories of ideal beauty, claiming that, while beauty was ordered by as-yet undefined laws, it was not an objective concept but was based on the infinite variety of nature. This would surely have struck a chord with Thompson, whose work aimed to reveal the fundamental laws that governed the extraordinary variety of nature's patterns.

Reviews for *On Growth and Form* appeared in publications as diverse as *Country Life*, the *Times Literary Supplement*, *Veterinary Review*, and *Engineering*. Cambridge University Press could afford to send out only a small number of review copies, and in November 1917 they wrote to Thompson to say that the *Journal of Decorative Art* had asked for a copy. Thompson replied to say: "It does not seem to me that it is a publication which is of much interest or importance to us," clearly suggesting that he had no idea of the kind of artistic significance his work would have.[42] It was only toward the end of his life that Thompson began a correspondence with the art critic Herbert Read (1893–1968), who arguably did more than anyone to raise awareness of *On Growth and Form* among painters, sculptors, and designers. In 1942, Read told Thompson, "I wonder if I ever thanked you for the enlightenment I got from your book '*On Growth and Form*': it helped me where perhaps you never intended it to help—in the understanding of art."[43]

As for Thompson's opinions of the modern abstract painters and sculptors of the 1930s and 1940s who were so inspired by his work, it is unlikely that he would have been a great enthusiast, though it does at least seem like he tried. In 1946, he wrote to Read praising his book *A Coat of Many Colours* and saying, "I envy you your knowledge of, and your sympathy for, a number of modern men whom I have had all too little patience to study and understand … Ben Nicholson [is] within my ken. But I have got along without Picasso, easily enough."[44]

Notes

1 Quoted in an appreciation in the *Dundee Advertiser* January 10, 1901—the same letter is reproduced in *George Dutch Davidson: A Memorial Volume* (Dundee: Graphic Arts Association, 1902) but with Thompson's name excised.

2 *The Piper o' Dundee*, March 28, 1888.

3 White regularly lent paintings to the Graphic Arts Association exhibitions, and his oriental collection was exhibited at the Victoria Galleries in 1899–1900.

4 Boardman, P., *The Worlds of Patrick Geddes* (London, Henley & Boston: Routledge & Kegan Paul Ltd, 1978), 78.

5 Some of the various and complicated causes of this are discussed in Macdonald, M., "Patrick Geddes—Science and Art in Dundee," in *The Artist & the Thinker—John Duncan & Patrick Geddes in Dundee*, edited by M. Jarron (Dundee: University of Dundee Museum Services, 2004), 13–29.

6 For full bibliographic details, see Meller, H., *Patrick Geddes: Social Evolutionist and City Planner* (London & New York: Routledge, 1990).

7 W. D. McKay, quoted in the Dundee Graphic Arts Association Annual Report 1890 (Dundee Central Library, Local History Centre).

8 For more on this period see Jarron, M., *Independent & Individualist: Art in Dundee 1867–1924* (Dundee: Abertay Historical Society, 2015).

9 Meller, op. cit., 70.

10 Boardman, op. cit., 129.
11 Geddes, P. (*c.*1893– 1894), "The Work of the Art School" annotated manuscript (University of Strathclyde Archives, T-GED January 5, 2018).
12 Ibid.
13 See Meller, op. cit., 63–4.
14 Geddes, P., "The Scots Renascence," *The Evergreen* (Spring 1895), 139.
15 Ibid., 4.
16 *The Piper o' Dundee*, August 7, 1889.
17 Madge was a close friend of Thompson's star student, Mary Lily Walker, and was married to W. R. Valentine, a life governor of UCD.
18 *Dundee Courier*, December 29, 1886.
19 Typed MS of address by D'Arcy Thompson (University of St Andrews Library Special Collections MS 45704). In 1924 Thompson was made an Honorary Member of Dundee Art Society.
20 Letter to D'Arcy Thompson January 25, 1896 (University of St Andrews Library Special Collections, MS 16359).
21 *British Association, Dundee 1912: Handbook and Guide to Dundee and District* (1912), Dundee: David Winter & Son.
22 *Illustrated Catalogue of a Loan Collection of Paintings, Watercolours & Engravings in the Victoria Art Galleries, Dundee on the occasion of the British Association Meeting* (1912), Glasgow: University Press.
23 Letter from Stanley Cursiter to D'Arcy Thompson as above, July 11, 1933 (University of St Andrews Library Special Collections MS 13031/1).
24 Invoice from Doig, Wilson & Wheatley January 1935 (University of St Andrews Library Special Collections MS 46483).
25 The commission is noted in Foggie's accounts book (held by the artist's family). The portraits are untraced.
26 From a review of the Dundee Art Society exhibition dated March 20, 1936 in volume four of Stewart Carmichael's scrapbooks (Dundee City Archives). The portrait is untraced.
27 Letter from D. S. Ewart to A. D. Peacock November 22, 1948 (University of St Andrews Library Special Collections MS 48539).
28 Letter from Henry Macbeth-Raeburn nd (*c.*1841) (University of St Andrews Library Special Collections MS 18222).
29 Letter from Laura Knight July 28, 1931 (University of St Andrews Library Special Collections MS 25404).
30 All letters in the University of St Andrews Library Special Collections—see index at https://www.st-andrews.ac.uk/media/special-collections/documents/Darcy%20Wentworth%20Thompson%20Index.pdf.
31 "Artistic Anatomy", *The College*, February 1890.
32 Obituary of Patrick Geddes in *The College,* June 1932.
33 Thompson, D. W., "Address Delivered at the Opening of Session 1903–4," *The College,* December 1903.
34 Letter from Patrick Geddes May 28, 1894 (University of St Andrews Library Special Collections, MS 16362).
35 Letter from Patrick Geddes April 7, 1896 (University of St Andrews Library Special Collections, MS 16363).
36 Geddes, P., "Art in Civic Progress and in Education," *Dundee Advertiser*, February 11, 1907.

37 Four of these models are still held in the D'Arcy Thompson Zoology Museum, University of Dundee.
38 Thompson, D. W., *On Growth and Form*, 2nd edn. (Cambridge: University Press, 1942), 751.
39 Thompson, D. W., *On Growth and Form* (Cambridge: University Press, 1917), 649.
40 Thompson, 1942, op. cit., 1080.
41 Thompson, 1917, op. cit., 740–1.
42 Letter from D'Arcy Thompson to Cambridge University Press November 15, 1917 (University of St Andrews Library Special Collections MS 42633).
43 Letter from Herbert Read nd (1942) (University of St Andrews Library Special Collections, MS 19370).
44 Letter to Herbert Read July 28, 1946 (University of St Andrews Library Special Collections, MS 45690).

5

D'Arcy Thompson and Dorothy Wrinch: A Friendship, 1918–1948

Marjorie Senechal

D'Arcy Thompson influenced many people he never met, but arguably none more deeply than one he knew well: the polymathic, polyhedral, poly-controversial Dorothy Wrinch. D'Arcy urged the brilliant but unfocussed young British mathematician to apply her talents and insights to problems of biological form, and she did. She also applied her belief in symmetry, simplicity, and beauty as a guide to truth—a belief they shared.

Dorothy Wrinch (1894–1976) was the first child of Hugh and Ada Wrinch. Her father, an engineer, managed the pumping station of the water-works in Rosario, Argentina, where Dorothy was born. When she was three, the family returned to their native England and he became head of the Surbiton branch of the Chelsea waterworks, near London. Her parents promptly enrolled their little Dot in the strict, Anglican, girls-only Surbiton High School. Her talents were already evident. "The tyke will become a mathematician," Hugh told Alice Proctor, the school's headmistress.

After Surbiton High, Dorothy entered Girton College, one of Cambridge University's two colleges for women. Enchanted by the free-thinking Bertrand Russell, she quickly abandoned her religious faith, but not the workaholism that Miss Proctor had instilled along with it. Dot graduated a Wrangler (first-class honors in mathematics) in 1916 and then … what? The stultifying careers of her female tutors did not attract her, nor did hand-crunching data in a lab. More study, then. She wrote to Bertrand Russell, asking to study logic. He took her on as a private student in London.

Fast forward thirty years to 1947. Formal logic had long since lost its grip on Dorothy. A battered but unbowed veteran of a bitter war over the model for protein structure she had proposed to the scientific community (more on this below), she now taught physics at Smith College in Massachusetts. Her new work on the mathematical theory of X-ray crystallography, *Fourier Transforms and Structure Factors*,[1] had been published to favorable reviews. In this book, her first (not counting her 1929 pseudonymous *The Retreat from Parenthood*[2]), she explained how X-rays scattered by the crystal reveal details of the crystal molecules.

I met Dorothy at Smith about twenty years later, and worked with her for the last ten years of her life as an unofficial post-doc. I made models and drew illustrations for a new book she was writing, *Whole Number Geometry and the Angstrom World.* Not a sequel to *Fourier Transforms*, but a prequel: the intricate symmetries of crystal structures, as they were then understood. She never mentioned, and I never guessed, its debt to D'Arcy Thompson.

Letters in the D'Arcy Thompson papers at the University of St. Andrews span the thirty-year gap from 1917 to 1947. In them, D'Arcy the person shines through, and D'Arcy as mentor and friend. They also show his encouragement and support for women. Dorothy is only one of D'Arcy's hundreds of correspondents. The fuller story awaits its teller.

* * * * *

In May 1918, Bertrand Russell was jailed for antiwar activities. Dorothy kept him in touch with the outside world by running errands and reporting their friends' goings-on, and in July she served as his eyes and ears at a joint session in London of the Aristotelian Society, the British Psychological Society, and the Mind Association. There, her path crossed with D'Arcy's, maybe not for the first time.

The main event of that joint session was a debate between two giants of science, D'Arcy Thompson and the physiologist John Scott Haldane.[3] The question burned brightly, then as now: "Are physical, biological, and psychological categories irreducible?" The hall was packed, and Dorothy was likely among the throng. Dr. Haldane, who spoke first, thought the answer was yes. "Fifty years ago many physiological processes which, from a physical and chemical standpoint, are now seen to be extremely complex and obscure, were regarded as quite simple," he explained. "There is a prevalent idea that the progress of chemistry, and particularly of physical chemistry, has helped toward an explanation of these processes. This is most certainly not the case."

D'Arcy's masterpiece, *On Growth and Form*, had been published the year before. In it he argued that such doubts were premature. Why look for more than physics and chemistry and mathematics, when biology still has so much to learn from these sciences? He called for help:

> While I have thought to shew the naturalist how a few mathematical concepts and dynamical principles may help and guide him, I have tried to shew the mathematician a field for his labor—a field which few have entered and no man has explored.

This field, he explained in the debate, spans all scales, from the skeletons in *On Growth and Form* to the invisible. "For my part look forward, in faith and hope, to the ultimate reduction of the phenomena of heredity to much simpler categories, to explanations based on mechanical lines … that the special science which deals with it has at least found, in Mendel, its Kepler, and only waits for its Newton."

To Dorothy, D'Arcy's words in that lecture hall were a trumpet call to glory. She was already a member of D'Arcy Thompson's wide scientific circle. In addition to logic with Russell, she was studying mathematical physics at University College, London, for her MSc (1920) and DSc (1921) degrees. Her advisor at UCL was John Nicholson of Oxford's Balliol College. They would marry in 1922.

Dorothy's first research problem concerned the skeletons of sponges. In *On Growth and Form* D'Arcy had shown that the skeletons in one particular species owed their shapes to the packing of cells of the sponge. In 1917, John Nicholson had published, with Arthur Dendy, the world's leading spongeologist, a paper on the skeleton of another sponge species, *Latrunculia*. This sponge is of interest today because it may play a role in curing pancreatic cancer.[4] But Nicholson and Dendy, a century earlier, were concerned with its skeleton's knobby form. Why did the knobs appear, and where? The stationary points in a vibrating string looked much like that; were the knobs formed by the motion of seawater? They showed that the analogy held experimentally, and Dorothy developed computational tools to nail it down.

* * * * *

The first letter in the Wrinch folder in D'Arcy's papers at St. Andrews is dated August 28, 1924. Dorothy wrote from Winchelsea, in Sussex:

> Dear Professor Thompson,
>
> I wonder if I might bother you with a question? I am working out some new wing profiles suitable I hope for aeroplanes and I want to find out if there is any bird whose wing profile is of a similar type. Aerodynamics is peppered with albatross wing sections and the Herring Gull etc. I want to find some type of wing profile sufficiently similar for the loan of the name of the bird to the section to be suitable. As so often happens in maths it's the maths itself which decides what may be worked and what may not! And I have managed to get these sections done not because I set out to do them, but because the matter in question allows the solution and now I want a nice name for them. Such is the powerlessness of the mathematician.
>
> I had hoped perhaps to see you at the Reading meeting. We should be absolutely delighted if you could ever find time to look us up when you are in the south, in London or in Oxford.
>
> Yours sincerely,
> *Dorothy Wrinch*

D'Arcy replied at length from Copenhagen on September 15. I quote his letter in full. As always with D'Arcy Thompson, the joy in reading his prose lies in not only what he says, but how he says it.

> My dear Dorothy Wrinch,
>
> Your letter has been forwarded to me here, and I snatch a moment to write you between two meetings. I shall be in London, I hope, by Friday, and shall stay there for the weekend, or a day or two longer. If you happen to be coming up to town about that time, do write me a line to the Athanaeum: I should love to meet you, and talk about Birds' wings, and other things.

We were disappointed that you did not come to Toronto: the Meeting was a great success, and the company on board the 'Caronia' was first class. We were still more sorry to be told that your good man had had an accident, and that this was the cause of your absence.[5]

As to your Birds' wings, I should like to take you to Leadenhall Market: where, any day of the week, you may find birds of many sorts, and study their wings to your heart's content. The curves you draw seem to me excellent curves, but not a bit like the profiles of any real bird's wing: - i.e., if I understand what you mean by a "profile", which I should rather call a section of the wing. The wing is always a curved surface: in other words (I suppose) it has something more to think of than mere stream-lines – it has to be stiff, to give resistance against the downward beat. The profile that you draw, - while it does not remind me in the least of the 'profile' of any bird' wing, - does remind me of the section of a bird's body, or (what is the same thing) of the horizontal section of a yacht. If I had to find a name for it, I think I should call it the yacht-curve, or the racing yacht-curve.

Just the same curve seems to me to be beautifully shown in horizontal section (i.e., on the water-line) of a swimming bird. You also see it in the vertical section of any ordinary flying bird: though here it is not (or is not necessarily) the same on the lower and upper half of the section. It looks to me as if the lower and upper half tended to differ in degree, - i.e. one may be one, and one another, of the various forms you have drawn in your letter to me.

Obviously whatever curve you find to be a suitable one ought to work also as a solid of revolution. This suggests to me an easy series of experiments. Make some lead weights of appropriate section, and let us sink them to the bottom of the sea, - a sounding line passing through the weight. In water of say 100 fathoms it will be very easy to time their fall with a stop-watch, and the differences are sure to be great enough to give one very significant results. We tried this some years ago with ordinary weights, which were to be used for releasing a catch on a `water-bottle' at the bottom of the sea. A sinker shaped approximately like your curves fell to the bottom in just about half the time taken by the ordinary cylindrical weight, – I have forgotten the actual data: but the method of experiment seems a good one, and much better than vertical movements in a confined tank.

There are heaps of questions I should like to ask you about stream-lines, and various bird and fish-forms: in fact, you are going to have a few such questions before long. There must, I suppose, be a simple answer to the question why so many birds' tails end in a long fork, - swallows, albatrosses, etc., etc.

I am snatching a few minutes to write you while there is a babble of talk going on all round, in several languages. I give it up, it is no good, and I cannot collect my thoughts.

Yours ever,
D'Arcy W. Thompson

When they met in London a few days later, D'Arcy asked her about the transformation of a gelatin cube as it shrinks and dries; he'd seen experiments of the colloid chemist Emil Hatschek. In a letter of September 17, he wrote to Dorothy (Figure 5.1),

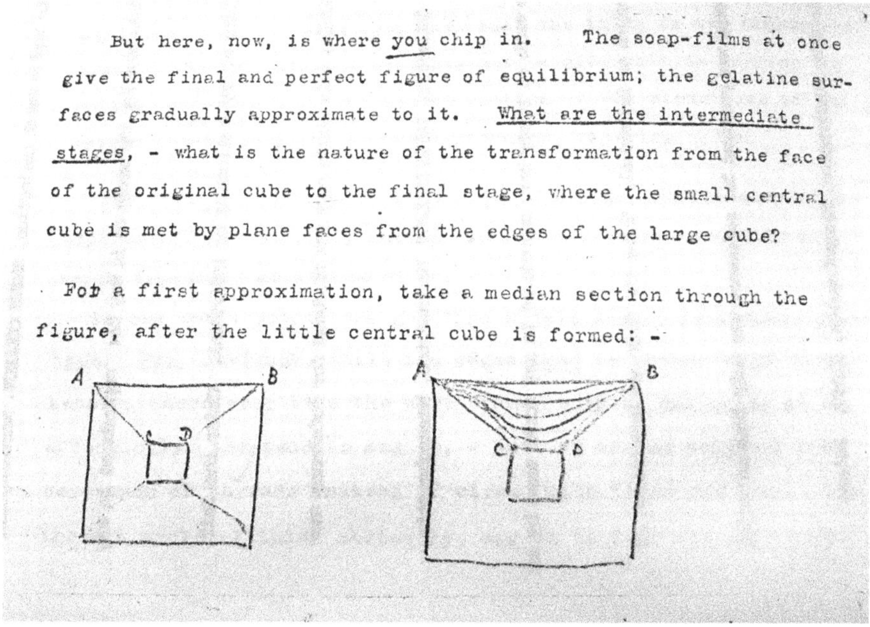

But here, now, is where you chip in. The soap-films at once give the final and perfect figure of equilibrium; the gelatine surfaces gradually approximate to it. What are the intermediate stages, - what is the nature of the transformation from the face of the original cube to the final stage, where the small central cube is met by plane faces from the edges of the large cube?

Fot a first approximation, take a median section through the figure, after the little central cube is formed: -

Figure 5.1 Letter: D'Arcy Thompson to Dorothy Wrinch, September 17, 1924, from D'Arcy Thompson papers. Courtesy of University of St. Andrews.

> The sides draw in, and a small hollow cube appears in the centre of the figure as in soap films on p. 337 of G. and F. But here, now, is where you chip in. The soap-films at once give the final and perfect figure of equilibrium; the gelatine surfaces gradually approximate to it. What are the intermediate stages … The actual forms produced in gelatine … are precise replicas of the typical cartilaginous vertebrae of a shark, or any other cartilaginous fish. But such vertebrae are not all exactly alike; the ends are certainly not always spherical; one has a lot to learn about what results would follow from slight asymmetry or non-homogeneity of the material.

It's the Schwartzian transformation, she told him.[6] D'Arcy replied on October 29:

> Hatschek is a beautiful experimenter, but mathematically he seems a bit of a duffer … the blessed ass can't see, apparently, that his rubber-membranes are separate surfaces, while the soap-films in Plateau's experiment, or the eight surfaces of the gelatine cube, are associated and connected surfaces.[7]

* * * * *

After her marriage, Dorothy had moved to Oxford, where she directed the mathematical studies at the university's five women's colleges. Her daughter Pamela was born in 1927. In 1929 Dorothy was the first woman to receive an Oxford DSc.

But the accident D'Arcy alluded to had permanent consequences: John's behavior became erratic and violent. In 1930 they separated legally (and later divorced); soon after the separation John was hospitalized, permanently as it turned out, for "lunacy."

Dorothy struggled financially. D'Arcy did his best to help.

27 May, 1931, Academic Registrar, University of London, to D'Arcy Thompson:

> Dear Sir,
> I write to inform you that Dorothy Maud Wrinch is applying for a grant of £100 from the Dixon Fund [to study the problems of cell division and the related problems in cytology] and has named you as a reference …

29 May, 1931, D'Arcy Thompson to the Academic Registrar:

> There are innumerable biological problems with a mathematical bearing; but the mathematics required is mostly of too high an order for the biologist. Nothing is more wanted in biology than that mathematicians of first-class standing should interest themselves in biological problems …
>
> It is precisely such problem as these which Dr. Wrinch wishes to investigate: especially the problem of Cell-division … Here it is for the mathematician to investigate the field of force within the diving egg (or other cell); the conditions of stability of its surface-equilibrium; and the point at which an unstable equilibrium breaks down, and is replaced by the new stability of the divided cell … I do not know of anyone better qualified – nor yet so well qualified – as Dr. Wrinch to undertake the task; and I strongly recommend that she be helped and encouraged to apply herself to it.

Cell division was the focus of several of Dorothy's unsuccessful grant proposals. She penciled comments in the margins of her copy of *On Growth and Form.* "But these wretched soap films are things of const. T:" she noted acidly on page 191 (Figure 5.2), "whereas bugs are made of materials in certain [shapes and T is a fn. of (?) etc. This is the real + fatal error of making such a stir abt s. films." (Here T is surface tension.)

Dorothy didn't receive the Dixon grant, or most of the others for which D'Arcy recommended her. This is not the place, nor is there room, to discuss the complicated reasons why. I note only that in the end the women's colleges of Cambridge and Oxford came through. In 1931 Dorothy took Pamela to Vienna. Her many postcards to D'Arcy (several in German) are semi-decipherable. Here's one:

Wien XVIII Scherdlstrasse 30, 16 Jan. 1932

> So many thanks for many things (α) the nice letter with the pleasant news on which the RS deserves congratulations (β) the charming (?) card which delights Pamela, (γ) the kindness in putting me on the (???). We are happy and flourishing. I think and pray (?) that I have got something about venation in leaves at last. I saw Prof. Przibram who talked enthusiastically about you. Will look up the Andrade

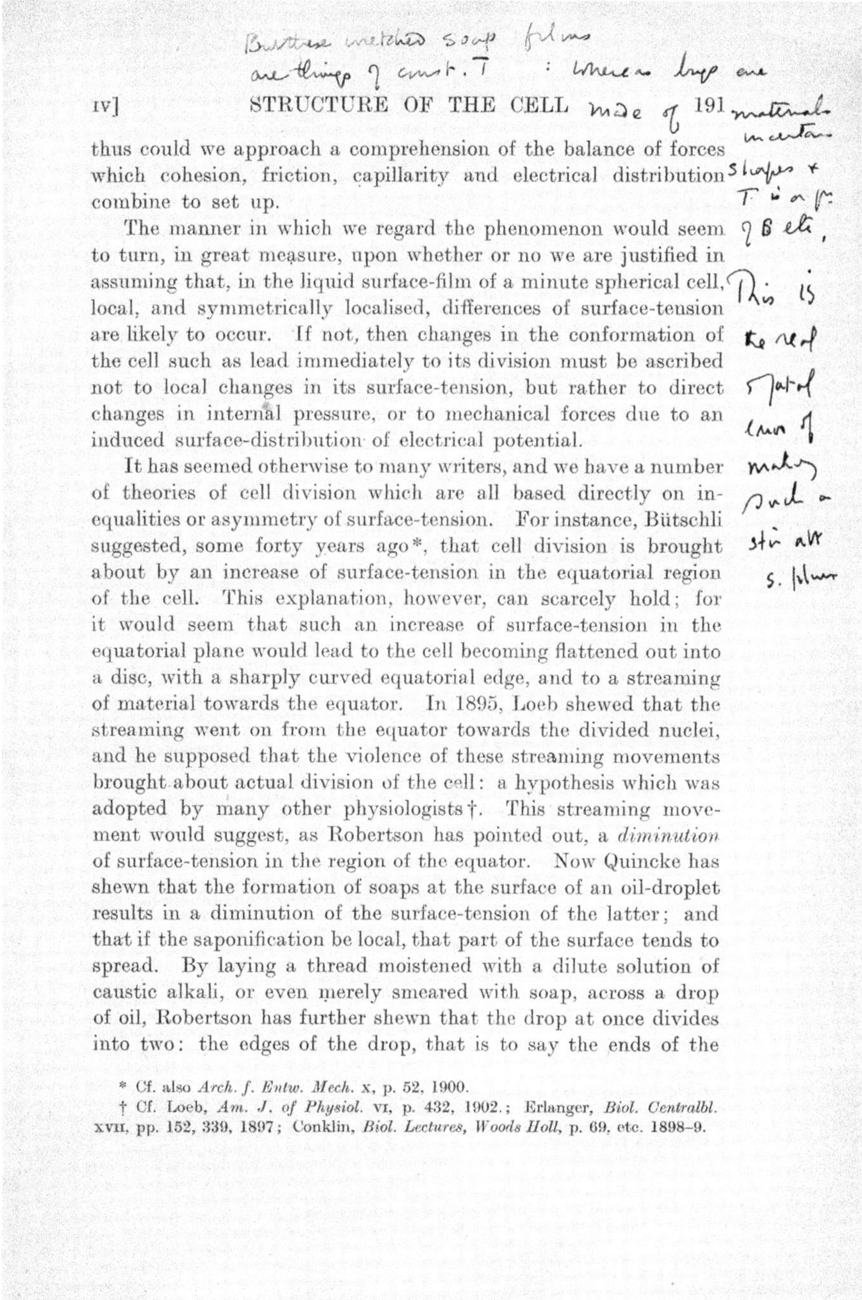

IV] STRUCTURE OF THE CELL 191

thus could we approach a comprehension of the balance of forces which cohesion, friction, capillarity and electrical distribution combine to set up.

The manner in which we regard the phenomenon would seem to turn, in great measure, upon whether or no we are justified in assuming that, in the liquid surface-film of a minute spherical cell, local, and symmetrically localised, differences of surface-tension are likely to occur. If not, then changes in the conformation of the cell such as lead immediately to its division must be ascribed not to local changes in its surface-tension, but rather to direct changes in internal pressure, or to mechanical forces due to an induced surface-distribution of electrical potential.

It has seemed otherwise to many writers, and we have a number of theories of cell division which are all based directly on inequalities or asymmetry of surface-tension. For instance, Bütschli suggested, some forty years ago*, that cell division is brought about by an increase of surface-tension in the equatorial region of the cell. This explanation, however, can scarcely hold; for it would seem that such an increase of surface-tension in the equatorial plane would lead to the cell becoming flattened out into a disc, with a sharply curved equatorial edge, and to a streaming of material towards the equator. In 1895, Loeb shewed that the streaming went on from the equator towards the divided nuclei, and he supposed that the violence of these streaming movements brought about actual division of the cell: a hypothesis which was adopted by many other physiologists†. This streaming movement would suggest, as Robertson has pointed out, a *diminution* of surface-tension in the region of the equator. Now Quincke has shewn that the formation of soaps at the surface of an oil-droplet results in a diminution of the surface-tension of the latter; and that if the saponification be local, that part of the surface tends to spread. By laying a thread moistened with a dilute solution of caustic alkali, or even merely smeared with soap, across a drop of oil, Robertson has further shewn that the drop at once divides into two: the edges of the drop, that is to say the ends of the

* Cf. also *Arch. f. Entw. Mech.* x, p. 52, 1900.

† Cf. Loeb, *Am. J. of Physiol.* VI, p. 432, 1902.; Erlanger, *Biol. Centralbl.* XVII, pp. 152, 339, 1897; Conklin, *Biol. Lectures, Woods Holl*, p. 69, etc. 1898–9.

Figure 5.2 Dorothy Wrinch's penciled comments on page 191 of her copy of *On Growth and Form*: "But these wretched soap films are things of const. T: whereas bugs are made of materials in certain shapes and T is a fn. of (?) etc. This is the real + fatal error of making such a stir abt s. films." Courtesy of Marjorie Senechal.

> paper. Am longing to read Lodge's *Advancing Science* after what you say of it. All my homage as ever and thanks. I return on April 15. But I love Wien.
>
> Dorothy Wrinch

In Vienna Dorothy visited the Vivarium, as the laboratory directed by the biologist Hans Przibram was known. She also met mathematicians Karl Menger and Olga Taussky and corresponded with British colleagues about forming a biology discussion group when she returned. *On Growth and Form* had posed the challenge:

> The biologist, as well as the philosopher, learns to recognize that the whole is not merely the sum of its parts. It is this, and much more than this. For it is not a bundle of parts but an organisation of parts, of parts in their mutual arrangement … This is no merely metaphysical conception, but is in biology the fundamental truth.

The group, later called the Theoretical Biology Club, would grapple with the details. What are the parts of a living whole, and why are they arranged as they are? How do atoms form molecules, molecules living cells, living cells tissues, tissues organs, organs organisms? At what level in this hierarchy should one begin to explore this?

Perhaps remembering D'Arcy's call to glory, Dorothy chose to start with the chromosome. She imagined it as a molecular fabric: proteins laid down parallel like a warp, nucleic acids threading under and over them like a woof. The two ends of each acid, hanging from this hypothetical loom, linked up in the living cell. Ergo, said Dorothy, the chromosome is a woven sheath. If she was not heredity's Newton, she was surely its Copernicus:

> I locate the genetic identify of a chromosome in its characteristic protein pattern. The genetic constitution of a chromosome is to reside not only in the nature of the [amino acid] residues of which it is composed, not only on the proportions of these different residues, but essentially and fundamentally in their linear arrangement in a sequence of sequences.[8]

The Rockefeller Foundation gave Dorothy a grant to continue her work on chromosomes, but her model predicted optical properties that differed sharply from those observed experimentally. She then switched her attention to proteins. Her failure had taught her, she said, that the widely accepted polypeptide chain model for proteins was probably wrong. Instead, she proposed, the chains loop into hexagons, and the hexagons into a lacey fabric.

In her first paper on proteins, she imagined these fabrics stacked in layers.[9] But experiments showed that some proteins were globular. She realized then that her lace could fold and unfold, like origami. That would also account for the proportions of the different amino acid residues, an accepted experimental fact (later proved false).[10]

Her model split the court of scientific opinion. Many resented her penchant for publicity; others resented that she was a woman. On the scientific sides, Nobel Prize winners lined up pro and con. Her model sent imaginations soaring. Niels Bohr, whose father had been a close colleague of John Scott Haldane, thought it might explain the difference between life and death. Figure 5.3 is a lantern slide of a model of her model that he had made in his lab. Albert St. Gyorgi foresaw designer drugs.

But Dorothy's model angered sticklers for chemical detail. Linus Pauling led the attack. He may as well have been attacking *On Growth and Form*. "Geometry is accidental. Symmetry is incidental," he declared. Dorothy dug in her heels. There's only one way to tell who's right, both sides agreed: crystallize the proteins, zap them with

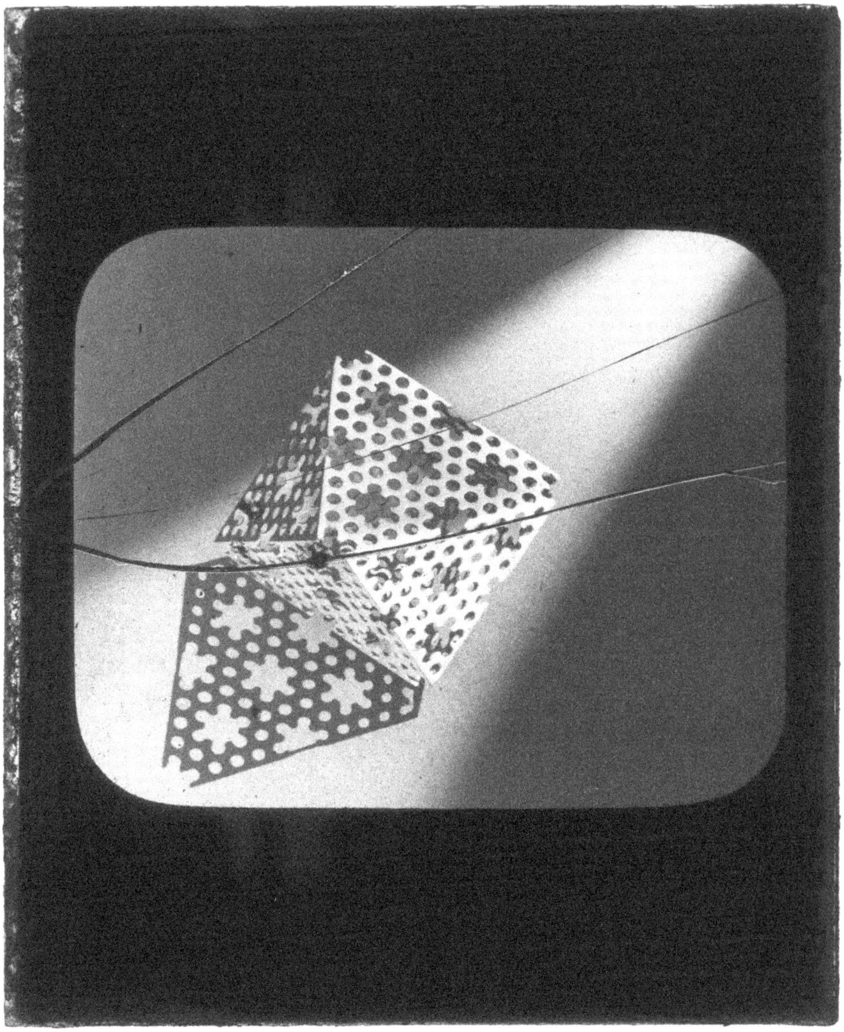

Figure 5.3 This lantern slide shows a photograph of the metal model of Dorothy Wrinch's protein structure model made in Niels Bohr's laboratory. (Photographer unknown.) Courtesy of Marjorie Senechal.

X-rays, and decipher the scattering patterns. Dorothy turned to crystals, X-rays, and deciphering.

D'Arcy supported her through it all. "One can hardly pass a fair judgment of D. W. without knowing her sad history. You doubtless know that she made an unhappy, even a tragic, marriage; from the consequences of which she still suffers acutely," he wrote to Somerville College, Oxford, in 1939, recommending her for a Lady Carlisle Fellowship. "What she wants is the chance of setting down to work, to work of a high class, without worry and anxiety for the next few years. I think it well worth while to help her to do so." The Fellowship was awarded unanimously.

But as war's shadows lengthened, Dorothy resigned the fellowship to emigrate, with Pamela, to America. There she met and married Amherst College biologist Otto Glaser and began teaching at Smith College in nearby Northampton.

Dearest D'Arcy,
Your letter to Otto Charles has just come apparently – he has sent it along to me as I am at the MBL at Woods Hole – so as I return it to him so that he can answer it, I am enclosing this note for him to send too. For though you may not know it, this wonderful and glorious person is MY SPOUSE. It is and has been the joy of joys for me since it happened entirely unexpectedly and wonderfully – right in the lab here – on Aug. 20 of last year. He is the being for whom I have been waiting all my life, Heart-head-hand all α^{+++}. So we have wonderful times and often talk of your work which, as the years go by, strikes me as more and more significant in the long range history of fundamental ideas. I believe you have been dead right in all your methods of thought (if a devoted admirer may say so) and I try daily to give to these ideas the chemical physical and geometrical presentations in protein terms to the best of my ability. Do you remember once you told me of some efforts to get Bragg Papa to see the relations of his ideas to your polyhedral pictures? – And the impossibility of getting him to understand? I still think this one of the most important bridges to build.

Would that we could talk together. Have you a little affection and congratulation to offer me in this Nunc Dimittis – Indian Summer-like new existence?[11]

Ever your most affectionate and admiring,

The University of St. Andrews
18th March, 1943

My very dear Dorothy,

Your letter (dated Feb. 3) and its enclosure, arrived this morning, and have given me extraordinary pleasure.

George Sarton told me, in a recent letter, that he had an article on G. and F. on the way; but he did not let the cat out of the bag that it was your doing. I can't help feeling that in every word you wrote you had it in mind to give an old man pleasure, and you have succeeded to the full. Besides, I think I deserved a little consolation, for some of my few reviewers have been a little hard on me. Two or three have been good, and one or two delightful – especially one in the "Lancet" four or five months ago; but Nature and now Science have each poured a little cold water down my back. Our old friend Harold Jeffreys meant well, but it seemed to me that he didn't quite know the difference between writing a review and correcting an Exam paper. McClung, in Science, has given me a regular trouncing! It doesn't worry me in the least; but it does so illustrate the kind of mind (of which I talk here and there in the book) which absolutely resents (as Bichat resented) the introduction of maths into biology.

I particularly like what you say about "over-mathematization," a real danger. R. A. Fisher is continually guilty of it. I'm quite ready to admit there are cases, a few cases, where all the refinements of K. P. and `Student' and Fisher himself may shew their usefulness; but when they will use these sledge-hammer methods when the tack-hammer or the drawing-pin would do as well, I've no patience with them. JBSH is by no means free from the same bad habit; he just loves to show off his mathematics. I remember a B.A. meeting, with poor old Forsyth in the Chair, when JBS, reading a paper, covered the board with very alarming diffl. equations. I said to Forsyth, as soon as he finished, "Aren't these equations very difficult?" "Sh." – said Forsyth – "I can't do 'em myself!"

... But we are none of us quite ourselves in these strange, sad times. One lives in a sort of chronic malaise, a dreadful lack of complete happiness, in which anxiety for what tomorrow, or the far future, may bring forth is only a small part of the trouble. Everything is somehow wrong, just everything. And, without being a gloomy pessimist, I have been a very, very anxious man all through. I was not only frequently in Germany before the war, but I was in Berlin only two or three months before it broke out; and I saw enough then to scare me stiff – including our own Ambassador!

And still, I keep wandering from the point; for the real point is, to congratulate you with all my heart – late in the day though it may be – on your most happy and fortunate marriage. God bless thee – and he will bless thee! – as I once heard a great man say in much the same circumstances. I never met your Otto, but I've known his work for many years. He has been generous in sending me his papers, and they have a box to themselves – you only have half-a-box – in my room in College.

You may be interested, you will certainly be a little amused, to hear that I am now writing – i.e., I am now working up long-accumulated notes – on the Greek Fishes, to make a companion volume to my Greek Birds. Nobody on earth wants a book on the Greek Fishes; it is highly probably that it will never even get printed. But you have no idea what a comfort it is to work at it. I go to it after supper, and work till the midnight news comes in. I forget all my own troubles, and my Wife's worries (which are worse than mine). The evening passes as quietly as my evenings have done for the last sixty years; and I am thankful to have found this quiet harmless drudgery to do.

I still do my College work, much as usual. I lecture as easily, and I think quite as well, as ever. But the day is much shorter than it used to be. And once thing I

> can't do – I can't scramble over sea-weedy rocks, and guddle[12] in the little pools along the shore!
>
> And so Pamela is 15+. She will presently be in her chrysalis stage, and then in no time her pretty wings will spread. Meanwhile, I can only imagine how fair, and sweet, and good she is.
>
> And now, at last. My best and warmest regards to both of you; my sincere and grateful thanks for your article; and an old man's love to you last of all.
>
> *D'Arcy W. Thompson*

* * * * *

Neither D'Arcy nor Dorothy fit the usual mold of a scientist, nor did their careers. Not in their own lifetimes, and not today. Their affinities stretched far beyond their shared bent for pattern and simplicity. Both had red hair. Both had wide interests and followed them, ignoring both real and imagined "boundaries," and paid a price for that. And both were kind.

Both were criticized for not using all the mathematics of their day: D'Arcy for ignoring calculus, Dorothy for group theory. These letters show they didn't think those tools were necessary for the work they wanted to do. Were they right?

On Growth and Form was, and is still, seen as a work of the imagination, not a treatise on mathematical biology. The same can be said of Dorothy's protein model. Yet without these amazing imaginations, where would those sciences be today?

* * * * *

Dorothy died in 1976, leaving her notes, models, and papers to Smith College. After reading them, I wrote about her struggles, and eventually a biography,[13] to which this essay is a postscript.

"A long time ago," Dorothy wrote in an undated note, "I wished to develop studies of the form of living organisms possibly on something approaching the lines laid down by D'Arcy Thompson in his beautiful work Growth and Form." She had set out that program in her review of the second edition of *On Growth and Form*:

> Once the discontinuities of the two realms are understood, the morphologist should be free to pass from the implications native to the molecular level to the explications that characterize his complex aggregates ... What D'Arcy Thompson envisages is not easy.

These few words contain worlds: the rigid world of crystals, with atoms or molecules repeating symmetrically in rows, rows stacked in planes, planes stacked in space; and the world of breathing, dying, wriggling living creatures. The question J. B. S. Haldane and D'Arcy Thompson debated in 1918, "Are physical, biological, and psychological categories irreducible?" would, she never doubted, one day be settled in the negative. It only remained to find that hidden passage.

D'Arcy himself suspected it wouldn't be easy. In *On Growth and Form*, on a page that Dorothy didn't annotate (p. 601), he wrote presciently, "In the molecular growth of a crystal, although one must of necessity assume that each molecule settles down in a position of minimum potential energy, we find it very hard indeed to explain precisely, even in simple cases and after all the labors of modern crystallographers, why this or that position is actually a place of minimum potential."

By "this or that position," D'Arcy meant a place in a stack. But the 2011 Nobel Prize in Chemistry was awarded for the discovery of unstacked crystals.[14] Today, crystallography is no longer just about crystals, it is "the science of structure … [and] the way in which large scale form is the expression of local force."[15]

Dorothy didn't see that coming. She saw certain inorganic crystal structures as simple prototypes of biological molecules, but she could not, and did not, imagine how they might come to life, or shape living beings. She delighted in symmetry, simplicity, and beauty. But Haldane wasn't altogether wrong. The closer one looks, the more complex things become.

Dorothy looked away.[16] *Whole Number Geometry and the Angstrom World* would be not only a prequel to X-ray crystallography, but also a preface to her sequel to *On Growth and Form*. An arch of a two-way bridge between the atomic and morphological levels of life. She never finished the book, and the bridge has yet to be built.

Acknowledgments

It is a pleasure to thank Moira Mackenzie, Maia Sheridan, and Mathew Jarron of the University of St. Andrews for their assistance over many years.

Notes

1 Wrinch, Dorothy, "Fourier Transforms and Structure Factors," *ASXRED* Monograph No. 2, American Society for X-ray and Electron Diffraction (1946).

2 Ayling, Jean, *The Retreat from Parenthood* (London: Kegan Paul Trench Trubner & Co., 1930).

3 *Proceedings of the Aristotelian Society*, n.s. 18 (1918) 419–61.

4 http://www.noaa.gov/news/noaa-discovery-of-green-deep-sea-sponge-shows-promise-for-cancer-research (accessed August 19, 2018).

5 The accident doomed her marriage, but that wasn't clear at the time.

6 Not to be confused with the modern Schwartzian transform, a sorting algorithm.

7 Emil Hatschek, colloid chemist, 1868–1944. "His services to colloid science were acknowledged when he was made the guest of honour at the Colloid Symposium at Ottawa in 1932, a distinction much appreciated by him," E. N. Da C. Andrade wrote in *Nature* after Hatschek's death (*Nature* 154, 46 (July 8, 1944) | doi:10.1038/154046a0). "His contribution at Ottawa was a paper on 'The Study of Gels by Physical Methods,' a subject to which he had devoted much attention."

8 Wrinch, D. M., "On the Molecular Structure of Chromosomes," *Protoplasma*, Vol. 25, No. 4 (1936), 550–69.

9 Wrinch, D. M., "On the Pattern of Proteins," *Proceedings of the Royal Society of London A*, Vol. 160 (1937), 59–86.

10 Wrinch, D. M., "The Cyclol Hypothesis and the 'Globular' Proteins," *Proceedings of the Royal Society of London A*, Vol. 161 (1937), 505–26.

11 *Nunc dimittis* is a canticle from the book of Luke. In the 1662 *The Book of Common Prayer*, it's translated as:Lord, now lettest thou thy servant depart in peace according to thy word.For mine eyes have seen thy salvation,Which thou hast prepared before the face of all people;To be a light to lighten the Gentiles and to be the glory of thy people Israel.

12 "Guddling" is a term used to describe attempts to catch fish with your bare hands—a rather messy and difficult task which would fit in with the widely accepted definition of something messy or confusing. www.scotsman.com/lifestyle/scottish-word-of-the-day-guddle-1-2915914.

13 Senechal, Marjorie, *I Died for Beauty: Dorothy Wrinch and the Cultures of Science* (Oxford: Oxford University Press, 2013).

14 https://www.nobelprize.org/nobel_prizes/chemistry/laureates/2011/press.html (accessed August 19, 2018).

15 Mackay, A. L., "Generalized Crystallography," *Izvj. Jugosl. centr. krist.* (Zagreb), Vol. 10, 15–36 (1975); posted online at http://www.cryst.bbk.ac.uk/surfaces/zagreb.html (accessed August 19, 2018).

16 Literally! When the first protein structure was determined, and it didn't match her model, she refused (at first) to look at the structure on display. See David Harker, "Colored Lattices," in *Structures of Matter and Patterns in Science*, edited by M. Senechal (Cambridge, MA: Schenkman Publishing Co., 1980), 11–29.

6

D'Arcy Thompson's Surrealism

Brandon Taylor

In truth, there are no more than a handful of instances in which the mental universe of D'Arcy Thompson can be associated with André Breton's project to liberate the human imagination. Too great is the distance between the patient researches of the Edinburgh zoologist, himself an old-school mechanist of sorts, and the visual researches of a Max Ernst or a Salvador Dalí, an Eileen Agar or a Yves Tanguy—even those of a semi-detached British Surrealist like Henry Moore. Yet a serious appeal was exerted by *On Growth and Form*—osmotically, perhaps, and with misunderstanding—upon artists who marveled at its diagrams or read parts of its stylish and erudite text. And this is not surprising. The interval between the book's first appearance in 1917 and its enlarged second edition in 1942 is the time of Surrealism, after all; and form—how it changes, how it should and could change—preoccupied them all. Thompson was not a Surrealist; but notoriously, he was entranced by the forms and modifications of nature, from water droplets to crystal structures to discharges of electricity to the shapes and proportions of the animal kingdom. In the first pages of *On Growth and Form* he remarks that "above all, in things both great and small, the naturalist is rightly impressed and finally engrossed by the peculiar beauty which is manifested in apparent fitness or 'adaptation.'"[1] Thompson is speculating here on why zoologists seemed reluctant to use mathematical methods to describe the forms of nature. Probably by "peculiar beauty" he meant "astounding" or "exceptional beauty," the kind that the mind cannot conceive of until face to face with it; but the dictionary reminds us that the word "peculiar" also means "individual," "private," "special," "singular," "odd"—even "strange." When André Breton in the *First Manifesto of Surrealism*, published in November 1924, reported his and Philippe Soupault's first experiments in "automatic" writing, he mentions "a very special picturesque quality … a strong comical effect" to be found among the fifty or so pages they produced by the end of the first day. They were "on the surface as strange to you as they are to anyone else, and naturally you are wary of them." Of those écritures Breton said that "poetically speaking, what strikes you about them above all is their extreme degree of immediate absurdity, the quality of this absurdity, upon closer scrutiny, being to give way to everything admissible, everything legitimate in the world … the actual functioning of thought."[2] Ten years later, for his article "La beauté sera convulsive" (Beauty will be convulsive), published in *Minotaure* in 1934, he provided a group of Brassai's photographs of crystal and coral—and prominently, a photo of an electric discharge, above a caption that read, "The image like that produced in automatic writing" (Figure 6.1).[3]

For Breton and his circle, the mutual alignment of nature and unconscious thinking remained a touchstone of the Surrealist path. Thompson, for his part, confesses that, to him, "that living matter influences and is influenced by mind is a mystery without a clue. Consciousness is not explained to my comprehension by all the nerve-paths and neurones of the physiologist; nor do I ask of physics how goodness shines in

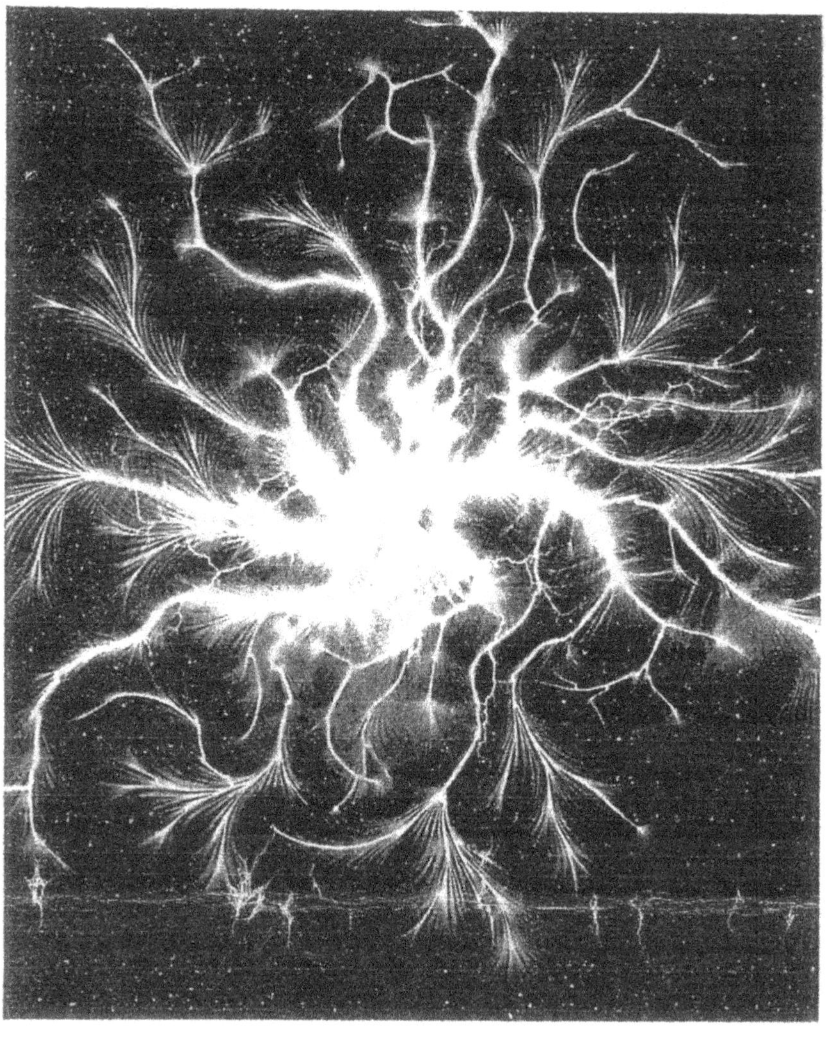

Figure 6.1 Photo of an electric discharge. From A. Breton, "La beauté sera convulsive," *Minotaure*, 5, Paris, 1934, p. 10.

one man's face, and evil betrays itself in another." There was a gap, in other words, between the quantifiable world of "outer" nature and the unfathomable "inner" world of imagination, temperament, and character. For Thompson, it was enough to try and fathom all "that is of the earth earthy," and to do so with "physical science ... our only teacher and guide." And there, in crystal and coral, in water splashes and electric discharges, "the harmony of the world is made manifest in Form and Number." He invokes Isaiah 40, verse 12 in his support. "He hath measured the waters in the hollow of His hand, and meted out Heaven with a span."[4]

Whether nature's strangeness can be accommodated within the rule of physical law is a question that Thompson and the Surrealists shared. Philosophical biology had been split, since Darwin, on the question whether nature was the inevitable outcome of physical law alone; whether the germination, growth, maturation and decay of the living parts of it—plants, animals, and humans—could be described as in some sense physical, or whether there would need to be a residue, a non-physical or supra-physical life-spark to distinguish its operations and appearances from those of inert matter. By his own admission Thompson was a mechanist; and mechanism in his morphological writings meant Newtonian physics and Greek geometry together, plus or minus some parts of the theory of "fluxions"—soon to become the calculus—for the capture of rates of growth and decay in systems whose measurements and proportions were known.

Mathematics was not in fact Thompson's forte. *On Growth and Form* contains only basic statistics, and little or nothing on modeling, topology, or the geometry of dimensions greater than three. The later development of those techniques is sometimes held to have rendered Thompson's findings irrelevant to biomathematics as practiced today. "At best a sideshow, at worst, nonsense" is how Ian Stewart recently summarized his main ones.[5] Even in 1917, Thompson's mechanism could seem dogmatic to some. His contemporary J. S. Haldane claimed that physiology and biology differ "not only in complexity but in kind" from physical and chemical phenomena, and demand different principles of explanation; that the functioning of the single cell was not a mechanical affair, and that the idea of "final cause" in the sense in which Aristotle defined it should not be too hastily abandoned. Thompson did not agree. "I am not willing to reject," he wrote in a review of Haldane's *The New Physiology and Other Addresses*, "as a working hypothesis of the physical universe, the concept of mechanical causation."[6] "Causes and effects thrust themselves on our curiosity," he had said in *On Growth and Form*, "and become the ultimate relations to which our contemplation extends," even if—here he qualifies himself—the known laws of physics are only "conditions *sine qua non*." Physical laws are not efficient causes. They do not make things work, but merely govern the physical description of them when they do.[7]

The reduction of nature's fluctuations and processes to basic physical laws was supported in *On Growth and Form* by some equally basic line diagrams of the shapes and outlines of living and material things. The book is known to have been seen, if not extensively read, by several artists in Europe in the 1920s. The Henry Moore scholar Christa Lichtenstern tells us that Moore, Barbara Hepworth, Paul Nash, and Herbert Read all owned copies at one time or another and that the book "passed rapidly from hand to hand in the Hampstead [London] studios of the day"; and that Moore's so-called *Transformation Drawings*, eight from 1932 and three more from 1935, show Moore patiently exploring the question of how, sculpturally, one natural form can suggest another, perhaps migrate visually and morphologically to and fro across the pair (Figure 6.2).[8] For sure, those drawings contain visible shadings and evidence of trial-and-error thinking—but as a group they seem engaged very differently from Thompson's argument in his final and celebrated chapter, "On the Theory of Transformations, or the Comparison of Related Forms." The concern of several of Moore's drawings is how to get holes into a sculpture, it might be said; how to sculpt the empty space first, and let the form come out as it may. Other drawings address the look of a bone when its disposition changes relative to the sculptor's angle of sight—or is shifted on the table in front of him, is tilted, or has the angle of lighting changed (Figure 6.3). The words "Transformation of Bones" that appear in the titles of some of Moore's works refer to visual correspondences—holes included—between bones, much magnified, and the human form. Thompson, for his part, is adamant that "transformation" in zoology or ethnography means a relation between things belonging to a single type, not the possible analogies between things belonging to different kinds. Seeing one kind of thing *in* or *as* another—for instance Moore's suggestion of a correspondence between

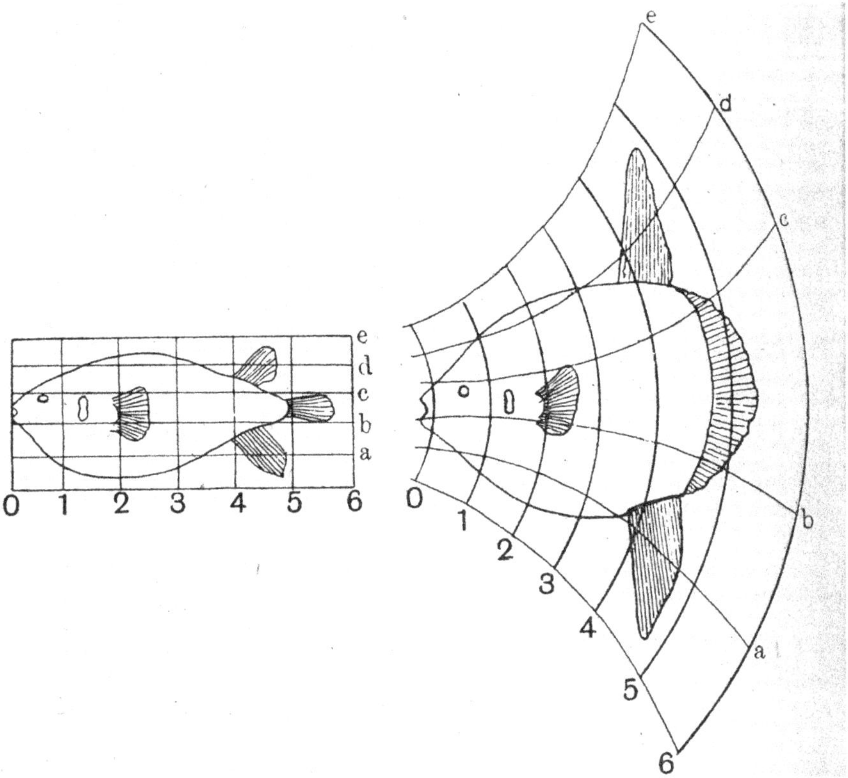

Figure 6.2 Diagrams of *diodon* and *orthagoriscus* from Thompson, *On Growth and Form* (1917), revised edition, Cambridge University Press, 1942, p. 1064.

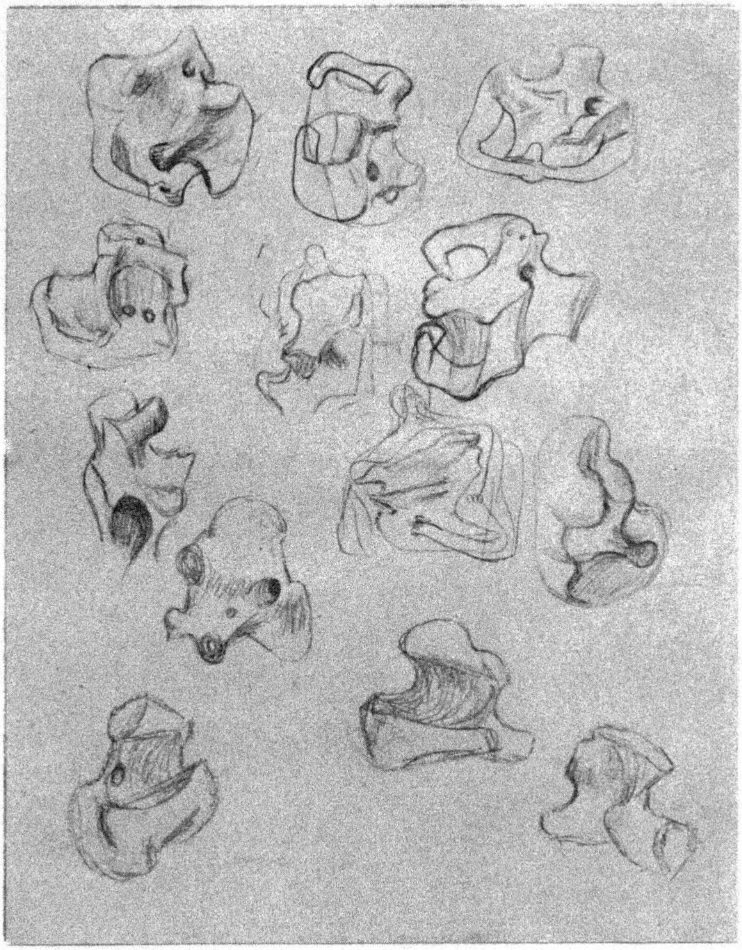

Figure 6.3 Henry Moore, "Ideas for Sculpture: Transformation of Bones," 1932, pencil drawing on paper. Reproduced by permission of the Henry Moore Foundation.

a bone and a "reclining" or "recumbent" female, or between a bone held upright and a standing or seated human figure: such was the method of Surrealism conducted as a parody of science.

Moore's relation to the diagrams and arguments of *On Growth and Form* may be put another way. The sculptor's interest in bones, rocks, and stones as formal structures is legendary, but symptomatic too of a certain necrophile regard. Most of Thompson's laboratory subjects had also been non-living entities. The majority of his "Transformation" diagrams are of shells, horns, tusks, and teeth, together with skulls or other bones such as the pelvis, the shoulder girdle, and the femur, harvested from animals that were no longer alive. His interest in inert body-parts whose livingness has long since departed is no less remarkable than Moore's, albeit differently motivated. Thompson measured creatures such as crustacea on condition that they were no longer inhabited; birds when they were no longer flying; skeletons when they were no longer functioning; skulls when they were no longer the seat of intelligence and cognition. It is a morphological range well-suited to the archaizing interests of British modernist British sculptors at a time when to carve in limestone, marble, or wood—to respond to materials "truthfully"—was as important as the morphology chosen. British sculpture found its modernity largely through its antiquity, that is to say; and it is unsurprising that those of Moore's own *Transformation Drawings* that propose an analogy between the shapes of a bird's bone and a standing figure, or between a human femur and a recumbent woman, eventuate in sculptures of the body's heaviness and passivity, seldom its alertness, its readiness to see and act. It is a quality that has no obvious counterpart in the work of Moore's contemporaries in mainland Europe, those such as Hans Arp, Wassily Kandinsky, or Paul Klee, whose much sharper vitalist impulses animated art-works of pulsing and frequently unstable energy. For them, a major image-type was that of the living fish; and by "fish" let us include sea-borne creatures, jellyfish, squid, crabs and mollusks, repellent and even repulsive organisms that thrived, if at all, in the rock-pools and deeper oceans, now made available to public consciousness thanks to the enlargements of underwater film and microphotography. In the deep-water fantasies of Yves Tanguy or the mobile chaoses whipped up in pen-and-ink by André Masson, or in the watery worlds of Gérard Vulliamy, Wols or the early Camille Bryen, this fishy world was a mainstay of the European Surrealist imagination, and one tinged with a kind of fascinated anxiety unfamiliar to Anglo-Saxon art.

D'Arcy Thompson was not uninterested in fish, of course. Starting with a study of Hydroida in 1882, his long list of publications contains studies of cuttlefish, dolphin, cod, haddock, the salinity and temperature of seawater, trawling patterns, sea-sponges, eels, Aegean lobsters, fish-names in ancient Egypt and Greece, sea-serpents and whales.[9] The list would even have a certain Surrealist incongruity about it, were it not for Thompson's patient and sober erudition; his dispassionate examination of sizes, dimensions, graphs of population growth and decrease; and his unswerving commitment to "Form and Number." Remorseless quantification was his guide, and no concession to mystery or inscrutability was allowed to interrupt it. Otherwise extraordinary phenomena such as the ganglion cell or medusa of various scales and ecologies were subject to schemes of measurement and observation that could be reconstructed in the laboratory, in the latter case by dropping liquid gelatin into coagulating fluids of fractionally differing densities. The resulting phenomena indicate, Thompson says in a typical passage, "at the very least, how certain simple organic forms might be naturally assumed by one fluid mass within another, when gravity, surface tension and fluid friction play their part, under balanced conditions of temperature, density and chemical composition."[10] Such passages give an account of sea-life, but with the consciousness of the observer missing.

It is Salvador Dalí to whom we must turn for a Surrealist invested in questions of morphology according to the latest science. Dalí is known to have been a keen student of Édouard Monod-Herzen's two-volume *Principes de Morphologie Générale*, published in 1927, in which, as might be expected, *On Growth and Form* is itself referenced with admiration. The book is "rich in the most suggestive insights," as Monod-Herzen generously puts it.[11] The *Principes*' own chapter on "Families of Form" reproduces no less than a dozen of the diagrams from Thompson's final chapter. For both authors, the human and anthropoid skull was of paramount interest. Both use the diagrams of horse and other skulls from Gerard Heilmann's *Vor Nuvaerende Viden om Fuglenes Afstamning* (Our Present Knowledge about the Origin of Birds), published in Copenhagen in 1916, and in doing so indicate the extent of wider scientific speculation on the relation between osteological specimens and types in the generation which Thompson and Monod-Herzen both revered. In the most complex transformations, a grid was superimposed over a given skull-shape and speculatively linked to a second shape through a succession of intermediate stages either of phylogenetic or ontogenetic change. Yet as both of the latter authors were careful to say, it was for the ethnographer to find fossil remains showing that different stages actually belonged to a succession within the same type, and was not a matter for geometers alone to decide.

Tellingly, Monod-Herzen refers to Thompson's "Transformations" as "anamorphoses," a term from the tradition of art that intersected from the sixteenth century onward with the many paradoxes of perspectival sight. Yet unlike Thompson, his sights were set ultimately on the morphologies of art. When Monod-Herzen turns in the second volume of the *Principes* to a discussion of facial types, he is inevitably faced not merely with facial measurement but with *tonicité*—muscle tone and quality—thence character, and thence more widely to what he calls "the aesthetic attitude" underlying various facial types as well as the seemingly transformative relations between them. For Monod-Herzen the work of art in its generality is "the tangible expression, the fixing or objectification, of a psycho-physiological complex that is not detached from but dominates a particular emotion, called 'aesthetic'."[12] He is aware that contemporary art shows evidence of "great perturbations, disarray, symptoms of reorganisation, of relative rather than absolute stability" following the devastation of 1914–1918, and that it was the job of the morphologist to understand those irregularities and hence prepare the ground for correcting them.[13] Quantification alone, he insists, is not enough to illuminate qualitative anamorphosis. To it must be added emotion and thought. "Geometry," concludes Monod-Herzen at the end of his second volume, "properly understood, can play an enormous role in the renewal of modern art."[14]

Given Dalí's interest in scientific thought, the speculation is that Monod-Herzen's book gave him the hint, around 1927 and in the years following, that *déformation*'s most enticing exemplifications were in geometry—that a certain route to psychic disorientation through anamorphosis was within his grasp. One of Dalí's interests was to demonstrate that modern science was Surrealist too, and that understanding that equivalence would help him paint paintings leading to "the collapse of reality" in the sense of the demoralization and confusion of the bourgeoisie. Anamorphosis and other tricks and devices of the geometry of vision would help "bring us back to the clear sources of masturbation, exhibitionism, criminality and love," as he put the matter bluntly to an audience in Barcelona in 1930—an audience ever-ready to be scandalized by "modern" art.[15]

By 1928 or early 1929 Dalí was aware that anamorphosis in the form practiced by Dürer, Jean-Francois Niceron, Hans Holbein, and others since the sixteenth century supplied some fertile possibilities for Surrealist art. The fun had started with perspective in its Renaissance form, which, when combined with a study of human proportions, had been impressively developed by Dürer, two of whose many facial diagrams, published in his *Vier Büchern von menschlicher Proportion* (Four Books on Human Proportion) of 1528, had appeared in the "Transformations" chapter of Thompson's *On Growth and Form* (Figure 6.4). Dürer had noticed how a regular grid drawn over a man's facial profile can be leant this way or that, producing other profiles, and, with them, other physiological types. In this way, facial angles (a line drawn from forehead to tip of nose, from earlobe to nostril) opened the way to morphological comparison of the commonalities of facial organization but also to the character or temperament-structure implied in each proportional variant. Whether through Monod-Herzen or Thompson or both, Dalí now had the cue he needed to examine more penetratingly, with lurching perspectives or extreme and paradoxical viewpoints, the puzzle-pictures and "secret" images of mid-sixteenth-century art in which forms can be hidden within each other with really consequential implication. The timing must have seemed providential. Dalí had already begun to understand that "the collapse of reality" could be accomplished by seeing one thing *in* or *as* another. Picasso had been his guide from around mid-1927 on. Now, in paintings of 1928 such as *The Bathers* and *Female Bather* (both in St Petersburg, Florida) he presented the fish-form as a human body-part or even a whole body: categories that in nature do not transform into one another but that a Surrealist could provocatively confuse. The slug-like droopings and sweatings of Dalí's beach figures can only be read as disgusting. Their swollen and elongated body-forms can only be read as signs of exaggerated,

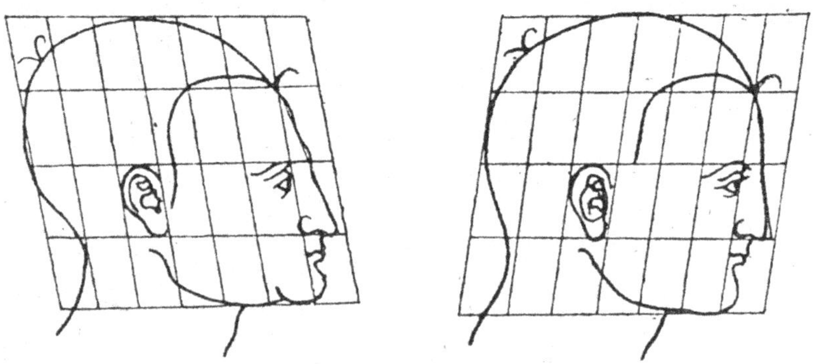

Figure 6.4 Facial analyses from A. Durer, *Vier Büchern von menschlicher Proportion*, 1528, reproduced in *On Growth and Form* (1917), revised edition, Cambridge University Press, 1942, p. 1054.

overheated desire. When Carl Einstein published Dalí's two paintings aligned with an anamorphic painting of St Anthony of Padua in the journal *Documents* in the spring of 1930, the transformative and disruptive power of anamorphosis was immediately clear (Figure 6.5). The St Anthony anamorph, dating from *c.* 1534, shows what looks like a bench with broad legs, stick-like tables of some sort below, and various abnormal forms in the higher part. Seen from a point to the far right and above, the image gels into that of Saint Anthony kneeling before the infant Jesus (an attribute of the saint, in some accounts), a chair, a book, and a cross.[16] Seen relatively miniaturized on the page in *Documents*, the bulging body-forms of Saint Anthony and the infant Jesus could be read straight, as actual bodily metamorphoses. The infant's cranial elongation would be seen as carnal, even sexual, too. Dalí's suite of etchings to Lautréamont's violent and misanthropic masterpiece *Les Chants de Maldoror*, completed in 1934, would provide a demonstration of how cranial and/or buttock-elongation could be taken to pictorial extremes.

Other anatomical and osteological forms—as well as soft food and bodily waste—now enter Dalí's devastating repertoire. In an essay titled "The New Colours of Spectral Sex-Appeal," published in *Minotaure* for February 1934 and therefore written in advance of the paintings and etchings of that year, Dalí gives a statement of two further anamorphic possibilities with which to disorient his fascinated detractors. Anamorphoses of the skull, following Holbein, now began to haunt his image-world.

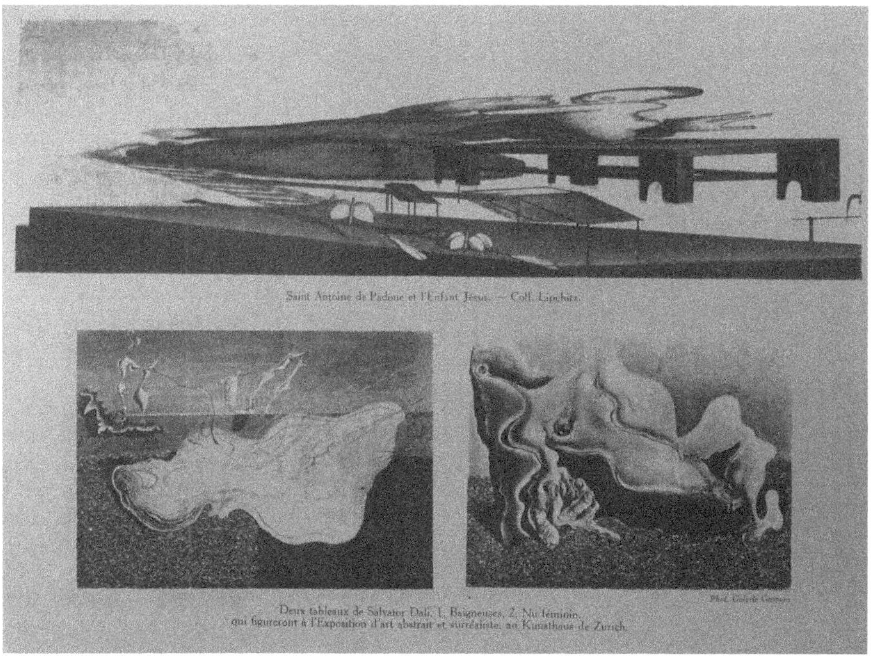

Figure 6.5 Illustration from C. Einstein, "St. Antoine de Padoue et L'Enfant Jesus," *Documents*, 4, 1930, 227.

He describes the painting that would become *Atmospheric Skull Sodomizing a Grand Piano* thus:

> A very large anamorphic skull, blindingly white, mad, sodomizing with its clear jaws a ruined grand piano, all this taking place in the bright spring sun behind an old tarred black boat, abandoned on a deserted and ill-equipped beach, on which hop along some sparrows, a herb omelette slipping dizzily on a polished marble ramp, etc.[17]

A second exemplification was streamlining—or blackheads. Such extrusions—Dalí now speaks about himself in the third person—are "aerodynamic beings-objects" that issue, when squeezed correctly from the nose, "from the very flesh of our thick and personal, non-Euclidean space-time," in becoming a "perverse, glandular, high-grade aerodynamism" continuous with "that super-gelatinous Art Nouveau and nutritive compressibility about which Salvador Dalí talks and instructs you tirelessly." By now, active anamorphic looking had become a component in the more general "paranoiac-critical" method that mandated reality as always multiple, always migrating, never self-identical or the same. The squeezing of blackheads is "nothing less than the ultra-concrete personalisation of what is most vital and lyrical in contemporary, scientific and artistic thought."[18] In Dalí's universe, the mind of the perceiver must itself function anamorphically in order that surprising new versions of objectivity will appear.

More generally, Thompson's "Transformation" diagrams were likely to have been regarded by artists in less frenetic and less destabilizing ways. For the scientist, diagrams such as Heilmann's equine skulls in which multiple instances of a type are arranged in sequence at least attested to the possibility of ontogenetic or phylogenetic transformation. On the other hand, certain paired diagrams—for instance *diodon* and *orthagoriscus*, or the rabbit's skull and that of the horse—were liable to raise questions even in the lay reader's mind. For such a reader might well ask what those diagrams are diagrams *of*. We do not see the drawings of *pomacanthus*, *polyprion* or *scarus* as individuals, after all, but as intended representatives of an aquatic type, delineated by line only and with color, texture, size, mass, and state of maturity all missing—let alone movement and the speeds and styles of motion thereof in the watery medium in which the creatures live. In such cases, the scientifically untrained and perhaps uncurious artist, looking at Thompson's diagrams in light of what they knew of Holbein and Niceron, might conclude that nature was stranger than art. Some might regard the diagrams as graphic inventions only—reminiscent of early Enlightenment efforts to draw nature as both individual *and* type, but with clarity as to terms like "specimen" and "average" missing.

We have now entered the territory of *On Growth and Form*'s relationship to the Surrealist journal *Documents* in 1929–1930. Carl Einstein, Michel Leiris, and Georges Bataille were no friends of scientific and mathematical reduction, in fact took every opportunity to ridicule it. They took every chance available to pillory the Enlightenment dream of defining nature's typicality or of subsuming its fabulous and apparently lawless regularities within predictive and retrodictive law. Thompson's own understanding of the concept of typicality becomes important here. He was not a Darwinian and carried out the majority of his descriptions of the animal world in plainly Aristotelian mode—with touches of Platonic metaphysics added when the case demanded it. It was Aristotle's *Historia Animalium* (of which Thompson published a translation in 1910) from which he took the assumption that animals belonging to the same genus (birds, for example, or fish) are identical "save only for a difference in the way of excess or defect" relative to a "normal" case.[19] But what *is* a "normal" case, of scaroid fish, cat or monkey? Thompson came to convince himself that Aristotle meant it mathematically—had intended to reflect a fascination with rational and irrational numbers widespread among mathematicians of his and previous times. What *is* a typical bird? "Just as we study the rational forms of an irrational number," Thompson says triumphantly about his method in an article of 1929, "and … draw nearer and nearer to the ideal thing, always failing to reach it by the little more or the little less; so we may survey the whole motley troop of feathered things, only to find each one of them falling short of perfection, deficient here, redundant there … and with their inevitable earthly faults and flaws." And then, "beyond them all we begin to see dimly a bird such as never was on sea or land, without blemish, whether of excess of defect: it is the ideal Bird."[20]

The statement exemplifies perfectly the rationalizing outlook that *Documents* wanted polemically to destroy. For Einstein and Bataille, and for Leiris too, a new metaphysics and therefore a new morphology would depend upon finding room for the individual case, each instance in its own way atypical while at the same time bearing no relation at all to a stabilizing norm, a norm that would diminish it by its capture in the net of natural law. Thompson subscribed to the concept of a natural kind not in one way but in two: making specimens typical within the diagram, but also showing himself prepared to cluster deviations probabilistically along a Gaussian error curve—itself "one of the most far-reaching, some say one of the most fundamental, of nature's laws."[21] Yet probability in the late 1920s and early 1930s was another territory on which mathematicians and Surrealists could both agree and disagree. Dalí himself chose Werner Heisenberg as his "father" in replacement of Sigmund Freud. Further, no less a mathematician than Hans Reichenbach made an appearance in *Documents* in 1929, explaining that the analogy of the machine for descriptions of nature no longer held good—given that observation and measurement both had become victims of unpredictability at the atomic level of scale. There, says Reichenbach, nature herself imposes limits deriving from its own inscrutability. Nature cannot be deterministically shown, and the residue of idealism implied in its metaphysics must be jettisoned too. The course of nature "should rather be compared to an endless game of dice."[22]

The abandonment of an objectively accessible observation point by scientists was congenial to those Surrealists who understood the implications of the new quantum physics, and who now hurried to derive appropriate conclusions for the conduct of bourgeois life. The young Gaston Bachelard, to take a pertinent case, in the early 1930s a philosopher of science, faced up directly to the physics of uncertainty and to the new kind of consciousness that would be required if quantum physics and relativity theory were—as they seemed to be—true. *Natura naturans* "is at work in us," he proposed in a phrase that beautifully enfolds consciousness of natural phenomena within the panoply of natural phenomena itself. "We can no longer think in any way but mathematically," he writes in an echo of Thompson's own deepest conviction. Yet unlike Thompson,

Bachelard wished not merely to record but to celebrate the "sudden animation that has been given to the soul by the creative syntheses of mathematical physics."[23]

The evidence is that, for Surrealists if not for artists everywhere, *On Growth and Form* exerted allure in ways that its author cannot have anticipated and may not have understood.[24] For American artists claiming to be building the foundations of the New York School in the early 1940s, *On Growth and Form* would assume real significance too, either in its 1917 edition or in the much-expanded version that appeared on both sides of the Atlantic in 1942. For artists of that milieu, the search was on for images of random, primitive, even violent life. Artists such as Gérôme Kamrowski, William Baziotes, and Jackson Pollock were already in thrall to aspects of European Surrealism and were getting to know the *emigré* Surrealist artists already in their city—Kurt Seligmann, Joan Miró, André Masson, Gordon Onslow Ford among them. The young Americans would frequent the American Museum of Natural History in search of form-languages that spoke to them of the development of life itself. The biology and mythology of both the very small and the very large were of interest to them, as a glance at their paintings of the early 1940s will quickly show.[25]

Aside from Thompson's book, a young artist interested in natural morphology also had access to a cornucopia of biological life from popular magazines in the form of microphotography, X-rays, cosmological charts, in graphic styles so diverse as to relativize almost to the point of extinction the idea of representation or else throw into abrupt relation ideas of style, verisimilitude, and historical truth. That at least was the context in which, in the Surrealist magazine *VVV* for 1943 we find two large *panoramagraphs*, as Kamrowski called them, in which Thompson's *diodon* and *orthagoriscus* diagrams once again appear, along with his skull-diagrams of a human and of a chimpanzee (Figure 6.6). Those quotations take their place amid meandering lines, images taken from Kandinsky, Miró, and astrology, a quotation from Breton about Thérèse d'Avila and further indexical signs, as evidence of how, in the midst of rich epistemological confusion, the scientific diagram could acquire an evidently mysterious face. Breton himself, a ringmaster of the New York community just as he had been for the Surrealists in Paris, described Kamrowski as having addressed "the cosmography of man's inner worlds" by keeping abreast of modern biology, specifically that of the Surrealist and polymath Pierre Mabille—a friend of Breton's—who in his 1949 book *Initiation à la Connaissance de l'Homme* (Initiation to the Understanding of Man) advanced the need for modern man to remain in contact with nature through bodily rhythms—"the outcropping of visceral feeling in peripheral conscious [translating itself] into mental forms, and thus into images."[26] Kamrowski's *panoramagraphs* take a long length of cloth, Breton tells us, and display in a single panoramic vision various relevant examples of the different kingdoms of nature as well as working tools and weapons from the history of man, "constructing a *thread* with the aim of stretching it to the farthest point of the physical world … before leading it back again towards the mental world, in order to take its mysterious workings by surprise." As late in Surrealism's history as 1950, in Breton's world and in Mabille's, biology's task was to go beyond the body's architectural stability and understand "the kernel … which supervises the maintenance of life" and in that way transcend the limits of "structural thought."[27] For Breton's "structural thought," here, read the Enlightenment project of quantificational morphology and natural law, a project that Thompson's book exemplified. By the same token Thompson had bequeathed to artists—it seems unknowingly—a set of visual principles and images that they had every freedom to exploit, but without any obligation to scientifically deploy. On the contrary, the world that Thompson had tried so hard to explain could now be celebrated for its strangeness, its otherness, its evocation of the fantastical within the real.

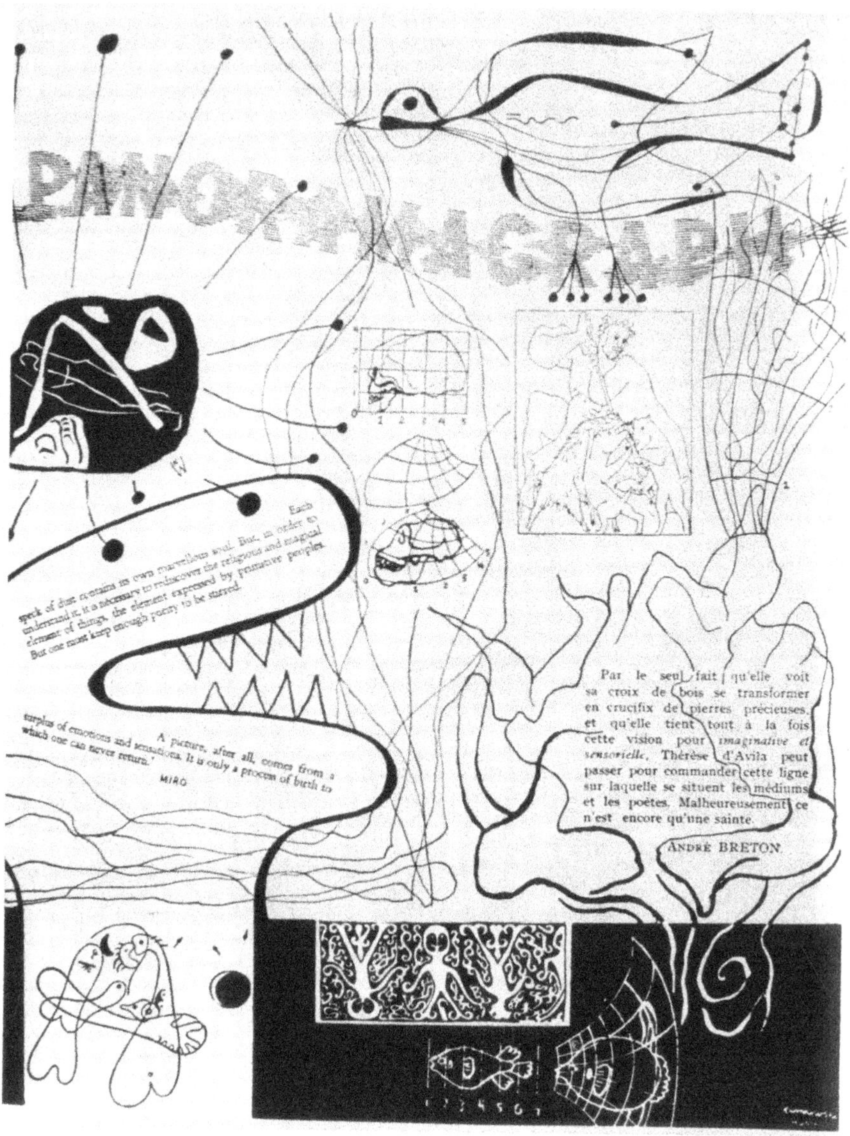

Figure 6.6 G. Kamrowski, *Panoramagraph* from *VVV*, New York, 1943.

The last word can be given to Dalí. In a book entitled *50 Secrets of Magic Craftsmanship*, published in New York in 1948, we find in an appendix a series of passages from the 1942 edition of *On Growth and Form*—a dozen long paragraphs in all—whose relation to the main text of Dalí's book is only partially clear. Dalí has reproduced Thompson's passages describing sea urchins, spider webs, water splashes, and gnomic forms—the latter being geometrical figures, as Thompson in his book explains through Aristotle, "which suffer no alteration (save magnitude) when they grow"—with a Surrealist mixture of mischief and delight.[28] *On Growth and Form* provided Dalí, as it provided others, with images of paradox, of wonder, of nature's perverse beauty in the multiple forms in which it could appear. We might agree that Thompson could never have predicted such a progeny. The Surrealist conscience may not have interested him. But by their mutual differences he and it were indelibly connected.

Notes

1 Thompson, D'Arcy, *On Growth and Form*, abridged edition, edited by John Tyler Bonner (Cambridge: Cambridge University Press, 1961/2014), 3. My emphasis.
2 Breton, A., *First Manifesto of Surrealism*, 1924; from the translation is C. Harrison and P. Wood (eds.), *Art in Theory: An Anthology of Changing Ideas 1900–2000* (Oxford: Blackwell, 2003), 451–2.
3 A. Breton, "La beauté sera convulsive," *Minotaure*, Vol. 5, (1934), 10.
4 Thompson, *On Growth and Form* (1961/2014), 8, 326–7.
5 Stewart, I., *The Mathematics of Life* (Basic Books, 2011), 9.
6 Haldane, J. S., *The New Physiology and Other Addresses*, C. Griffin, Edinburgh 1919, cited by Thompson in his review, "The New Physiology," *Mind*, Vol. 28 (1919), 362.
7 Thompson, *On Growth and Form* (1967/2014), 3. Disagreement with Julian Huxley over the mathematical expression of *relative* rates of growth only added to Thompson's reputation as an erudite but maverick figure in the field. See the respectful yet critical remarks in Huxley, *Problems of Relative Growth* (London: Methuen and Co., Ltd, 1932) chapter 1.
8 Lichtenstern, C., *Henry Moore: Work—Theory—Impact* (London: Royal Academy of Arts, 2008), 407; as well as "Henry Moore and Surrealism," *Burlington Magazine*, November 1981, 651–2.
9 For the list, see "A List of the Published Writings of D'Arcy Wentworth Thompson," in *Essays on Growth and Form Presented to D'Arcy Wentworth Thompson*, edited by W. E. Le Gros Clark and P. B. Medawar. (Oxford: Clarendon Press, 1945), 386–400.
10 Thompson, D'Arcy, *On Growth and Form* (1967/2014), 74.
11 Monod-Herzen, E., *Principes de Morphologie Générale* (Paris: Gauthier-Vilars,1927), Vol. I, 33.
12 Monod-Herzen, E., *Principes* (Paris, 1927), Vol. 2, 115.
13 Specifically, it is the evidence of *déformation*, more specifically *hypertonicité*, found in the works of the Hungarian artist Gyula Zilzer in the aftermath of 1914–1918 where such symptoms are most evident.
14 Monod-Herzen, E., *Principes* (Paris, 1927), Vol. 2, 173.
15 Dalí, S., "Posicío moral del surrealisme" [The moral position of Surrealism], *Hèlix*, Barcelona, March 22, 1930; in H. Finkelstein (ed.), *The Collected Writings of Salvador Dalí* (Cambridge: Cambridge University Press, 1961/2014), 272.

16 The St Anthony panel, a "puzzle-picture" in woodcut by the German artist Erhard Schön, and a fine anamorphosis of Charles V, all from the collection of Jacques Lipschitz in Paris, were displayed in the landmark exhibition *Fantastic Art, Dada, Surrealism* in New York in 1936.

17 Dalí, S., "Les nouvelles couleurs du sex appeal spectral"; *Minotaure*, February 5, 1934; in Finkelstein (ed.), *The Collected Writings*, 205–6.

18 Dalí, S., "Apparitions aérodynamiques des 'Êtres-objets,'" *Minotaure*, 6, Winter 1934–1935; Finkelstein, *Collected Writings*, 210, 209, 208.

19 Thompson, "Excess and Defect: Or the Little More and the Little Less," *Mind*, 38 (1929), 43.

20 Thompson, "Excess and Defect: Or the Little More and the Little Less," 55.

21 Thompson, *On Growth and Form* (London and New York edition, 1942), 120.

22 Reichenbach, H., "Crise de la causalité," *Documents*, Vol. 2 (1929), 108.

23 Bachelard, G., *Le Nouvel Esprit Scientifique* (Presses Universitaire de France, Paris 1934; in translation by Arthur Goldhammer as *The New Scientific Spirit* (Boston: Beacon Press, 1985), 132.

24 The interest taken in *On Growth and Form* by Naum Gabo, not a Surrealist, is well documented in M. Hammer and C. Lodder, *Constructing Modernity: The Art and Career of Naum Gabo* (New Haven: Yale University Press, 2000), 385–8.

25 A small canvas painted collectively by all three of them, known as *Untitled* and dated 1940–41, seems richly suggestive of the techniques and image-types they found there. It is reproduced in *Abstract Expressionism*, ed. D. Anfam (London: Royal Academy of Arts, 2017), 140.

26 Breton, A., "Gérôme Kamrowski" (1950), referring to P. Mabille, *Initiation à la Connaissance de l'Homme*, Paris 1949, in Breton, *Surrealism and Painting*, trans. S. Watson-Taylor (London: Macdonald, 1972), 226.

27 Breton, A., "Gérôme Kamrowski," op cit, 229.

28 It is the passage from *On Growth and Form*, 1942, 759–62, reproduced in Dalí, S., *50 Secrets of Magic Craftsmanship* (New York: Dial Press, 1948), 187.

7

Structures of Light as "An Ethnologist's Jewels": D'Arcy Thompson, The Independent Group, and Montage

Assimina Kaniari

During the early 1950s, The Independent Group (IG) drew heavily on imagery from D'Arcy Thompson's book *On Growth and Form* (1917), which was first published in 1917, comprising a wealth of drawings and photographs depicting complex forms from nature and art as illustrations to developmental phenomena embodied in pictures and in graphic form, as, in Martin Kemp's words, "geometries of growth."[1] This becomes apparent in two of the major exhibitions The Independent Group organized at the Institute of Contemporary Arts (ICA) in London titled Growth and Form (1951), after the title of Thompson's book, and *Parallel of Life and Art* (1953). In what follows, I discuss the migration of images from Thompson's book and against the backdrop of their original context as illustration into the format of display as an exhibition context, giving special attention to the idea of montage. I argue that, underpinning both Thompson's image strategies as regards book illustration in *On Growth and Form* and the IG's 1950s exhibitions at the ICA, there is the same attention to and affinity with the idea of image montage, which demanded, by consequence, the active participation of the observer or the reader.

Paths of Montage between Scientific Illustration and Avant-Garde Art Display

Indeed, in the 1953 exhibition *Parallel of Life and Art,* both artists and viewers are cast as "editors."[2] And in the 1917 edition of Thompson's book, illustrations are often composite, constructed as picture montages of pictures, which demanded the powers of association of the reader. They appeared to the eye as a juxtaposition of forms in the same page spread, as illustrations edited of diverse sources, including science journals, prior to being printed in the book as "readable" images.[3] Photographic stills from science journals as images already accessible on a mass scale were prominent examples among both Thompson's illustrations and the IG's 1950s exhibitions, with Arthur Worthington's splash a notable example from Thompson's book.[4]

Worthington's image was printed across all consecutive editions of Thompson's 1917 book, including the new edition of 1942.[5] Thompson's illustration of Worthington's "phases of a splash" had been reproduced from the physicist and photographer's original 1908 publication "A Study in Splashes" from a science journal.[6] (Figure 7.1 *Phases of a Splash.* Illustration from Thompson's *On Growth and Form* (1917) 1942 (Fig. 115), after Worthington.) Thompson's second edition of 1942, for example, included also an illustration captioned "An instantaneous photograph of a 'splash,'" "from Harold

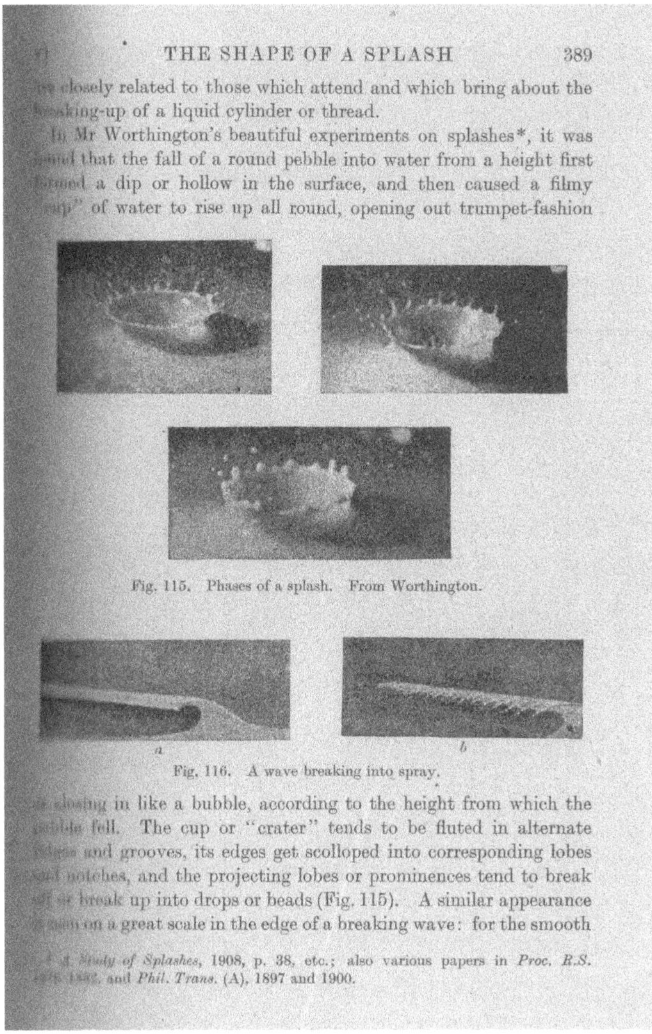

THE SHAPE OF A SPLASH 389

closely related to those which attend and which bring about the ...king-up of a liquid cylinder or thread.

In Mr Worthington's beautiful experiments on splashes*, it was ... that the fall of a round pebble into water from a height first ... a dip or hollow in the surface, and then caused a filmy ... of water to rise up all round, opening out trumpet-fashion

Fig. 115. Phases of a splash. From Worthington.

a b

Fig. 116. A wave breaking into spray.

... in like a bubble, according to the height from which the ... fell. The cup or "crater" tends to be fluted in alternate ... and grooves, its edges get scolloped into corresponding lobes ... notches, and the projecting lobes or prominences tend to break ... break up into drops or beads (Fig. 115). A similar appearance ... a great scale in the edge of a breaking wave: for the smooth

* *A Study of Splashes*, 1908, p. 38, etc.; also various papers in *Proc. R.S.* ... and *Phil. Trans.* (A), 1897 and 1900.

Figure 7.1 *Phases of a Splash.* Illustration from Thompson's *On Growth and Form* (1917) 1942 (Fig. 115), after Worthington.

E. Edgerton, Massachusetts Technical Institution" reproduced as a single page spread photographic plate opposite the title page.[7] Like Worthington's photographic montages of "Phases of a Splash" and Edgerton's blown up portrait of "An instantaneous photograph of a 'splash,'" Thompson's own illustrations were constructed as a pictures montage comparing diagrams of forms to other kinds of pictures emphasizing visual analogy across nature and art and often animate and inanimate matter. One such example is Thompson's diagrammatic, linear drawings of medusas juxtaposed (as in montage) to photographic stills of splashes.[8] Thompson's fascination with the possibilities of montage is evident in the 1942 edition of his book while describing "Mr. Worthington's beautiful experiments on splashes" as a quasi-cinematic temporal sequence embodied in image montage: "the fall of a round pebble into water from a height" "first formed a dip or hollow in the surface, and then caused a filmy 'cup' of water to rise up all round, opening out trumpet-fashion or closing in like a bubble."[9]

In the new edition of 1942, Thompson reproduced Edgerton's print of an "instantaneous" splash which, by contrast to Worthington's 1908 printed forms of splash imagery, comprised an enlarged image which read as a portrait of a splash. (Already by 1939 Edgerton had produced a popular book and a film on splash imagery.[10]) According to Martin Kemp, comparisons between "stilled splashes" and medusas allowed Thompson to intuit analogies between nature and art, both considered as the location of dynamic phenomena where physical forces in action impacted matter being responsible for "natural" and artificial form.[11] (Figure 7.2 Thompson's analogies of splashes and medusas from chapter V "The forms of cells.")

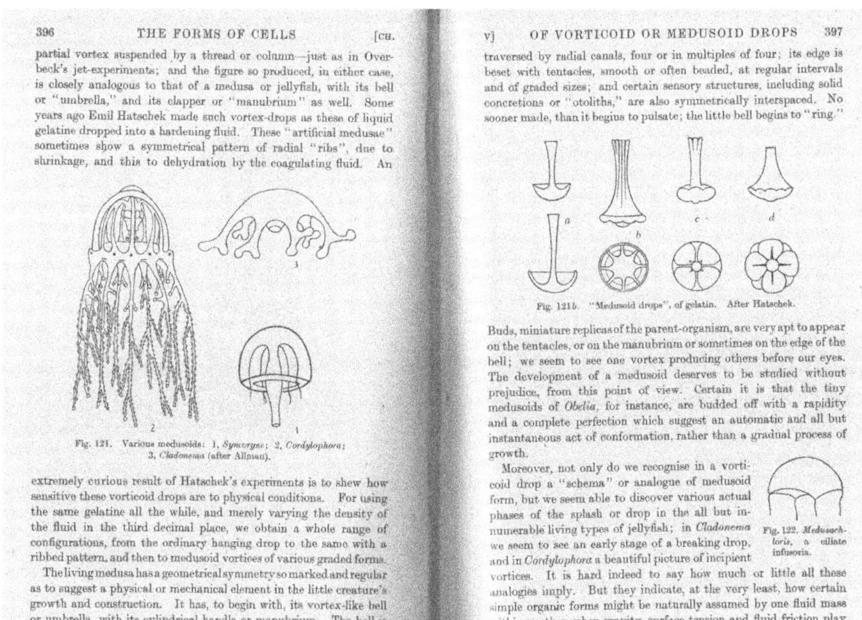

396 THE FORMS OF CELLS [CH.

partial vortex suspended by a thread or column—just as in Overbeck's jet-experiments; and the figure so produced, in either case, is closely analogous to that of a medusa or jellyfish, with its bell or "umbrella," and its clapper or "manubrium" as well. Some years ago Emil Hatschek made such vortex-drops as these of liquid gelatine dropped into a hardening fluid. These "artificial medusae" sometimes show a symmetrical pattern of radial "ribs", due to shrinkage, and this to dehydration by the coagulating fluid. An

Fig. 121. Various medusoids: 1, *Syncoryne*; 2, *Cordylophora*; 3, *Cladonema* (after Allman).

extremely curious result of Hatschek's experiments is to shew how sensitive these vorticoid drops are to physical conditions. For using the same gelatine all the while, and merely varying the density of the fluid in the third decimal place, we obtain a whole range of configurations, from the ordinary hanging drop to the same with a ribbed pattern, and then to medusoid vortices of various graded forms.

The living medusa has a geometrical symmetry so marked and regular as to suggest a physical or mechanical element in the little creature's growth and construction. It has, to begin with, its vortex-like bell or umbrella, with its cylindrical handle or manubrium. The bell is

v] OF VORTICOID OR MEDUSOID DROPS 397

traversed by radial canals, four or in multiples of four; its edge is beset with tentacles, smooth or often beaded, at regular intervals and of graded sizes; and certain sensory structures, including solid concretions or "otoliths," are also symmetrically interspaced. No sooner made, than it begins to pulsate; the little bell begins to "ring."

Fig. 121*b*. "Medusoid drops", of gelatin. After Hatschek.

Buds, miniature replicas of the parent-organism, are very apt to appear on the tentacles, or on the manubrium or sometimes on the edge of the bell; we seem to see one vortex producing others before our eyes. The development of a medusoid deserves to be studied without prejudice, from this point of view. Certain it is that the tiny medusoids of *Obelia*, for instance, are budded off with a rapidity and a complete perfection which suggest an automatic and all but instantaneous act of conformation, rather than a gradual process of growth.

Moreover, not only do we recognise in a vorticoid drop a "schema" or analogue of medusoid form, but we seem able to discover various actual phases of the splash or drop in the all but innumerable living types of jellyfish; in *Cladonema* we seem to see an early stage of a breaking drop, and in *Cordylophora* a beautiful picture of incipient vortices. It is hard indeed to say how much or little all these analogies imply. But they indicate, at the very least, how certain simple organic forms might be naturally assumed by one fluid mass within another, when gravity, surface tension and fluid friction play

Fig. 122. *Medusachloris*, a ciliate infusoria.

Figure 7.2 *The Forms of Cells.* Illustration from chapter V of Thompson's *On Growth and Form* (1917) 1942, drawing visual analogies between the forms of splashes and medusas.

And just as Thompson's strategy of illustration, suggested "parallels" by way of visual association captured as printed image juxtapositions, visual juxtaposition as a metaphor for "parallel" underpinned the exhibition strategy of the IG in the 1950s at the ICA. In the 1953 exhibition, the role of montage becomes visible in their own words describing the process of exhibition making as a form of image editing.[12]

Image-Montages of Art and Life: Between Thompson's Illustrations and the IG's 1950s Exhibition Practice at ICA

A 1953 note on The *Parallel of Life and Art* exhibition, kept among Nigel Henderson's papers at Tate Archive, suggested that "Technical inventions such as the photographic enlarger, aerial photography, and the high speed flash … have given us new tools, with which to expand our field of vision beyond the limits imposed on previous generations."[13] "The common visual denominator" among the images was the camera[14] and the visitor was referred to as "the observer."[15] The camera allowed selected images on display to be viewed as independent from their context, acting "as recorder of nature objects, works of art, architecture and technics; as reporter of human events … and, as scientific investigator extending the visual scale and range, by use of enlargements, X rays, wide angle lens, high speed and aerial photography."[16] Participating artists were listed in the following order: "Nigel Henderson, Photographer, Eduardo Paolozzi, Sculptor, R.S. Jenkins, Engineer" and "Alison and Peter Smithson, Architects."[17] Yet, as the 1953 Memorandum of the *Parallel of Life and Art* makes clear, they conceived of their role not in terms of a contemporary curator, but rather as "editors of the material," which is connected in turn to the printed or technically reproduced image; for example, the exhibition is defined as "an exhibition of documents, principally photographs and diagrams."[18]

Such an interest in experimental practice and photography goes back to 1951 but even earlier in the context of Nigel Henderson's experimental photographic practices in printing images, both examples having a strong resonance with Thompson's book as a source of inspiration. Nigel Henderson's photographic practice, in addition to the two ICA exhibitions of the 1950s and in the context of what he described *c.* 1950 as "stressed photographs," may be recast as perhaps the earliest articulation of Thompson's impact on 1950s avant-garde art practices of the IG. At a formal level, and also by way of artistic research practices in a post-Bauhaus context, his photographs echo Thompson's illustrations of complex forms and imagery of "geometries of growth" in nature and art as expressions of dynamic phenomena produced by forces acting upon matter.[19]

Nigel Henderson's "Stressed Photographs" of the 1950s

Indeed, Henderson's "stressed photographs" recall the topology of Thompson's examples of illustrations describing phenomena of deformation but also the fluidity

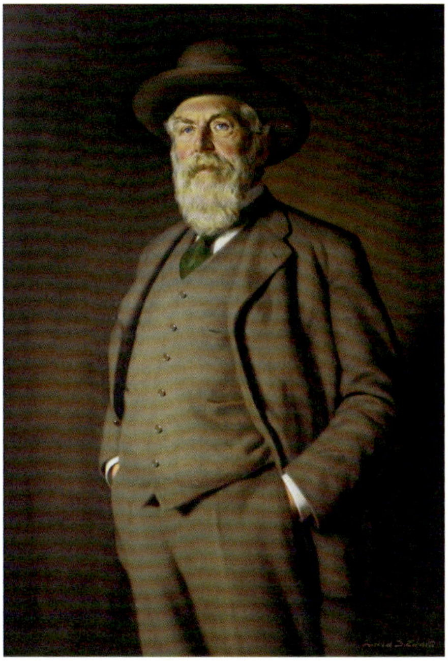

Plate 1 David S. Ewart, Professor Sir D'Arcy Wentworth Thompson, 1938/50. Courtesy of University of Dundee Museum Services.

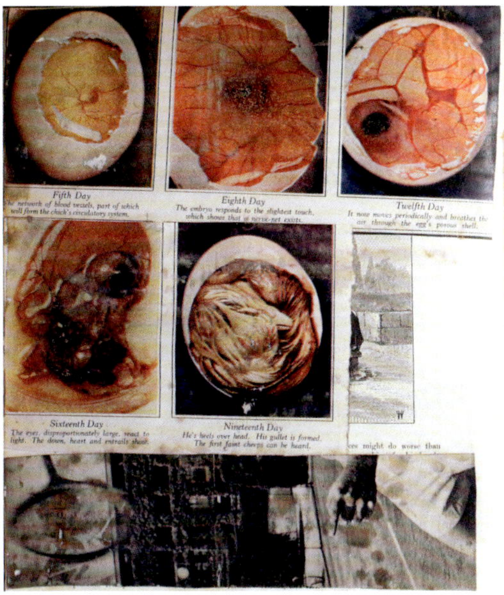

Plate 2 Henderson, Nigel, *Page from Scrapbook Showing Photographs, with Captions, of a Chicken Embryo Cycle*, c. 1952; Nigel Henderson 1917–1985; Image # M01616. Source: © Tate, London 2018.

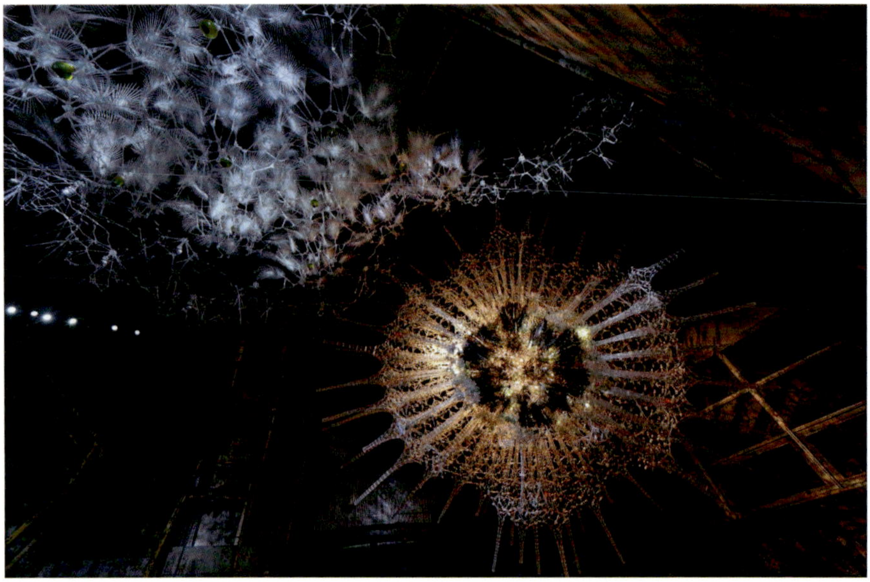

Plate 3 Philip Beesley, Partial view of Astrocyte, by Living Architecture Systems/Philip Beesley Architect Inc. et al., at DX EDIT festival, Unilever Factory, Toronto October 2018.

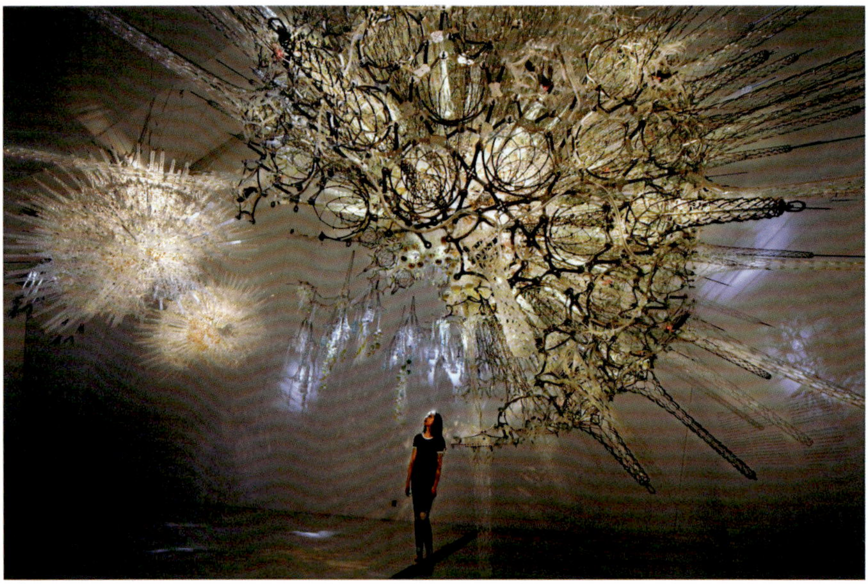

Plate 4 Philip Beesley, Partial view of Noosphere by Living Architecture Systems/Philip Beesley Architect Inc., 4DSOUND et al., at Transforming Fashion Exhibition, Royal Ontario Museum, Ontario, Canada, June–October 2018.

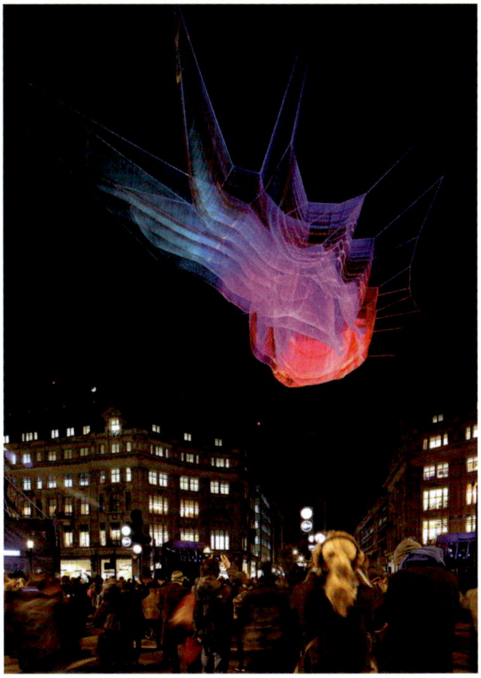

Plate 5 Janet Echelman, *1.8*, 2016, Initially sited in London, colored lighting, Wi FI, and interactive computer programming. Fibers are braided with nylon and Ultra high molecular weight polyethylene, Net: L 100′ × W 45′ × D 20′. Installation: L 180′ × W 180′ ft. × H 70′. Photo: Ema Peter. Courtesy of Janet Echelman.

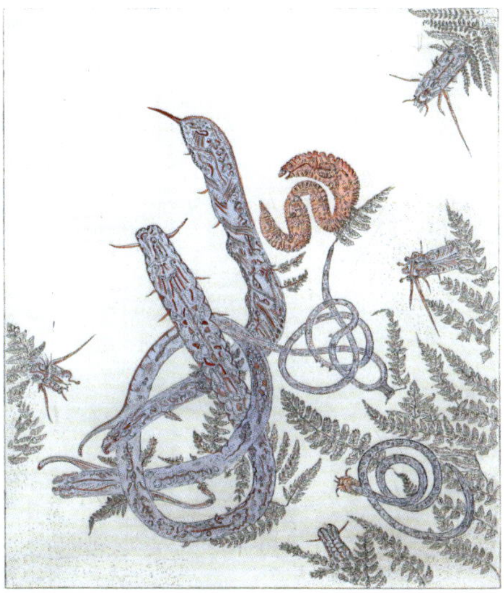

Plate 6 Gemma Anderson, A copper etching of a nematode drawn from specimens at the Natural History Museum, 2012, 18.5 × 16.5 inches. Courtesy of Gemma Anderson.

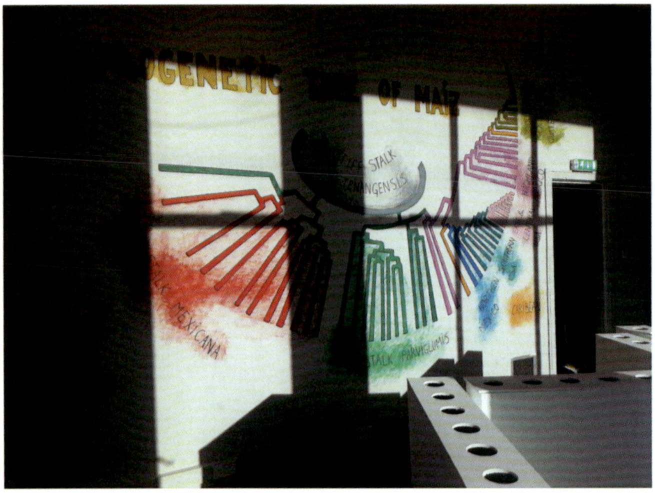

Plate 7 Marta de Menezes in collaboration with Maria Antonia Gonzalez Valerio, *Origin of Species—Post Evolution—Maiz*, 2017–. Installation including a drawn Phylogenetic Tree of Maiz. Dimensions variable. Acknowledgments: Dr. Nelson Saibo, Principal Investigator @Plant Gene Regulation Laboratory, ITQB, Portugal. Courtesy of Marta de Menezes.

Plate 8 Tauba Auerbach, *Chiral Fret (Meander)/Extrusion/Ghost*, 2015. Woven canvas on wooden stretcher. Credit: @Tauba Auerbach. Courtesy of the Paula Cooper Gallery, New York. Photo: Steven Probert.

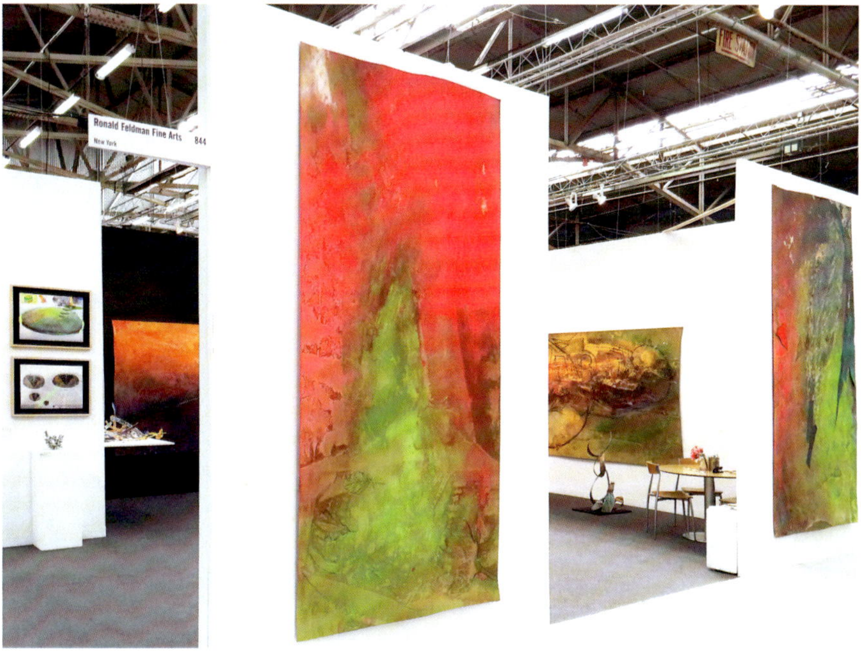

Plate 9 Todd Siler, "NanoWorld," Ronald Feldman Fine Arts, at The Armory Show (March 5–9, 2014), New York, NY. Courtesy of Ronald Feldman Fine Arts, New York, NY, and www.ArtNanoInnovations.com.

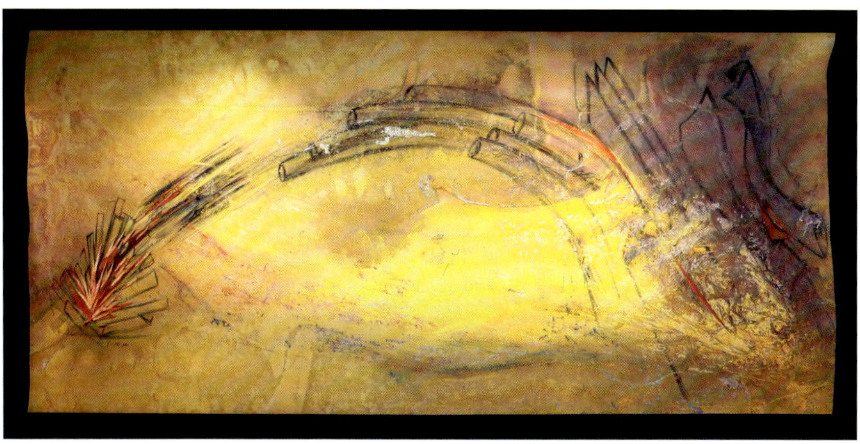

Plate 10 Todd Siler, *Metaphorming Four Building Blocks of the NanoWorld* (2011–2012), mixed media on synthetic canvas with collage elements. Courtesy of ArtNano Innovations.

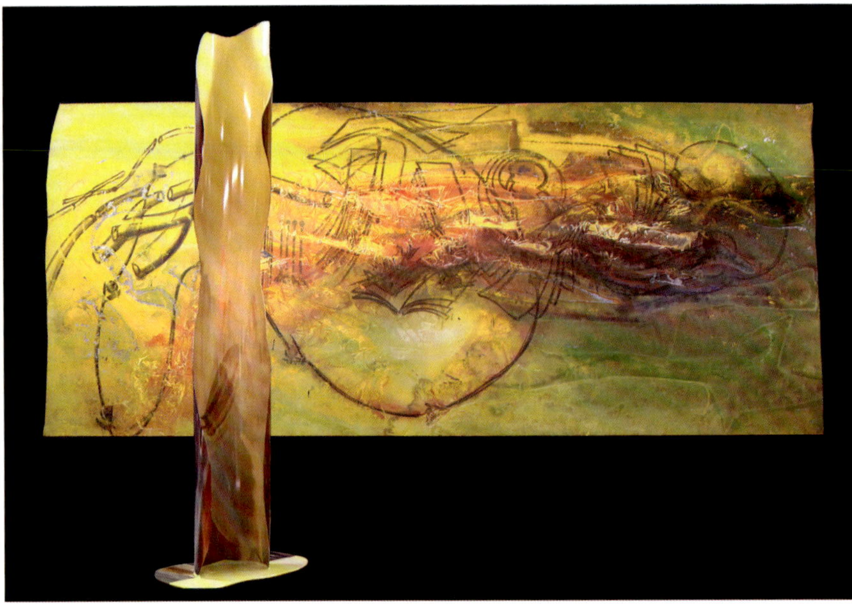

Plate 11 Todd Siler, *Envisioning Minds & Nature Forming Nanomaterials (1 to 100 nm) That Form All Materials* (2011–2013), mixed media on synthetic canvas with freestanding *Photosculpture: Synthesizing Nature's Nanotubes #1* (2011); Painting: Gift of Edwin & Barbara Prober to the Luskin Conference Center at UCLA, Los Angeles, 2015; Sculpture: Gift of Ronald & Frayda Feldman. Courtesy of the Luskin Conference Center at UCLA, Los Angeles, 2015.

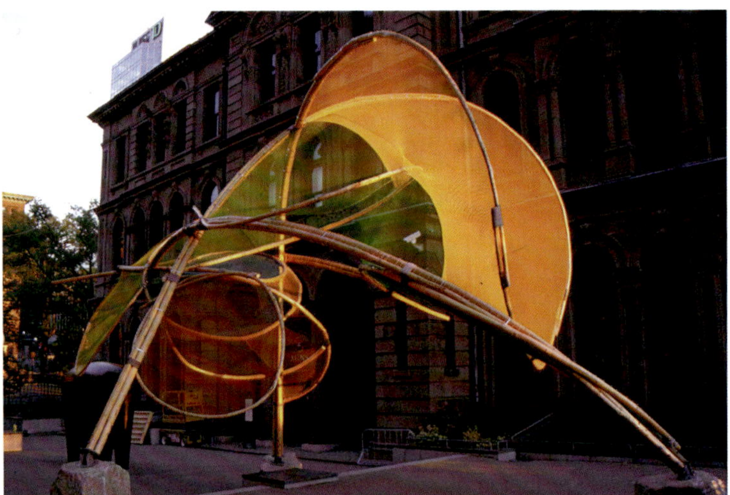

Plate 12 Sarah Bonnemaison, *Hummingbird*, 2000. Dry laminated wood and hand-dyed nets. Courtesy of the Art Gallery of Nova Scotia, Halifax.

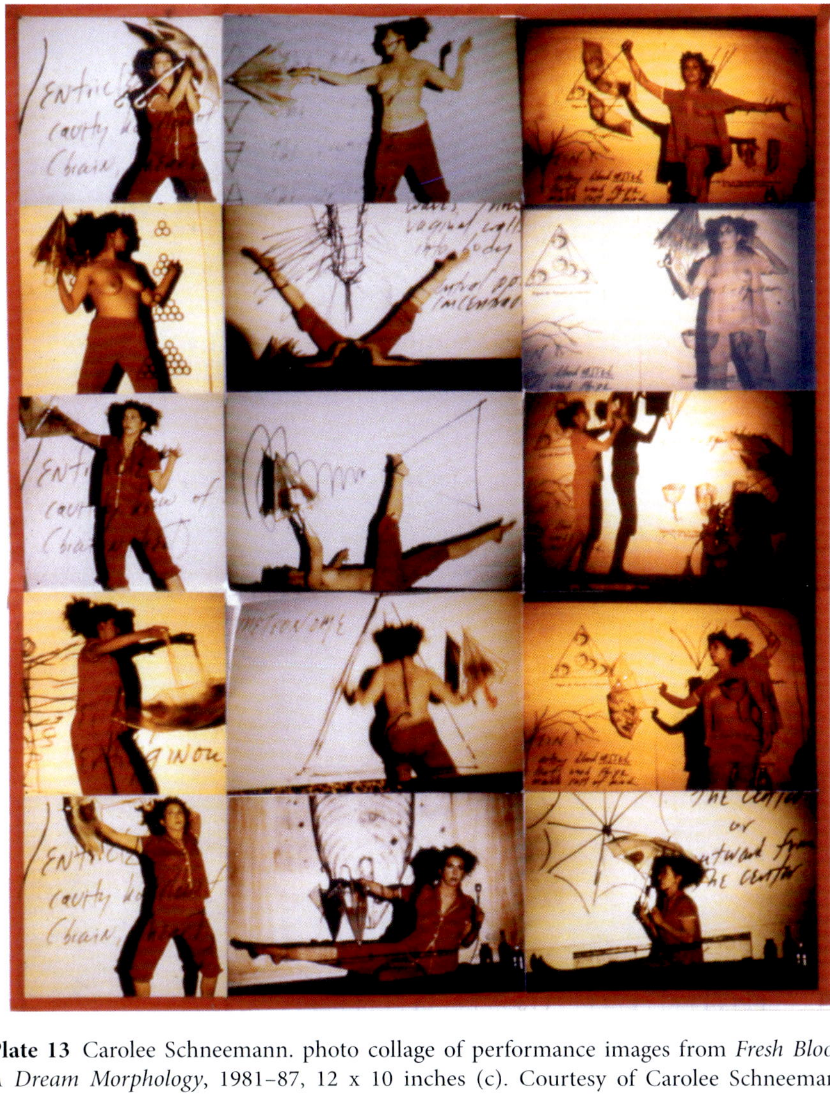

Plate 13 Carolee Schneemann. photo collage of performance images from *Fresh Blood: A Dream Morphology*, 1981–87, 12 x 10 inches (c). Courtesy of Carolee Schneemann Foundation

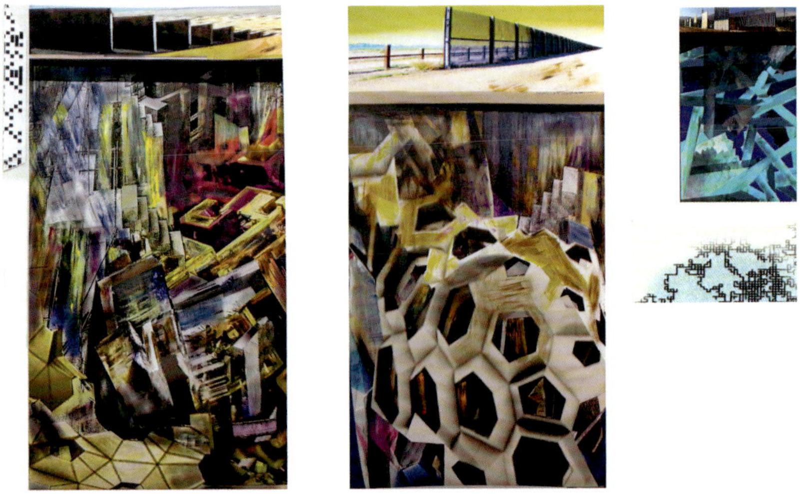

Plate 14 Ellen K. Levy, Installation: *Crossing Borders: Mexico*, 2018. Each of the 2 large vertical works is 80 × 38 inches. Acrylic and gel over archival print. Courtesy of Ellen K. Levy.

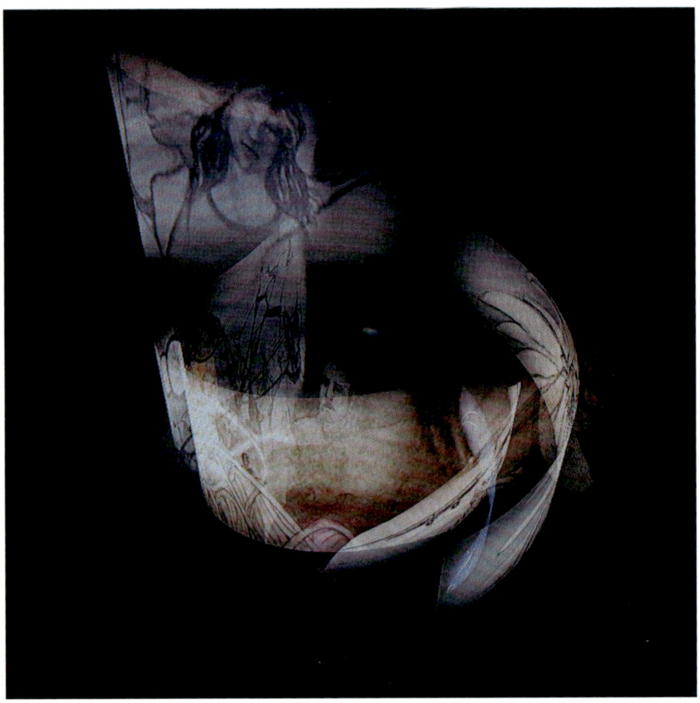

Plate 15 Meredith Tromble with Dawn Sumner, *Dream Vortex* (detail), 2014, Virtual Reality Installation. Courtesy of Meredith Tromble and Dawn Sumner.

of the stilled splashes. (Figure 7.3 Nigel Henderson, "Stressed photograph" *c.* 1950, **TATE.**) And Henderson's working methods underpinning his "stressed photographs" mirror Thompson's explanation of complex phenomena in nature and art giving rise to the forms he uses as illustrations in the book.[20] Henderson describes, for example, the alterations he produced on the images as the result of the "printing process" or as a game of observation across scale allowed by the use of an image enlarger.

In order to produce "stressed" photographs, he applied stresses directly and literally on the printing paper and used an image enlarger in order to isolate and enlarge his portraits. Henderson explains:

> I noticed … that when I had an actual negative that interested me (let's say a boy on a bicycle) I could sometimes enrich the impact of the image by slanting the paper under the enlarger projecting lens … If I pleated the paper horizontally I could create a pattern of stress which further animated the situation by putting the wheels and frame "through it" as it were and creating an identification with the boys' efforts and the tension of the wheels and frame in a somewhat "Futurist" way.[21]

And as portraits of people, Henderson's "stressed photographs" are and act also as ethnographic records of human action. The same can be argued for the 1953 exhibition and the role of experimentation in this context.

Experimental art practice in the context of the 1953 exhibition aimed to produce images that suggested parallels of life and art as evidence of "a single cultural whole."[22] And three years prior to the exhibition, Henderson's *c.* 1950 portraits had already

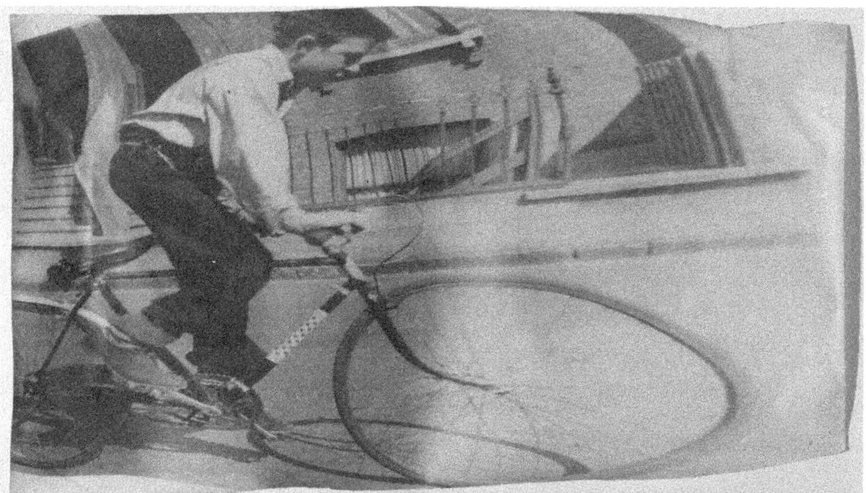

Figure 7.3 Henderson, Nigel, *Stressed Photograph c.* 1950. Nigel Henderson 1917–85; Image # P79309; Photograph, gelatin silver print on paper, mounted on board. *Source*: © Tate, London, 2018.

articulated in the context of photographic practice the key idea underpinning the 1953 exhibition's rationale. Henderson's "stressed photographs" allow for analogies between art and life to emerge as a social and visual fact and as the product of the artist's creative modifications of both technique and material reality (and perhaps social life and social reality as well). In this light, the parallel of art-and-life as the object of the 1953 exhibition, in addition to its references to Thompson's image montages between nature and art in the context of illustration, may be understood as an extension of Henderson's own experimental avant-garde photographic practice (and perhaps then as an art work embodied in an image, similar to Thompson's idea of illustration as the product of image montage) re-captured in an ethnographic light.

In this, Henderson was not alone. Going back to the 1951 Growth and Form exhibition, a letter from Eduardo Paolozzi to Henderson kept in the Henderson papers at the Tate Archives is a record of a sketch by Paolozzi, across half the first page, which comprised an attempted description of an image enlarger.[23] The same letter suggests Henderson's and Paolozzi's early involvement in the exhibition Growth and Form (1951), before it eventually fell under Richard Hamilton's purview.[24] In addition to Henderson's experimental practice, could it be that the emphasis on different scales and pattern may be credited also to Paolozzi's input?

Pictures on display (often photographic prints of photo-enlargements) included technical images exhibited as products of a number of scientific devices and inventions past and present (not in chronological order) which made accessible structures invisible to the naked eye. Among the exhibits of the ICA Growth and Form exhibition there was, for example, an image of a micro-photograph of molybdenum oxide smoke, reproduced as a cover image in the 1951 October issue of the *Architectural Review* devoted to the ICA Growth and Form exhibition.[25] And while Growth and Form (1951) at ICA is often credited to IG artist Richard Hamilton, and some of his drawings at TATE have also drawn inspiration from Thompson's 1917 book,[26] a photogram by Henderson in the Tate archive suggests the close affinities between his earlier experimental photographic practice and the perception of matter as a game of observation across scales enacted in Thompson's book prior to 1951.

In Nigel Henderson's own archive at the Tate, there is also a page from a science journal where his interest in the biological clearly becomes manifest as a possible influence on his work and portraits. It depicts an embryo in its various stages. (Color Plate 2. Nigel Henderson, **TATE**.) Just as Thompson's own analogies between animate and inanimate matter if captured as an image (through the mediation of the lens, as also in the example of Haeckel's drawing of radiolarian supposedly observed under the microscope), Henderson worked with an expanded field of imagery and forms as a source of inspiration for his photographic practice.[27]

Inorganic form, stilled as a lens-based image, such as the splash, appears in retrospect as Thompson's privileged instrument of observation and (quite in a photographic sense) as a structure of light; projected over other images and forms (as in through a lens), producing image overlays. The latter produced, as Kemp suggested in the context of Thompson's comparisons of splashes to medusas, "intuitive analogies."[28] Deformation, Thompson's privileged phenomenon used in the book to account for the impact of physical forces on living matter and on form, in retrospect and as a visual image, reads almost as a "photo-enlargement." And it is hard to distinguish whether Henderson's stressed photographs were literal translations in the context of art practice of Thompson's biological explanations or if Thompson conceived of deformation photographically.

His image montages may also be conceived as literally montages of thought processes (and perhaps disciplines) brought together through experimental art practice and via the image of montage printed in the book as an illustration. Indeed, in the 1917 book, the biologist observes form, very much as the editors do in the 1953 exhibition, as a camera. And here as well, the biologist-as-a-camera functions as the "common visual denominator" for illustrations of art and life to come together in one narrative in the book (and as an "illustration").[29] The same vision (recast as a form of ethnographic observation) materializes in the thought and thinking about visual analogy which Claude Lévi-Strauss articulates in the context of an anthropological-fiction (indebted as he writes in retrospect to Thompson and his 1917 book).[30] In this context, the image of the splash, and by way of his debt to Thompson, is recast into "an ethnologist's jewel."[31] Like Thompson's biological image and art forms' analogies of 1917, Lévi-Strauss's ethnographic analogies similarly compare splash imagery to artifacts suggesting diverse types of social order across different cultural and historical contexts.

In an ethnographic act of montage, Lévi-Strauss edits Thompson's splash into an anthropological sequence/narrative of a "nature" as "culture" nexus and argument. In doing so, he projects the form of the splash, enlarged and delineated over the pattern of an early modern design of a count's crown. Ethnographic explanation, just as Thompson's "biological" one in 1917, resembles the processes of montage: images as illustrations of social phenomena are produced literally in his thinking (just like Henderson's "stressed photographs" and in a materialist light) as the products of the "printing process" itself (that is, as crystallizations of the avant-garde art idea of form).

With thanks to Martin Kemp and David Hockney for inspirational discussions that helped develop ideas in this essay. Many thanks to Ellen Levy and Charissa Terranova for commenting and editing drafts of this paper.

Notes

1 For Thompson's "geometries of growth" see Kemp, Martin, "Growth and Form," in *Seen Unseen. Art, Science and Intuition from Leonardo to the Hubble Telescope*, edited by Martin Kemp (Oxford, NY: Oxford University Press, 2006), 200–38. Thompson, D'Arcy Wentworth, *On Growth and Form* (Cambridge: Cambridge University Press, 1917; 1942). For a discussion on the impact of Thompson's writing on Nigel Henderson and The Independent Group in the context of post-Bauhaus, design theory and the writing of László Moholy-Nagy, see Assimina, Kaniari "D'Arcy Thompson's '*On Growth and Form*' and the Concept of Dynamic Form in Postwar Avant-Garde Art Theory," *Interdisciplinary Science Reviews*, Vol. 38, No. 1 (2013), 63–73, DOI: 10.1179/0308018813Z.00000000035. See also, Assimina Kaniari, "Uma

genealogia da nano-art através da escala: práticas fotograficas de Nigel Henderson no arquivo da Tate Gallery" ["A genealogy of nano-art across 3 scale: Nigel Henderson's avant garde photographic practices in the Tate Archive" in Portuguese], Translation Marta De Menezes, *Nada,* No 13, 2009, 40–51. For a reading of Nigel Henderson's photographic practice and early photograms against the work of Moholy-Nagy, see Walsh, Victoria and Henderson, Nigel, *Parallel of Life and Art* (London: Thames & Hudson, 2001).

2 Tate Archive, Henderson Papers, 9211-5-1-1.

3 Thompson summarizes his argument in Chapter 1, "Introductory" and section titled "Of Life itself" in the new edition of 1942, as follows: "The terms Growth and Form, which make up the title of this book, are to be understood, as I need hardly say, in their relation to the study of organisms. We want to see how, in some cases at least, the forms of living things, and of the parts of living things, can be explained by physical considerations and to realise that in general no organic forms exist save such as are in conformity with physical and mathematical laws." Thompson, D'Arcy Wentworth, *On Growth and Form* (Cambridge: Cambridge University Press, 1942), 15.

4 For a discussion on images already printed in scientific journals and popular science books reproduced in Thompson's first and consecutive editions, see Kemp, Martin, "Stilled Splashes," *Visualizations: The Nature Book of Art and Science* (Oxford, New York: Oxford University Press, 2000), 78–9. For The IG's references to scientific imagery as a form of mass culture in the 1953 ICA exhibition see Tate archive, Henderson Papers, 9211-5-1-2.

5 Thompson, *On Growth and Form* (1942), 389.

6 Ibid. Worthington's "Spark photographs of splashes" are among the permanent collection of the Science and Media Museum at Bradford, dated, circa 1900, and Arthur Mason Worthington, FRS, is indexed as "Physicist specialising in fluid mechanics and splashes" and a "pioneer of high speed photography." https://collection.sciencemuseum.org.uk/people/cp107560/arthur-mason-worthington Kemp describes Worthington's splash imagery published in 1908 and reproduced by Thompson in 1917 on display in the museum, as "photographic montages." Kemp, "Stilled splashes," 78. Bradford is also the home town of David Hockney whose iconic painting of a splash titled *A Bigger Splash* (1967) is considered a landmark of 1960s Pop. Hockney painted a whole series of images of water and swimming pools during the 1960s, many of which drew on comparisons with photographs, and consistently throughout this work approached the painting of water as a "formal" problem. See, for example, David Hockney, *Pictures by David Hockney*. Selected and edited by Nikos Stangos (New York: Harry N. Abrams, 1979). The form of Hockney's pictorial splash a photographically enlarged blow up.

7 In the first (1917) edition of the book, Thompson reproduced, for example, FRS Arthur Worthington's photographic imagery from "A study of splashes 1908" (Kemp, "Stilled splashes," 79) and Worthington's stills survived across consecutive editions and in the New Edition of 1942. See, for example, Thompson, *On Growth and Form*, 389. In the 1942 New Edition, Thompson reproduced as a full-page photographic plate MIT professor Harold Edgerton's image of a splash dating from *c.* 1930s, produced by Edgerton's "perfected stroboscopic system" (Kemp, "Stilled splashes," 79) that could "deliver 3000 images per second" (ibid.). In the 1942 New Edition of his book, Thompson describes the photographic plate reproduced opposite his title page as follows: "An instantaneous photograph of a 'splash' of milk.

From Harold E. Edgerton, Massachusetts Technical Institution." Thompson, *On Growth and Form*.

8 See, for example, Thompson's illustrations in chapter V "The forms of cells" from the 1942 edition of Growth and Form, between pages 389 and 397. On these comparisons see also Kemp, "Stilled splashes," 79.

9 In the 1942 edition, Thompson described Worthington's experiments using language that appears to retrace the experiment as an almost cinematic sequence of events (following the idea of montage). Referring to "Mr Worthington's beautiful experiments on splashes," Thompson describes the experiment producing the pattern of the splash as an almost cinematic sequence: "the fall of a round pebble into water from a height" "first formed a dip or hollow in the surface, and then caused a filmy 'cup' of water to rise up all round, opening out trumpet-fashion or closing in like a bubble." See Thompson, *On Growth and Form*, 389.

10 See Kemp, "Stilled splashes," 79.

11 Martin Kemp, "Stilled Splashes," 79.

12 Tate Archive, Henderson Papers, 9211-5-1-1.

13 Tate archive, Henderson Papers, 9211-5-1-2.

14 Ibid.

15 Henderson Papers, 9211-5-1-1.

16 Tate archive, Henderson Papers, 9211-5-1-2.

17 Tate Archive, Henderson Papers, 9211-5-1-1.

18 Ibid.

19 Tate catalogue gives a definition of Henderson's "*Stressed Photograph*," with reference number P79309, "one of several photographs Henderson took of boys on bicycles." See "Nigel Henderson, Stressed Photograph. C. 1950." https://www.tate.org.uk/art/artworks/henderson-stressed-photograph-p79309. The image, not on display, is described as "Photograph, gelatin silver print on paper, mounted on board" and the dimensions of the support are 305 x 507 mm. The distortion apparent in the image is connected to Henderson's experimental treatment of the early 1950s and the series are dated "from around 1950, not long after he had taken up photography." Ibid. According to the catalogue, distortion is produced by both Henderson's modifications of the substrate, the printing paper, and by alterations on the scale of the image while enlarging it. Ibid. Referring to content it notes that "most of these works, as this one, deal with street scenes from London's East End, where Henderson was living in the early post-war years" (Henderson having transformed the bathroom at his home in Bethnal Green into a darkroom, so as to develop experimental photographic techniques). Ibid. "During the printing process this image has been distorted (or 'stressed'), so that the scene seems warped, as if viewed in a distorting mirror." Ibid.

20 For Henderson's "Stressed photographs" at TATE, see, for example, https://www.tate.org.uk/art/artworks/henderson-stressed-photograph-p79309.

21 Henderson, Nigel quoted in *Nigel Henderson: Photographs of Bethnal Green 1949–1952* (Nottingham: Midland Group, 1978), 5, cited from the Tate catalogue. https://www.tate.org.uk/art/artworks/henderson-stressed-photograph-p79309.

22 Tate archive, Henderson Papers, 9211-5-1-2.

23 Ibid.

24 Tate archive, Henderson papers, 9211-2-23.

25 *The Architectural Review*, 110 (October 1951).

26 For Hamilton's role in Growth and Form 1951 see Isabelle Moffat, "'A Horror of Abstract Thought': Postwar Britain and Hamilton's 1951 Growth and Form Exhibtion," *October,* Vol. 94 (Fall 2000), 89–112. See also the text in the Tate catalogue under "REAPERS AND GROWTH AND FORM,1949/1951" in connection to displays in Room 1 and 2 at Tate Modern, following the reconstruction led by Victoria Walsh with Elena Cripp. https://www.tate.org.uk/whats-on/tate-modern/exhibition/richard-hamilton/richard-hamilton-room-guide/richard-hamilton-room-1 "The etchings *Variations on the theme of Reaper* were made while Hamilton was studying at the Slade in 1949, and exhibited the following year at the London gallery Gimpel Fils … This meeting of the natural and mechanical also underlay the exhibition Growth and Form, organised by Hamilton at the ICA for the 1951 Festival of Britain, and reconstructed for the first time in Room 1. Inspired by D'Arcy Wentworth Thompson's book *On Growth and Form* 1917, Hamilton brought together a range of organic and scientific materials, demonstrations, and photographs, making use of the most up-to-date imaging technologies. One of the 'benefits' of the exhibition, Hamilton wrote, was 'the influence it may have upon design trends'. By using grid-based structures to look at form in nature, the exhibition challenged the apparent opposition between geometric and organic approaches to contemporary architecture and design. With its strange jumps of scale, Hamilton's exhibition was infused with a Surrealist sensibility. Another link to Surrealism was the anomalous portrait photograph listed simply as 'Head' and provided courtesy of Jean Painlevé, an important Surrealist filmmaker." Ibid. For a discussion on Hamilton's Reaper and mechanization in the context of Sigfried Giedion, see Papapetros, Spryros, "Soil, land, reaper. Revisiting Giedion's 'Documents on mechanized life'," in *Richard Hamilton and Sigfried Giedion: Reaper*, edited by Carson Chan, Fredi Fischli, Niels Olsen, Linda Schädler. (Zürich: Graphische Sammlung der ETH Zürich, 2017), 215–39.

27 See, for example, Thompson, *On Growth and Form* (1942), 15. For a similar idea against the foundations of Art History and Abby Warburg's image juxtapositions see Papapetros, Spryros, *On the Animation of the Inorganic. Art, Architecture and the Extension of Life* (Chicago and London: The University of Chicago Press, 2012). In many ways, Thompson might be seen also to perform an act of "animation," perhaps in the sense discussed by Papapetros apropos Warburg's method as a process that inscribes agency to the image; visual and printed form as illustration being grasped as the expression of dynamic phenomena. See Papapetros, *On the Animation of the Inorganic. Art, Architecture and the Extension of Life*.

28 In his comparisons, Thompson appears to bring together examples of pictures that depict aspects of both the animate and inanimate world. Thompson "characteristically intuits morphological analogies with medusoids and hydroid polyps," Kemp writes, for example (Kemp, "Stilled Images," 78), using a term which he further develops into a thesis about art methodology in his own influential writing and comparisons between structures in nature and art. See Kemp, Martin, *Structural Intuitions. Seeing Shapes in Art and Science* (Charlottesville and London: The University of Virginia Press, 2016).

29 Such as, for examples, splash stills in comparison with medusas. The reference to the camera as a "common visual denominator" for images on display comes from The IG and the 1953 exhibition at ICA. See Tate archive, Henderson Papers, 9211-5-1-2.

30 In 1991 essay, he pronounces Thompson's book a great intellectual achievement and compares the form of Edgerton's splash, against the title page of Thompson's 1952

edition of the book, to the pattern of a count's crown from the early modern period considered as "the ethnologist's jewel." Claude Lévi-Strauss, "The Ethnologist's Jewels." In *We Are All Cannibals. And Other Essays.* With a foreword by Maurice Olender. Translated by Jane Marie Todd (New York: Columbia University Press), 57–63, 58. [In translation from the Italian, originally published in *La Repubblíca*, September 10, 1991 as 'Ma perchè ci mettano I gioielli?]. Jean Petitot has already suggested that morphology may be seen to have impacted structuralism and the 1991 fascination with Thompson's book may be accounted in this light. Could it be that it was Thompson's avant garde techniques of editing in the form of image montage that perhaps inspired Lévi-Strauss's notion of ethnographic analogy (articulated in the essay) in the first place? Petitot Jean, "Morphology and Structural Aesthetics from Goethe to Lévi-Strauss," in *The Cambridge Companion to Lévi-Strauss*, edited by Boris Wiseman (Cambridge: Cambridge University Press, 2009), 275–95. https://doi.org/10.1017/CCOL9780521846301.015. In retrospect, his 1991 essay reads as a defense of human creativity cast as a "natural" process growing in art (and culture) from within nature and not from outside in the context and as empirical observation. Given that the form of the stilled splash that the artisan unwittingly imitated and that the pattern of the crown prefigures the form of the stilled splash, human creativity, and imitation (and by extension art) are a part of nature; the artist having contemplated nature from within, he concludes. Lévi-Strauss, "The Ethnologist's Jewels," 58.

31 Lévi-Strauss, "The Ethnologist's Jewels," 59.

References

The Architectural Review, 110 (October).

Henderson, Nigel (1950). "Stressed Photograph," TATE. https://www.tate.org.uk/art/artworks/henderson-stressed-photograph-p79309.

Henderson, Nigel (1978). *Nigel Henderson: Photographs of Bethnal Green 1949–1952* (Nottingham: Midland Group).

Hockney, David (1979). *Pictures by David Hockney*. Selected and edited by Nikos Stangos (New York: Harry N. Abrams).

Kaniari, Assimina (2013). "D'Arcy Thompson's *On Growth and Form* and the Concept of Dynamic Form in Postwar Avant-Garde Art Theory." *Interdisciplinary Science Reviews*, 38(1), 63–73, DOI: 10.1179/0308018813Z.00000000035.

Kaniari, Assimina (2009). "Uma genealogia da nano-art através da escala: práticas fotograficas de Nigel Henderson no arquivo da Tate Gallery" ["A genealogy of nano-art across 3 scale: Nigel Henderson's avant garde photographic practices in the Tate Archive" in Portuguese], Translation Marta De Menezes, *Nada*, No. 13, 40–51.

Kemp, Martin (2006). "Growth and Form." In Martin Kemp (ed.), *Seen Unseen. Art, Science and Intuition from Leonardo to the Hubble Telescope* (Oxford, NY: Oxford University), 200–38.

Kemp, Martin (2000). "Stilled Splashes." In *Visualizations: The Nature Book of Art and Science* (Oxford, NY: Oxford University Press), 78–9.

Kemp, Martin (2006). *Structural Intuitions. Seeing Shapes in Art and Science* (Charlottesville and London: The University of Virginia Press).

Lévi-Strauss, Claude. "The Ethnologist's Jewels." In *We Are All Cannibals. And Other Essays*. With a foreword by Maurice Olender. Translated by Jane Marie Todd (New

York: Columbia University Press), 57–63. [In translation from the Italian, originally published in *La Repubblíca*, September 10, 1991 as 'Ma perchè ci mettano I gioielli?].

Moffat, Isabelle (2000). " 'A Horror of Abstract Thought': Postwar Britain and Hamilton's 1951 Growth and Form Exhibtion." *October*, 94 (Fall), 89–112.

Papapetros, Spyros (2012). *On the Animation of the Inorganic. Art, Architecture and the Extension of Life* (Chicago and London: The University of Chicago Press).

Papapetros, Spryros (2017). "Soil, Land, Reaper. Revisiting Giedion's 'Documents on Mechanized Life.'" In Carson Chan, Fredi Fischli, Niels Olsen, Linda Schädler (eds.), *Richard Hamilton and Sigfried Giedion: Reaper* (Zürich: Graphische Sammlung der ETH Zürich), 215–39.

Petitot, Jean (2009). "Morphology and Structural Aesthetics from Goethe to Lévi-Strauss." In Boris Wiseman (ed.), *The Cambridge Companion to Lévi-Strauss* (Cambridge: Cambridge University Press), 275–95. https://doi.org/10.1017/CCOL9780521846301.

"REAPERS AND GROWTH AND FORM,1949/1951," https://www.tate.org.uk/whats-on/tate-modern/exhibition/richard-hamilton/richard-hamilton-room-guide/richard-hamilton-room-1.

Tate archive, Henderson Papers, 9211-5-1-1.

Tate archive, Henderson Papers, 9211-5-1-2.

Tate archive, Henderson papers, 9211- 2-23.

Thompson, D'Arcy Wentworth (1942). *On Growth and Form* (Cambridge: Cambridge University Press).

Walsh, Victoria and Henderson, Nigel (2001). *Parallel of Life and Art* (London: Thames & Hudson).

8

Exhibition as Extended Organism: The Evolutionary Agency of Richard Hamilton's Growth and Form

Charissa N. Terranova, PhD

> It opened my eyes to the idea that the world is as it is because it must follow certain mathematical principles. [Thompson] describes phenomena, like the spirals on a cauliflower, so that you see it has to be this way, because time and the activity of growth are related. It is a beautiful notion; it hangs in the mind as an explanation of life itself. Organisms start as a single cell, and then this cell splits into two, then two into four and then 8 and 16 and 32 and so on, till it has achieved the number of desired cells as a group. Then it rotates and assumes another form when a concavity appears, in much the way as a thumb pushed into a ball of clay makes a pot. The ball becomes a cylindrical tube. Then excrescences emerge from the sides of this tube and you end up with a creature with a mouth, an anus and four legs, or two legs and two arms. That this procedure should create something as complex as a human being is magical. That it should produce this extraordinary object at the end is one of the mysteries of existence.
>
> Richard Hamilton[1]

British artist and designer Richard Hamilton was fascinated with the open-ended empiricism of D'Arcy Wentworth Thompson's *On Growth and Form*, as the quote above illustrates. This chapter explores how Thompson's non-reductionist take on biological morphogenesis made science inviting and openly reachable to a group of mid-twentieth-century London artists, which included Hamilton. In particular, it focuses on Growth and Form, an exhibition based on Thompson's book curated by Hamilton at the Institute of Contemporary Arts, London, in 1951, just four years after its founding. Many scholars have written about this event according to the translation of Thompson's scientific work into an imagistic art exhibition, focusing on an array of themes concerning the fusion of art-and-science.[2] None of these has evaluated the import of Thompson's ideas into art in terms of evolutionary theory. What follows is an evaluation of evolutionary theory, art, and culture through the lens of Hamilton's eponymous show, Growth and Form, in three parts: starting with the exhibition itself and its ecological aesthetics of distributed consciousness; followed by an examination of Thompson's book and Hamilton's exhibition in light of evolutionary theory; and concluding with an evaluation of the evolutionary agency of Hamilton's exhibition conceived as an "extended organism," or a physiological extension of the human brain–body synthesis.

Man in His Proper Environment: Growth and Form at the Institute of Contemporary Arts, London, 1951

> This exhibition is devoted to the description of life microbiologically, micromineralogically; where life is mathematically and cosmically ordered … the authors of this exhibition have been able to use equally the magnificent tools of science, instruments of measurement, and the apparatus of photography and of the cinema … the lighting which shows us things it was not possible for our ancestors to see.
>
> Le Corbusier[3]

In London for the tenth meeting of the International Congress of Modern Architecture (CIAM), the Swiss-French modern architect Le Corbusier introduced Richard Hamilton's exhibition on July 4, 1951, to a small crowd at the show's vernissage at the ICA on Dover Street in London.[4] Thompson's monistic comparisons between manmade design and organic morphology made *On Growth and Form* popular among architects. Yet, beyond architecture, Le Corbusier was moved by Hamilton's exhibition as it presented a distinct paradigm of the "new" and avant-garde across artistic media. "I found myself in the presence of this new conception which was not madonnas or religion or anti-religion, the revolution or anti-revolution," Le Corbusier exclaimed, "but a conception which showed *man in his proper environment* (my emphasis)."[5] He described the show in immersive terms, according to both the array of technologies Hamilton deployed to translate Thompson's text into an exhibition and its profoundly body-based and kinetic organization. If Thompson's appeal in architecture seems logical, within art it is prima facie counterintuitive. The book's imagistic presence, however, proves otherwise, making it penetrable and prescient. In the mid-twentieth century, Thompson's image-rich book (its second edition of 1942 includes over 550 photographs, diagrams, and drawings) corroborated the image-saturated post–Second World War visual culture created by film, television, photo-based publications like *Life* magazine and *Paris Match*, and art-centric publications such as György Kepes's *Language of Vision* (1944), André Malraux's *Le Musée imaginaire* (1944), and Lászlò Moholy-Nagy's *Vision in Motion* (1947).

Beyond Thompson's thinking about structure, it was his ideas concerning growth in time—that biological morphologies must be understood as changing through lived temporality and real-time systems—that took hold in the art world of mid-century London. In the milieu of the ICA during the first half of the 1950s, Thompson's book inspired the making of "organic art," a strain of art related to, while distinct from, the biomorphic sculptural practices of 1930s British artists such as Barbara Hepworth and Henry Moore in its appropriation of process rather than appearance. Mid-century art works inspired by zoologist Thompson's *On Growth and Form* were biological in logic and development but not always in appearance. David Thistlewood describes this Thompson-inspired practice in terms of intuitive spontaneity and moving-image technology that was further reinforced by philosophers Henri Bergson's *élan vital*, Alfred North Whitehead's "philosophy of organism," and biophysicist Lancelot Law Whyte's hybrid art-and-science philosophy. Above all, Thistlewood's organic art names a kinetic process-based approach within art rather than inert formalism. Rather than a thing, organic art is a series of actions. It is "a way of working in which images or forms would appear to develop, and to gain their ultimate characteristics, with little specific pre-intention on the part of the artist, during the process of creative activity [as well as] chronological successions, [which] constituted 'time-lapse' recordings of their evolving creative awarenesses," particularly in the work of Richard Hamilton and Victor Pasmore.[6]

Hamilton was introduced to Thompson's book by friend and artist Nigel Henderson whom he met at the Slade School of Fine Art in London in the late 1940s.[7] It is very telling that, in these years, Henderson introduced Hamilton to two seminal though distinct works: Thompson's *On Growth and Form* (1917) and Marcel Duchamp's *Green Box* (1934). If the former is a quasi-linear scientific tome with literary luminosity, then the latter is a discursive romp in fragmented thinking about avant-garde art. Henderson took Hamilton to artist, historian, and co-founder of the ICA Roland Penrose's for tea and to see his library, which contained one of the 300 copies of Duchamp's *Green Box,* an amalgam of explanatory notes connected to Duchamp's famously difficult but paradigm-shifting work of mixed-media dada art, *The Bride Stripped Bare by Her Bachelors Even* (1915–1923).[8] In all likelihood Hamilton's consumption of one influenced and shaped the other.

Together, Thompson's book and Duchamp's box constitute a sense of "bio-dada," an inventive and imaginative importation of the natural sciences into art by way of a nonsensical and poetic dadaist sensibility. With bio-dada, artists use nonsense and poetry as tools to playfully explain biological data to a lay audience but also in order to educate them about or critique the greater natural sciences. If Hamilton's would-be bio-dada goal in the 1951 exhibition Growth and Form was heuristic and design-oriented, a matter of teaching the London public about organic processes through good exhibition design, then the bio-dada of twenty-first century's bioart is often intended to educate a public about biology while also performing a Foucauldian biopolitical critique of Big Pharma and biotechnology.

While an instance of bio-dada today, in 1951 the exhibition's intentions were different. It was originally intended to be part of the Festival of Britain in London, a massive celebration of postwar renewal and a "display of civic optimism that would mark both the centenary of the Great Exhibition of 1851 and the beginning of a new era of increased productivity and security for Britain."[9] However, Herbert Read, co-founder with Penrose of the ICA, prohibited the exhibition from being part of the Festival, arguing that it was too far afield from its main theme of 100 years of British achievement.[10] The exhibition was thus installed at the Dover Street location of the ICA, organized solely by Hamilton with the help of other Independent Group artists.

Already operating on a small budget from the ICA, Hamilton received little in the way of funding from outside industry in support of the exhibition. While Metal Box Co. Ltd. donated £50, other interested businesses donated services and materials. Metal Box and Carlton Artists provided photographic services and another company, Rank, donated film projectors.[11]

Hamilton's show unfolded in a mixture of media, including projected moving images, microphotographs, specimen displays, and X-rays (Figure 8.1). "Shapes and forms move[d] on screens on the ceilings or on the floor," one reviewer wrote.[12] A film showing crystal formation was projected overhead on the ceiling while on another screen viewers experienced a sea urchin's eggs dividing. Hamilton divided Growth and Form into seven succinct parts intended to bring home the processual kernel of Thompson's thesis on physical forces and form. These were time as a dimension of form; forms of cells; cell groupings; skeleton structure; related forms; form and

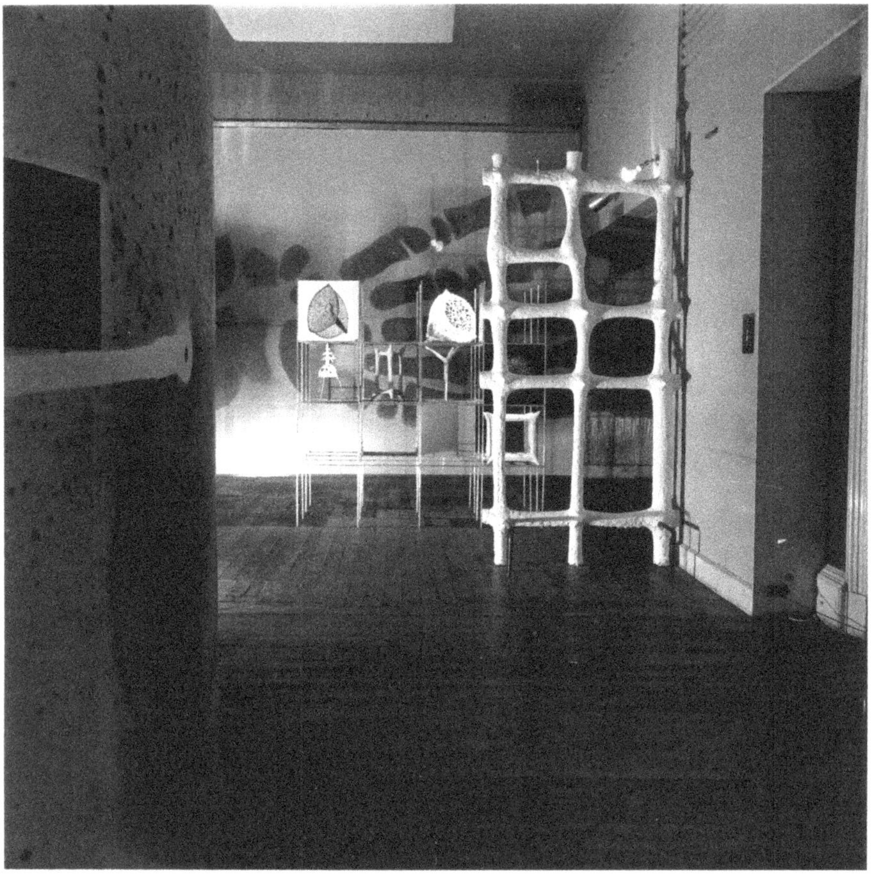

Figure 8.1 Nigel Henderson, Installation shot of Richard Hamilton's work (title unknown) in the exhibition "Growth & Form" at the ICA, 1951. *Source:* Black and white photograph from black and white negative. © Tate, London, 2018.

mechanical efficiency; and the formal realization of pure mathematics.[13] Viewers meandered through the small space of the exhibition in which Thompson's laws of form came to life as a living light-and-object experience—a manifestation of the scientific visualization as a distributed image and lived event. According to Le Corbusier, it created an all-over and enveloping "environment [in which the viewers'] eyes see, and which his senses appreciate … that is discoverable by tools which man has been able to make himself."[14] Pushing the text and diagrams of Thompson's original book into the realm of expanded, extruded, and immersive full-body phenomenon, the image here goes from being two-dimensional print on paper to a four-dimensional living, time-based, and phenomenological event.

Hamilton's choice of celluloid arose out of the London bohemian art scene. Setting this bohemia apart, it was decidedly transdisciplinary, driven to cross-pollinate art and science. Some fifteen years later, Nigel Henderson recalled the hybrid interaction between artists and scientists. "I seem to remember an evening of scientific films," Henderson reminisced, "when Eduardo and Roland and I went to Guy's Hospital where there was a pathologist interested in the ICA."[15] Their decision to interact with scientists shows not simply an openness to other modes of knowledge and expertise, but also a tweaked take on Duchamp's anti-retinal art and thus another instance of bio-dada. Here the anti-retinal—that which goes against the conventions of beauty and painterly formalism—arrived as instructional science films. "And one of us promoted," he continued, "the notion that it might be interesting to have an evening of technical films at which the scientist would talk about the meaning from his narrow discipline and we would make comments on it from another viewpoint, trying to establish why we liked looking at technical films."[16] By the end of the decade, Hamilton and Henderson had largely moved on from science, focusing instead on the techniques of popular advertising, thus efflorescing the earliest forms of pop art. The penchant to bring scientific films into the realm of art nonetheless continued, taking place elsewhere in London, such as The Common Room. The Common Room, a subterranean space specifically devoted to art-and-science mash-ups, opened in 1957. Located in the basement of the avant-garde Polish filmmakers Stefan and Franciszka Themerson's Gaberbocchus Press office, The Common Room hosted "82 evening events that included programmes of experimental and scientific films and lectures, mainly about art and science, and various topics in between."[17]

Despite its inimitable invention, Hamilton's show in 1951 was not a roaring success. The exhibition traffic for Growth and Form was low; there were only 1,140 visitors. It also lost money, a sum of £223, which the ICA subsumed.[18] Yet, this does not detract from its major significance not simply to the annals of history but in contemporary transdisciplinary art, science, and design practices. That said, there is a major aspect of the show that often goes overlooked: this is its mistaken status solely as an exhibition rather than a prototype for contemporary installation and conceptual art. In its own moment, it was not in any way a conventional art exhibition. It did not show the work of one artist or a group of artists, as was the status quo *c.* 1950. Instead it was an intermedia installation of Hamilton's ideas about Thompson's book: it was truly an immersive experience rooted in Hamilton's imagination and design expertise. Hamilton deployed a battery of technologies in order to encapsulate the processual morphogenesis of form as a means to translate Thompson's book *On Growth and Form*. The installation was equal parts visual experimentation, heuristic experience, and aesthetic reverie into new territories of thought and phenomenology. It did not emerge out of the studio practices of a group of painters or sculptors, but through the Bauhaus-inflected design prowess and singular approach of Hamilton, an artist often singly connected to British pop art. The 1951 installation sets Hamilton in the full light of his manifold skills and expansive breadth of his imagination, revealing him to be a fearless polymath working across fields.

Yet, none of this meant success for the show. In many ways, Hamilton's Growth and Form has experienced greater fame and recognition in its long afterlife. In addition to the exhibition's reinstallation at London's Tate Modern and Madrid's Museo Nacional Centro de Arte Reina Sofía in 2014 as part of a Richard Hamilton retrospective, the exhibition today materializes the collective will to unite art and science within the ever-expanding field of new media art. It jibes well, perhaps better today than in 1951, with the vast and variable bodies of knowledge made accessible through the internet and the commonality of algorithms within everyday life. Though almost seventy years old, the exhibition in its original incarnation preternaturally tuned into systems-based aesthetics, eco-criticism, and the distributed ecological image made possible by digital technologies in the twenty-first century. As gallery-goers walked through and amid the projected and moving images of Thompson's ideas, they experienced scientific visualization as a matter of extended mind: science as moving-image consciousness; mind not so much as brain-borne congelation unique to humans but a matter of technological extensions, site specificity, and environmental connectedness within a world of form, both living and non-living.

Culture, Science, and a Capacious Evolutionary Theory: From *On Growth and Form* to Growth and Form

> Morphology then is not only a study of material things and of the forms of material things, but has its dynamical aspect … And here it is well worth while to remark that, in dealing with the facts of embryology or the phenomena of inheritance, the common language of the books seems to deal too much with the *material* elements concerned, as the causes of development, of variation or of hereditary transmission … we must most carefully realise in the outset that the spermatozoon, the nucleus, the chromosomes or the germ-plasm can never *act* as matter alone, but only as seats of energy and as centres of force.
>
> D'Arcy Wentworth Thompson[19]

> Art—or, to use a more exact phrase, aesthetic experience—is an essential factor in human evolution, and, indeed, a factor on which *homo sapiens* has depended for the development of his highest cognitive faculties.
>
> Herbert Read[20]

If, in its own moment, critics received Hamilton's Growth and Form as a pioneering though eccentric experiment in curating, scholars today see it as archetypal for art-and-science hybrid practice.[21] Missing in its contemporary celebration is an analysis of how the two together, Thompson's book given form in the exhibition, inform and activate in new ways Charles Darwin's theory of evolution and, more precisely, Neo-Darwinism. While there is little evidence of exactly how Hamilton felt about evolutionary theory, Thompson himself took a specific stand. I argue here that, even while his exhibition was not intentionally a foray into evolutionary theory, Hamilton situates himself in the landscape of Thompson's own ideas about evolution, thus incorporating them through translating his ideas to London's art-loving public.To clarify, *On Growth and Form* is not a book about evolutionary theory per se. Yet Thompson's thesis on physical forces, growth, and form does have very significant repercussions on evolutionary theory, in particular Darwinist evolution that gives primacy to natural selection and genetics above all other forces, be they epigenetic, linguistic, or cultural. Thompson knew this, inasmuch as the quote above reveals: there is more to morphogenesis than "hereditary transmission." In contemporary terms, life is greater than the genome. Writing in the early days of genetics and molecular biology, one can assume that Thompson understood that his book constituted something of a counterforce to rising gene-centric evolutionary theory, otherwise known as Neo-Darwinism or the Modern Synthesis. While not quite synonymous, the terms "Neo-Darwinism" and "Modern Synthesis" describe the revival at the end of the nineteenth century of Darwin's theory of natural selection and the incorporation of Mendelian inheritance or genetics into the theory of evolution, both of which became central engines of population genetics. In short, these concepts quickly allayed biology's "physics envy" by providing modes of quantification to the all-too-difficult-to-quantify complexity of induction, gene action, and phenotypic expression. For Thompson, the problem with Darwin's theory, or more precisely its codification as Neo-Darwinism by the 1930s, was the unquestioned centrality of natural selection and its functional means across deep time. Thompson thought that, with natural selection as *the* causal force of morphological change over gradual time, Darwin's theory of evolution missed other causalities—such as physical forces—in real, lived time.

The late evolutionary biologist Stephen Jay Gould argues that Thompson bases his critique "of natural selection directly upon his own idiosyncratic theory of form."[22] Yet, let it be clear: Thompson does not reject wholesale Darwin's theory, but rather positions himself in adjacency, in a place that is critical of what Gould more generally calls "Darwinian functionalism," or Neo-Darwinism by another name. Gould's Darwinian functionalism is the dominant idea "that adaptation drives evolution as organisms change to secure better fit to their environments."[23] Thompson agreed with Darwin, and even celebrated the preeminent existence of adaptation. Where Thompson departed from Darwin was on the idea that physical adaptations are predetermined for a specific function—that they arise "'for' a given utility by a functionalist mechanism like natural selection."[24] Unlike Darwin, who argued that changes in body morphology unfold gradually across deep time for functional reasons, namely, according to natural selection, Thompson held out space in his theory of form for temporal ruptures and leaps. "Our geometrical analogies weigh heavily against Darwin's conception of endless small continuous variations," Thompson forthrightly states.[25] "They help," he continues, "to show that discontinuous variations are a natural thing, that 'mutations'—or sudden changes, greater or less—are bound to have taken place, and new 'types' to have arisen now and then."[26]

While for Gould, Thompson's critique of Darwinism and Neo-Darwinism made him an outlier in the "century of the gene," for historian of science Maurizio Esposito Thompson's position was typical of a certain early twentieth-century camp of thinkers. Thompson was part of a rich, broad, and diverse community of scientists developing "organismal biology," which included Berthold Hatschek, Franz Hofmeister, Hans Przibram, Joseph Woodger, Edward Stuart Russell, Ludwig Von Bertalanffy, Paul A. Weiss, and Albert Dalcq. As organismal biologists, these scientists took a holist approach to organisms and life itself. While "holist" very generally refers to interdisciplinary activities between art and science, in this specific context I use the word "holist" as it refers to Thompson's view of the genes as one source among others, albeit a very important one, contributing to biological development and organismic morphology. Holism propounds the interconnection of forces and forms within nature. Esposito argues in these terms, "Thompson believed that Evolution had nothing to do with transmission and selection of discrete characters [such as genes] because there was nothing purely discrete in the organic world."[27] The organismal, organicist, and holist position within biology of a century ago has a newfound, if not greater, purchase today than it did then, broaching contemporary categories of evolutionary theory that argue for an "extended synthesis" over the Modern Synthesis, and evolution in multiple dimensions, including the epigenetic, behavioral, and symbolic, in addition to the genetic.[28] These takes on evolution are more capacious than those of Neo-Darwinism and the Modern Synthesis, opening up room to understand cultural expression and the arts as both the result of and, reciprocally, the catalyst of evolutionary forces. It is precisely its functional feedback loop—culture acting upon evolution and vice versa—that makes the reductionism typical of twentieth-century natural science in urgent need of revision.

That scientists working on evolutionary theory have historically had little to say about art did not keep ICA co-founder Read from devoting energy to the topic. On April 10, 1951, just three months prior to the opening of Hamilton's Growth and Form at the ICA, Read gave the Conway Memorial Lecture in London on the topic of art and evolution. Titled "Art and the Evolution of Man," his talk was a tour de force in art-and-science imagination. Read was keenly aware of the ancillary place given to art in evolutionary theory, launching his lecture with an invocation of art's "non-existent" status within "Dr. Julian Huxley's 'modern synthesis' of evolution."[29] Read set out to remedy this problem by non-reductively integrating art into evolutionary theory. Yet Read devoted his lecture not to the evolution of art-as-artefact but of human kind's capacity to "think" and "process" art, or her "*maximum aesthetic sensibility*—combined with other factors which may be social or ideological" (Read's emphasis).[30] Read describes an evolutionary foundation of complexity and multiple vectors, a nonlinear ecological network including the seemingly indeterminable forces of culture far distinct from the Neo-Darwinism of Huxley's Modern Synthesis.

The main thrust of Read's talk concerns consciousness and art, how "perceptual imagery or symbolism" are part of "biological urgency," and thus are material evidence of the evolution of high cognitive capabilities in humankind. For Read "biological organization and mental activity," Thistlewood explains, "are processes of closely similar, or even identical, character."[31] Art is a biological need because of the human mind's inherent complexity and its reciprocal thirst for complexity. Prehistoric humans made images as records of events, or "memory-images," and also for ritual acts bearing "the desire … to utter, to give out a strongly felt emotion or desire by representing."[32] In this paradigm of evolution, the biological serves not as simple impetus, say, as the catalyst of intuition or need, but a physical model of process—art as a network of desire unfolding in real time.[33]

While aware of Huxley's *Evolution: The Modern Synthesis* (1942), Read's thinking about evolution was more directly molded by the philosophies of Henri Bergson and Alfred North Whitehead, whose ideas were markedly less mechanistic and "remorselessly sequential."[34] Lancelot Law Whyte, who wrote the Foreword to the published version of Read's lecture, adds anthropologist Susanne Langer and embryologist Albert Dalcq to this list of influences. Read's ideas, according to Whyte, resonate in particular with Dalcq's "analogy between the process of aesthetic creation and the formative organising processes of organic growth and differentiation," thereby reinforcing the processual take on consciousness and aesthetics within Read's lecture.[35] I focus here on Dalcq, for his ideas serve as a linchpin connecting Hamilton to Thompson's ideas about an expanded take on natural evolution, in particular by way of his participation in the 1951 symposium connected to Growth and Form titled *Aspects of Form*.

Like Thompson, the Belgian embryologist Dalcq was an organismal biologist and holist. Dalcq was also an avid aficionado of Thompson's. He wrote a letter to the Secretary of the Royal Society of Edinburgh in 1939 requesting a portrait of Thompson for his lab in Brussels.[36] Written eleven years later, Dalcq's essay for *Aspects of Form*, titled "Form and Modern Embryology," focuses on embryological development as a mode of form-making amenable to art-making. Bringing one formalism to another, the changeful aspects of embryological development turn the inertia of conventional artistic formalism on its head. At the same time, Dalcq's subtle vitalism introduces a new take on consciousness to artistic form. If living matter has inherent time-based consciousness, so too might the inorganic form after which it is modeled. The essay begins with a soupçon of related vitalist metaphysics, with Dalcq arguing that "form is both deeply material and highly spiritual."[37] Establishing organic form with agency, Dalcq continued, "form is never trivial or indifferent; it is the magic of the world."[38] Dalcq argues that natural form is autopoetic, that "life does not imply any kind of matter or energy which is not present," and that the human mind is drawn to form not so much by the inherent organization of form but out of the mind's need to organize the encompassing world.[39]

Dalcq addresses evolutionary theory in the closing pages of the essay, proposing a three-pronged critique of Neo-Darwinism because it "is not capable of providing a complete interpretation of the whole process of evolution."[40] First he argues for an agency-based consideration of adaptation. Second, he demands "parallelism between ontogenesis and evolution," or a heightening of the importance of development in lived-time (ontogenesis) over the gradual shifts unfolding across deep history (phylogenesis). And third, rather than reducing species differentiation to the chromosomes in the nucleus, namely, the genes, Dalcq calls for a "causal embryology" that looks to the interaction between the cytoplasm and the nucleus coupled with careful analysis of "morphogenetic regulation." This approach offers a means of analyzing evolution according to the cybernetic feedback between various agencies: the agency of the organism and environment, nucleus as well as cytoplasm. As such, Dalcq was a purveyor of a more capacious evolutionary theory in which art and culture serve as vectors of influence in the ongoing morphogenesis of life itself. By way of Thompson, Dalcq, and Read, we can surmise that Hamilton's Growth and Form bears an evolutionary agency for humankind, functioning in particular as a mode of design-based physiological extension of the human organism.

Design and the Extended Organism of Hamilton's Growth and Form

> He appears, in a roundabout way, committed to the late Bauhaus attitudes of the Institute of Design, founded by Moholy-Nagy. Anyone who saw—and to see was to become involved in—Hamilton's Growth and Form exhibition at the I.C.A., will remember his Bauhaus display tactics.
>
> Lawrence Alloway[41]

> Growth and form [*sic*] seemed an ideal subject for another involvement of that time, exhibition design. By the turn of the century the "exhibition" was beginning to be understood as a form in its own right with unique properties.
>
> Richard Hamilton[42]

One might argue ultimately that Hamilton's adaptation of Thompson's *On Growth and Form* was more a matter of design than art. In addition to Alloway's Hamilton-Bauhaus connection, one need only look to Bauhaus designer, artist, pedagogue, and biocentric László Moholy-Nagy's 1929 exhibition in Stuttgart titled *Film und Foto* where metal structural components organized scientific photographs much in the way of Hamilton's Growth and Form.[43] Second-generation Bauhaus artist, impresario, and fellow Hungarian biocentric György Kepes further reprised this quasi-industrial mode of exhibiting scientific photos in his 1951 exhibition, *The New Landscape*, about technology and vision at MIT's Hayden Gallery. Yet, what does such an argument garner? Perhaps declaring the show a design project makes it more persuasively about process and thus a better approximation of the processes of growth and form-making central to Thompson's book. I conclude something other, namely, that the exhibition functioned as an extension of human consciousness out into the world, and as such, the designed props of the exhibition—the projected moving images, microphotographs, specimen displays, X-rays and their armatures—were vectors of a collective consciousness specific to the event itself. As such, these vectors turned scientific visualization into full-blown means of intellectual metabolism. That is to say, Hamilton deployed a design-based expression of scientific concepts to make an exhibition that was physiological in nature. Growth and Form was a mode of mentally combusting the new world of scientific and technological romanticism in 1951—a world that was equal parts hopeful overture and uncanny closure. Like other designed forms, such as a ranch house, termite mound, beaver's lodge, or bowerbird's nest, Hamilton's show marked the externalization of internal mammalian physiology. It extended a site-specific sense of human consciousness, one bound to the exhibition space and Thompson's tome, and thus made fleeting and ephemeral, for beyond the exhibition this collective consciousness dissipated only to form differently under other circumstances. In turn, Hamilton's exhibition as extended organism offered an externalized and space-time-specific sense of mind that moves beyond the idea of consciousness as strictly brain-borne and separate from the body. Consciousness from this perspective is shared, extended across bodies, and nomadic.

Here I expand physiologist and biologist J. Scott Turner's thesis on homeostasis and evolution to the realm of human design and art exhibitions. Based on decades of research into the "extended organism," or the physiology of animal-built structures, Turner argues that homeostasis, beyond natural selection, is key to a correct and full understanding of natural evolution.[44] Building on nineteenth-century physiologist Claude Bernard's work, homeostasis refers to an organism's adaptational modulation within an environment by way of built structures and the "relentless striving of living systems for persistence and self-sustenance."[45] Like Thompson's physical forces, Turner's ideas about homeostasis bring real-time and immediate activities within the environment to bear upon evolution. From this perspective, multiple vectors in real time are agents within the complex ecological network that is evolution itself. Agents and material constitute a feedback loop in which positions change and shift; one moment one is an agent acting on material, and in the next material being acted upon by an agent. "Homeostasis does not derive from natural selection," Turner argues; "it is homeostasis that drives selection, and there is nothing natural about it. *What drives the course of evolution is not the soulless lottery of the gene pool, but life's striving for persistence* (my emphasis)."[46] Like the thinking of Thompson and Dalcq, Turner's take on evolution is capacious, bearing space for art and cultural events like Hamilton's Growth and Form. As such, Hamilton's exhibition must be seen as an evolutionary agent of equilibrium for humans active mid-last century. It did not provide the reverie of escape but the reverie of going deeper within; it gave a prescient path to understanding a world in which science and technology would help reveal the multiple vectors of evolution, a world of life, death, change, and artistic invention beyond fitness, competition, genetics, and natural selection.

Notes

1 Olbrist, Hans Ulrich, "Pop Daddy: The Great Richard Hamilton on His Early Exhibitions," an interview with Richard Hamilton (April 1, 2003), https://www.tate.org.uk/context-comment/articles/pop-daddy-richard-hamilton-early-exhibition (accessed July 16, 2018).

2 See footnote 24.

3 Le Corbusier, Opening Speech, Growth and Form, Institute of Contemporary arts, July 4, 1951, Tate Gallery Archives, 955.1.12.26 33/36.

4 Morgan, Emmanuelle, "Le Corbusier Parle … 1951," *Twentieth Century Architecture*, Festival of Britain, No. 5 (2001), 8. See also Sean Keller, *Automatic Architecture: Motivating Form After Modernism* (Chicago: University of Chicago Press, 2018) 35.

5 Le Corbusier, Opening Speech, Growth and Form, Institute of Contemporary arts, July 4, 1951, Tate Gallery Archives, 955.1.12.26 33/36.

6 Thistlewood, David, "Organic Art and the Popularization of Scientific Philosophy," *British Journal of Aesthetics*, Vol. 22, No. 4 (1982), 311.

7 Hamilton, Richard, *Collected Words 1953–1982* (London: Thames and Hudson, 1982), 11.

8 Hamilton, *Collected Words 1953–1982*, 11.

9 Jolivette, Catherine, "Representations of Atomic Power at the Festival of Britain," in *British Art in the Nuclear Age*, edited by Catherine Jolivette (New York: Routledge Press, 2014), 104.

10 See Jolivette, "Representations of Atomic Power at the Festival of Britain," 104 and Massey, 42–4.

11 Massey, 43.

12 Moffat, Isabelle, "'A Horror of Abstract Thought': Postwar Britain and Hamilton's 'Growth and Form' Exhibition," *October*, Vol. 94, The Independent Group (Autumn 2001) 101.

13 Growth and Form Exhibition Archive, Tate Britain, 955.1.12.26 4/36.

14 Le Corbusier, Opening Speech, Growth and Form, Institute of Contemporary Arts, July 4, 1951, Tate Gallery Archives, 955.1.12.26 33/36.

15 Nigel Henderson Archive, Tate Britain, TGA955.1.14.6 Nigel Henderson interview August 17, 1976 with Dorothy Morland.

16 Nigel Henderson Archive, Tate Britain, TGA955.1.14.6 Nigel Henderson interview August 17, 1976 with Dorothy Morland.

17 Reichardt, Jasia, "Gaberbocchus Press and the Common Room," *Interdisciplinary Science Reviews*, Vol. 42, Nos. 1–2, 30–41 (2017), 34.

18 Massey, 44.

19 Thompson, 20.

20 Read, Herbert, *Art and the Evolution of Man* (London: Freedom Press, 1951), 12.

21 See Moffat, Isabelle, "The Physiology of Thought: Neuroplasticity in Lancelot Law Whyte's Model of 'Unity Thought' and Richard Hamilton's 1951 Growth and Form Exhibition," in *Habitus in Habitat II: Other Sides of Cognition*, edited by Sabine Flach and Jan Söffner, eds. (New York: Peter Lang, 2010), 185–98; Jolivette, Catherine, *Landscape, Art, and Identity in 1950s Britain* (New York: Routledge, 2009); Massey, Anne, *The Independent Group: Modernism and Mass Culture in Britain, 1945–1959* (Manchester: University of Manchester Press, 2009); and Juler, Edward, "A Bridge between Science and Art? The Artistic Reception of *On Growth and Form* in Interwar Britain, c. 1930–1942," *Interdisciplinary Science Reviews*, Vol. 38, No. 1 (March 2013), 35–48.

22 Gould, Stephen Jay, *The Structure of Evolutionary Theory* (Cambridge, MA: Harvard University Press, 2002), 1203.

23 Gould, 1179–80.

24 Gould, 1200–1.

25 Thompson quoted in Gould, 1202.

26 Thompson quoted in Gould, 1202.

27 Esposito, Maurizio, "Problematic 'Idiosyncrasies': Rediscovering the Historical Context of D'Arcy Wentworth Thompson's Science of Form," *Science in Context*, Vol. 27, No. 1 (2014), 100.
28 See Pigliucci, Massimo and Müller, Gerd B. (eds.), *Evolution: The Extended Synthesis* (Cambridge, MA: MIT Press, 2010) and Jablonka, Eva and Lamb, Marion J., *Evolution in Four Dimensions: Genetic, Epigenetic, Behavioral, and Symbolic Variation in the History of Life* (Cambridge, MA: MIT Press, 2005).
29 Read, Herbert, *Art and the Evolution of Man*, 11.
30 Read, *Art and the Evolution of Man*, 11.
31 Thistlewood, David, *Herbert Read: Formlessness and Form: An Introduction to His Aesthetics* (London: Routledge & Kegan Paul, 1984), 138.
32 Read quoting Jane Harrison's *Ancient Art and Ritual* (1913), *Art and the Evolution of Man*, 26.
33 Thistlewood, David, *Herbert Read: Formlessness and Form*, 138.
34 Thistlewood, *Herbert Read: Formlessness and Form*, 37.
35 Whyte, Lancelot Law, Foreword, *Art and the Evolution of Man* (London: Freedom Press, 1951), ix.
36 Esposito, 82.
37 Dalcq, Albert M., "Form and Modern Embryology," in *Aspects of Form*, ed. Lancelot Law Whyte (Bloomington, IN: First Midland Book Co, 1961 [1951]).
38 Dalcq, 91.
39 Dalcq, 93.
40 Dalcq, 101.
41 Alloway, Lawrence, "Re Vision," *Art News and Review*, January 22, 1955, unpaginated.
42 Hamilton, *Collected Words 1953–1982*, 10.
43 Botar, Oliver, "László Moholy-Nagy's' New Vision and the Aestheticization of Scientific Photography in Weimar Germany," *Science in Context*, Vol. 17, No. 4 (2004), 525–56.
44 Turner, J. Scott, *The Extended Organism: The Physiology of Animal-Built Structures* (Cambridge, MA: Harvard University Press, 2000).
45 Turner, J. Scott, *Purpose & Desire: What Makes Something "Alive" and Why Modern Darwinism Has Failed to Explain It* (New York: Harper One, 2017), 292.
46 Turner, *Purpose & Desire*, 292.

9

The Invisible Motives of Growth and Form

Caroline O'Donnell

Typologies and Portmanteaus

Eighteenth-century treatises on architecture owed much of their organization to the classification systems of plant and animal encyclopedias of the previous decades. In collections such as *Histoire Naturelle* (1749), or *Species Plantarum* (1753), plants and animals were organized and tabulated, whether by evolutionary lineage or reproductive organs, in terms of their formal typologies.[1] Many architecture books that followed showed buildings organized and tabulated in the same way. Julien David Leroy, for example, in *Les Ruines des plus beaux monuments de la Grèce*, first published in 1758, shows a series of different yet similar temples in plan. Leroy's innovation here—new to architecture but commonplace in the sciences—was to compare like-types from different times in one table.[2]

Building on Leroy's work, Jean-Nicolas-Louis Durand organized and classified a large number of buildings in several books. Some examples are *Recueil et parallèle des édifices de tout genre, anciens et modernes: remarquables par leur beauté, par leur grandeur, ou par leur singularité, et dessinés sur une même échelle* by J. N. L. Durand. pub. l'Imprimerie de Gillé fils (1799 or 1800) and *Précis des leçons d'architecture données à l'École royale polytechnique* by J. N. L. Durand. pub. Chez l'auteur (1809). Through this scientific classification system, they anticipated that some theory of architecture would emerge beyond the presentation of architecture for historical and analytical purposes. The expectation was that "out of an ordering of the variety of buildings of the past will come theoretical principles, which may be applied in designing new buildings, of new forms, to answer new programmes and new circumstances."[3] Durand believed that this systematization was essential to uncover the underlying truths of architecture without which one could not design, writing: "A man who plans a career as a playwright does not learn how to do this or that tragedy; a musician this or that opera; a painter this or that painting. Before composing, in whatever genre, one must know what one composes with."[4]

In order to achieve these overarching theories, and, as in the natural science treatise, differences and variations were eliminated in order to deduce the basic model of the type. Architectural types were stripped down to their bare essentials in order that the designer could build up more complex forms from a solid conceptual base.

In his *Recueil*, Durand organized buildings by program in two groups: stylistic programs (Egyptian temples, Roman palaces, Moorish details, etc.) and functional programs (theaters, markets, hospitals, etc.) (Figure 9.1). In some cases, however, programmatic classification was replaced by formal properties: round temples, for example, are "a type corresponding to a simple, geometric form, from which more elaborate forms can be derived."[5] This shift toward shape and geometry opens up the possibility that type is not only stylistic or programmatic but also formal, and potentially fundamentally so: or, as Giulio Carlo Argan writes in *On the Typology of Architecture*, "the initial geometric move is the type."[6] Crucially, Durand defined architecture as the combination of elements: "Architecture is the art of composing and executing all public and private buildings."[7] To design, then, was to compose, collage, or combine some known forms elements according to certain rules.

This composition of elements related back, in many ways, to the thinking about species in the early nineteenth century, before the revolutionary theories of Lamarck, Darwin, and Wallace transformed the discipline of biology (and, with it, the world). A giraffe, for example, was a *cameleopardis*: a portmanteau of the words camel (in form) and leopard (in material) that refers to the comprehension of the new species as being a combination of two established species. All invention, it seemed, whether designed or natural, emerged from a combinatorial logic.

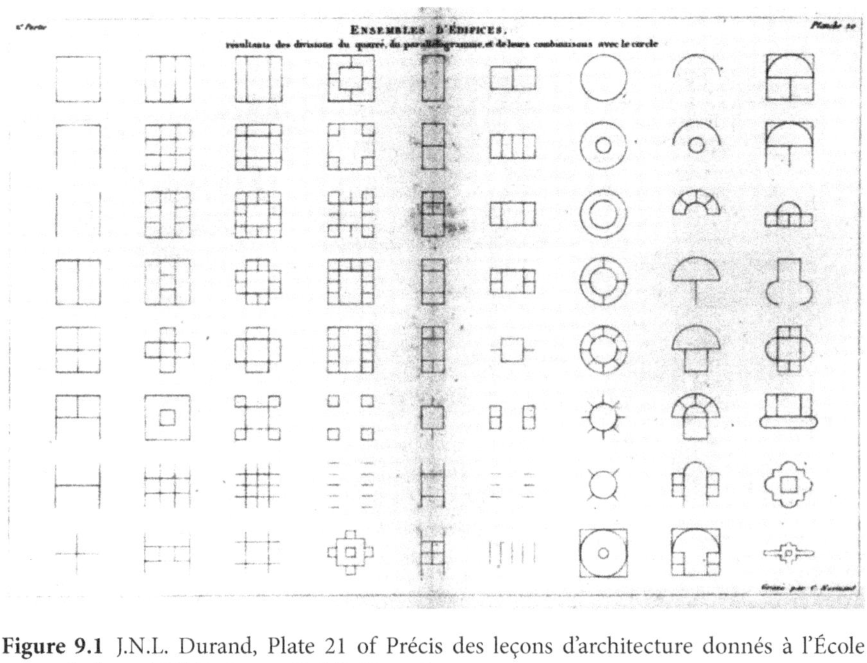

Figure 9.1 J.N.L. Durand, Plate 21 of Précis des leçons d'architecture donnés à l'École polytechnique, 1802/5. *Source:* Public Domain.

Evolution and Growth

Though the malleability of species has its roots in antiquity, it was not until the nineteenth century that Jean-Baptiste Lamarck (*Recherches sur l'Organisation des Corps Vivants*, 1802) and then Charles Darwin (*On the Origin of Species*, 1859) proposed the first theories of evolution. Darwin's summary of his theory is clear in its dependence on two factors—variation and fitness: "Variations, however slight and from whatever cause proceeding, if they be in any degree profitable to the individuals of a species, in their infinitely complex relations to other organic beings and to their physical conditions of life, will tend to the preservation of such individuals, and will generally be inherited by the offspring."[8] Though in many ways a critic of Darwin (with specific arguments in support of macromutation as well as in favor of a greater emphasis on part-to-whole relationships, and force over lineage[9]), D'Arcy Thompson's renowned diagrams serve in some ways as an illustration of the possible mathematical rules underlying variations within species (though they do not account for and in fact are at odds with Darwin's theory of evolution). These diagrams famously superimpose a Cartesian grid on a series of organisms and generate, through a series of stretching, compressing, skewing, shearing, and radial operations, a range of possible variations within a type. While morphometrics developed into a useful scientific inquiry in itself, Thompson's hope was that the diagram might reveal "some guidance as to the 'law of growth' or play of forces by which the transformation has been effected"[10] (Figure 9.2). What is important to note is that Thompson was interested in the underlying mathematics of translation much more so than the animals themselves, and the diagrams suggest that while a great number of variations are *possible* along any spectrum of grid deformation, not all of these species necessarily exist in the world. A lengthy quote from Stephen J.

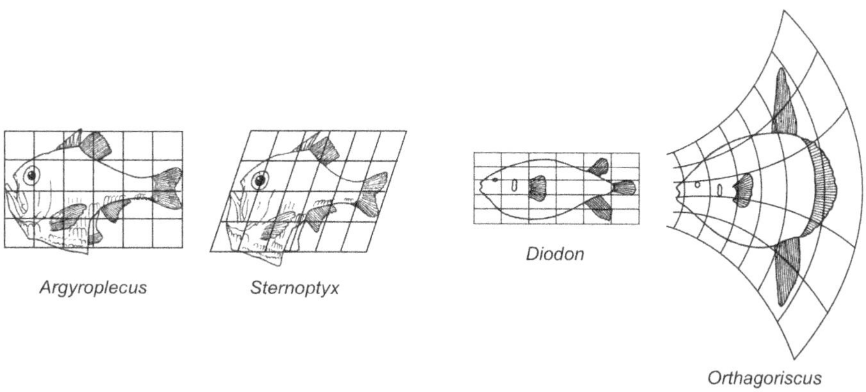

Figure 9.2 Evolution of body form. The transformation of Cartesian coordinates from (left) (A) the body plan of the fish *Argyropelecus Olfersi* to (B) the body plan of the fish *Sternoptyx diaphana* (Fig. 517 and 518, 1062) and (right) from (A) the body plan of the fish *Diodon* to (B) the closely related fish *Orthagoricus* (Fig. 525 and 526, 1064). *Source:* D'Arcy Wentworth Thompson, edited by John Tyler Bonner, *On Growth and Form*. Copyright © 1961. Reprinted with the permission of Cambridge University Press.

Gould illustrates this point. Referring to Thompson's last chapter of transformations (1037), Gould writes:

> D'Arcy Thompson was interested in the deformed net, not primarily in the animal that it generated. He saw that net as a diagram of forces; and just as the trabeculae of the stressed femur reflected the forces responsible for their deposition, so would the deformed net depict the forces that could transform one animal to another. Since these forces might produce a form directly, the deformed net is no mere framework for description; it may be a display of efficient causes. If "diverse and dissimilar fishes can be referred as a whole to identical functions of very different coordinate systems, this fact will of itself constitute a proof that variation has proceeded on definite and orderly lines, that a comprehensive 'law of growth' has pervaded the whole structure in its integrity, and that some more or less simple and recognizable system of forces has been in control.[11]

In the physical world, there is not an even spread of forms across a deformed grid of possibilities: in fact, some forms proliferate and others do not exist at all. This lack of attention to the actual world was the source of much criticism of Thompson: British ecologist G. Evelyn Hutchinson, for example, wrote that Thompson constructed a "floating mathematics for morphology, unanchored for the time being to physical science."[12] The motivating forces implied in Thompson's diagrams are not completely "unanchored" since, on several occasions, Thompson describes the causal factors. Thompson does describe in several instances what these motivating causes of the force might be, "from simple imbibition of water to the complicated results of the chemistry of nutrition."[13] He mentions specifically, at various points throughout the book, gravity, temperature, light, and water. For example, he notes that "the gravitational field is part of the complex field of force by which the form of the organism is influenced and determined";[14] or in plant stems: "anomalies may be such as arise intrinsically from structural peculiarities in the stem itself, or externally to it by reason of unequal illumination or through various other localized forces."[15] In noting the positive effect on ultraviolet light on plants, Thompson notes that "It is a physiological problem, and as such it shews how plant-life is adapted, on the whole, to just such rays as the sun sends; but it also shews the morphologist how the secondary effects of climate may so introduce growth as to modify both size and form."[16]

However, though described textually, the motives remain invisible in Thompson's diagrams. The deformations end at the extent of the gird: arrows, annotations, or objects that represent the motives of the deformation are not included. And this is all well and good, because, as Gould notes, Thompson's interest is only in the net, not so much in the animal itself, and presumably, even less so in the motivators of the transforming force.[17]

When architects adopt the diagrams into their discipline, the result is that the grid is overlaid on formal typologies, and while both the grid and the objects are much discussed, the invisible motives rarely enter into the discussion. Strangely enough, these elements are the same for the architect and the biologist: gravity, light, water, temperature, and so on.

Architectural Adoption

It is not surprising that the translation from organic form to non-living material is so enthusiastically taken on by architects. The majority of Thompson's chapters cite correspondences between organic form and the forms of non-living material that have been acted upon by physical forces.[18]

> The waves of the sea, the little ripples on the shore, the sweeping curve of the sandy bay between the headlands, the outline of the hills, the shape of the clouds, all these are so many riddles of form, so many problems of morphology, and all of them the physicist can more or less easily read and adequately solve, solving them by reference to their antecedent phenomena, in the material system of mechanical forces to which they belong, and to which we interpret them as being due … Nor is it otherwise with the material forms of living things. Cell and tissue, shell and bone, leaf and flower, are so many portions of matter, and it is in obedience to the laws of physics that their particles have been moved, moulded and conformed.[19]

Moreover, Thompson's use of geometry over equations allowed the arguments to be graphically delivered, therefore visualizing otherwise complex differential calculations and tapping into an architectural tradition of borrowing from scientific categorizations. Perhaps most thrillingly, by overlaying the grid over the organism and using grid deformations and their implied forces to explain formal differentiation within a species, Thompson not only described existing forms but predicted new forms that could occur within a type. Here we come full circle to Durand's search for the elements of architecture, the building blocks that could be used to combine and build up new forms within the type. Only now, these new forms are arrived at not through combinatorial methods, but through smooth morphology. Architectural type became species.

For the first time, formal variation in architecture was motivated. There were forces outside the form that were pushing and pulling. For the time being they were invisible and, in most cases, unknown, but the overlay of the grid and its visible deformation attested to their very real and powerful presence.

This shift from static typological thinking to species thinking is well documented in the family of forms that is known as the Palladian Villas. Beginning with Wittkower's diagram of 11 of Palladio's villas that first appeared in his 1949 work *Architectural Principals in the Age of Humanism,* a pattern was sought that would link these variations into one type. Wittkower redrew eleven of Palladio's villas,[20] in search of this "geometric formula" that might give structure to the differences. A parallel study by Wittkower's student Colin Rowe refers to Wittkower's "pattern" as the "ideal"[21] in the seminal text *Mathematics of the Ideal Villa*. Rowe's analytic formalism—as made clear by the inclusion of "mathematics" in the title—argues that the comparable proportions of modernist and renaissance form are a transcendent language, a geometrical language in which the mathematical harmony of nature, of music, and the universe was an underlying law of architecture.[22]

While both of these influential studies discussed variations, they immediately closed down the possibilities by imposing an ideal grid. These analyses, as well as Rowe's later bricolage techniques, can be aligned here with Durand and the scientific encyclopedias. That is to say, in the 1940s and 1950s, architecture (or at the very least the architecture of formalism) continued to be a composition of disparate and static forms

Greg Lynn later returns to the mathematical roots of Rowe's argument and rebuilds it with crucial tweaks. Lynn argues: "Interest in diversity, difference and discontinuity do not preclude formal and mathematical thought" and instead proposes an "alternative mathematics of forms; a formalism that is not reducible to ideal types but is in essence freely differentiated."[23] Lynn observes that architecture is, like an organism, a body acted upon by exterior forces, and notes that the metaphor of the body has been used frequently in architecture from Virtuvius to Le Corbusier, but also that that body has been a static and average one. Instead he proposes that we might look to transformations which, "rather than reducing the differential variations between elements to a static type, ... employed a continuous, differentiated system of transformation." This more dynamic method of diagramming allows us to think about type platonically, as a single pattern from which all others are imperfect copies, but as a gradient of ever-changing form. Of Thompson's deformations specifically, Lynn writes:

> particular information influences and transforms a general grid, making geometry more compliant to the matter it describes. The enlargement of a fish's eye, for example, is registered in the deformation of its accompanying grid. The dimensional fluctuation became, for Thompson, an indication of light level and water depth influencing that particular species. In this manner, the type or spatial organism is no longer seen as a static whole separate from external forces, but rather as a sensibility continuously transforming through its internalization of outside events. But within the pact of the organism and the geometric language with which it is exactly described, these fluid characteristics are generally reduced to fixed principles.[24]

Lynn's thinking in the 1990s, it is important to note, is built not only upon a formalist education but also on decades of computational thinking that, fortunately for our purposes, also engaged the villas as a typology. In *The Palladian Grammar*[25] (1978), by G. Stiny and W. J. Mitchell, the villa variations were explored as a potential example of *Shape Grammars,* which "use replacement rules to change objects of a certain kind-strings or shapes in to new objects of the same kind,"[26] an explanation (albeit a description of a linguistic phenomenon) that might as easily be applied to Thompson's transformations.

Where Wittkower set up a static template, George Stiny and Mitchell proposed a set of instructions. Still using the Palladian villas as their source material, this methodology became digital in 1985, when George Hersey and Richard Freedman developed computational rules in order to produce all "Possible Palladian Villas."[27] Whereas their predecessors' first operation was to set out a grid, their program,

Planmaker, began by randomly selecting the dimensions of a starting perimeter rectangle and subsequently operated by "splitting" the repeated subdivision of the original rectangle until certain conditions of configuration, number, and proportions were generated. After several iterations, Hersey and Freedman discovered that while the rules *can* generate new Palladian plans, they can also generate some "ridiculous" "foibles," that "could and should eventually be programmed out of existence."[28] Stiny, along with another colleague, James Gips, had already noted that, given the potential infinite number of possibilities, such an editing operation must occur. They wrote that "a mechanism, a selection rule, is required to determine which of the shapes in the language defined by the shape grammar is used. The selection rule acts as a halting criterion for the shape generation process."[29]

This selection rule is potentially aligned with the exterior context and the relationship between the organism and that environment. Again, the contextual influence is implied, but not drawn. It is not unexpected that the influence of the natural environment is neglected, since it was not drawn in Wittkower's originals. It is not until Lynn notes the analogy between the villa transformations and Thompson's transformations that the first inklings of the world beyond the grid enter into the discussion.

Lynn first identifies Wittkower's collection of similar forms as a "brood," a "species."[30] Soon thereafter, in a critique of Wittkower's fixed set of formal examples, Lynn notes that "the prejudice toward fixed orders is achieved at the cost of repressing local differences of program, structure, form, and culture."[31] And that these conditions are exacerbated only in the subsequent studies that serve only as an "extension of a previously delineated and closed set of potential forms whose characteristics can be stated in advance through an ideal mathematics."[32]

In response to Wittkower's set of eleven variations of a single type then, Lynn asks: "might there be another way to respect particularities and differences without 'returning our inquiry' to universal types?" one characterized by "more pliant systems of description"[33] and to one that includes "a system of local affiliations outside itself?"[34] In a direct analogy to Thompson, these variations are now motivated by their context, according to the text, though, as in Thompson's diagrams, no visual representation of that context is present in either example. His project, *Embryonic House*, explores a basic form controlled by twelve control points to which forces are applied to create a series of possible houses (Figure 9.3). Architecture was not the fish or crustacean trapped within the grid, but the grid itself, and the computer provided the means to visualize the force which enabled architecture breaking out of the box.

In later projects, *Stranded Sears Tower*, contextual features (the river, the city's gridded structure, and various transportation lines) are mentioned in the text as transformative forces, but not drawn. In the *Cardiff Bay Opera House Competition*, the text notes that site alignments are made with existing buildings and infrastructure.[35]

In his shifting the disciplinary thinking away from a fixed type and toward a spectrum of variations as motivated by Thompson, Lynn comes to the realization that "the type or spatial organism is no longer seen as a static whole separate from external forces, but rather as a sensibility continuously transforming through its internalization

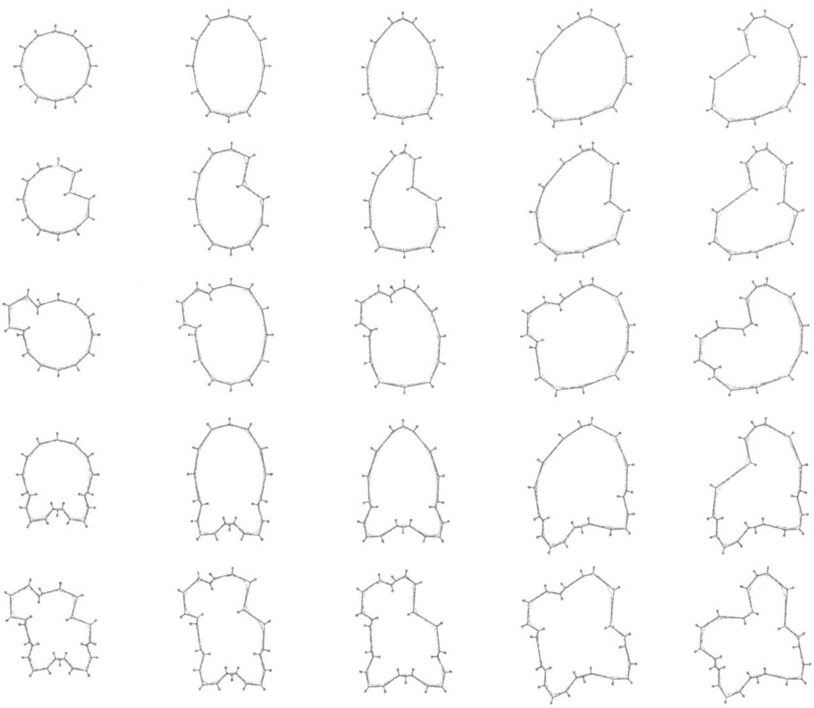

Figure 9.3 Greg Lynn, *Embryological House*, 1998. Courtesy of Greg Lynn/FORM.

of outside events."[36] Furthermore, while both Hersey's and Lynn's projects propose a computationally generated range of possibilities in a Thompsonian manner, the crucial difference between the two versions is that Hersey's digital operations are finding possible non-existent forms within a known typology. In Lynn's they are producing new forms *in a new typology*.

In the subsequent digital revolution in architecture that owes much to Lynn's work in linking computation, architectural typological thinking, and Thompson's morphological approach, the shift in evolutionary thinking has been mirrored by a contemporary trend toward the production of variation in architecture. As scripts became capable of generating varied options, evolutionary terminology has tiptoed into the language of architecture: words such as *species*, *iteration*, *generation*, *variation*, and *mutation* have by now become commonplace.[37]

These advances, however, have often eliminated the consideration of the underlying issues of these evolutionary transformations.[38] The result is that the discipline of architecture continues to act formally or sustainably but has not yet managed to elegantly create either a theory or a practice in which the form of the object is in direct functional alignment with an environmental motivator.

Architectural Force and Fitness

Thompson's diagrams, depicting species through continuous and simple geometric transformations, render the biological development of physical forms graphically legible.[39] The grid, which represents the forces of deformation across species, suggests an underlying ordering system in the natural world. "Force," as understood by Thompson, is "the appropriate term for our conception of the causes by which these forms and changes of form are brought about."[40] Thompson also uses the word "strain." In describing the process of overlaying and deforming the net, he writes: "we obtain a new figure which represents the old figure under a more or less homogeneous strain."[41]

Where Thompson's "force" described all possible forces that are "theoretically imaginable,"[42] acting along a mathematical hypothetical range, Darwin's "fitness" is more ambiguous. Stephen J. Gould notes that while Darwin never specifically defines the term, and that "the survival of the fittest" was in fact a term coined by Herbert Spencer, for Darwin, Gold writes, fitness was "a property of form, a measure of good design that did not entail survival, a priori."[43]

The ambiguity of the term "fit" in the English language becomes an issue here, as fit can mean "in good health" or "having the right shape for." Likely Darwin's notion of fitness is somewhere between the two, reproductive fitness being a consequence of both common meanings. The terms "force" and "fit" certainly have an overlap if we consider the invisible arrow pulling the organism into a particular deformation in its representation of an external context. This invisible arrow is the relationship between the animal and its environment. For these arrows to yield a form in the physical world, they must provide a true relationship between the organisms and their environments. This positive relationship is famously illustrated via the variations of finch types in the Galápagos Islands, each having adapted to its specific environmental niche. This so-called "adaptive radiation" is defined as a process in which organisms diversify rapidly into a multitude of new forms, particularly when a change in the environment makes new resources available, creates new challenges, and opens environmental niches.[44] The birds, it appeared, had diversified their claws, beaks, and other characteristics in order to better survive in their specific niches, which included ground or tree habitats and related foot morphology, and the related food sources, whether insect or fruit.

We might imagine that these birds, if they were to be mapped in a Thompsonian manner, would (perhaps), represent twelve instances along a gradient of possible instances (though likely the grid would need to become more complex, as variations are unlikely to be smooth). What is it about the twelve contexts[45] that motivate the bird types becoming real rather than just possible? And likewise, what is it about the contexts where no physical example exists, that preclude or demotivates the forms' successes? (Figure 9.4).

Thompson notes that once the deformation has been represented, it is easy to "postulate the direction and magnitude of the force capable of effecting the required transformation."[46] He does not, as I have noted, include this information graphically. And when Lynn criticizes formalist typological thinking and opens the doors to a Thompsonian thinking (and drawing) applied to architecture, the same language of

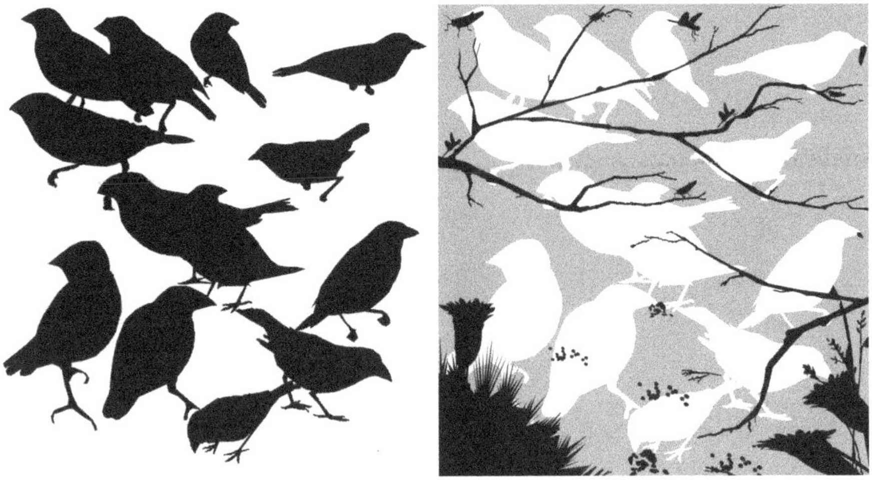

Figure 9.4 Darwin's Galapagos Finches. Based on a drawing in "Biological Science: Molecules to Man," Houghton Mifflin Co., 1963. CODA, 2012. Courtesy of Caroline O'Donnell.

representation comes with it. If we return to the source and imagine that Thompson *had* included graphical arrows or objects that visualized the forces and that the subsequent adoption of Thompsonian morphological transformations into architecture had included the presence of a motivating context, what would be the consequences for architecture?

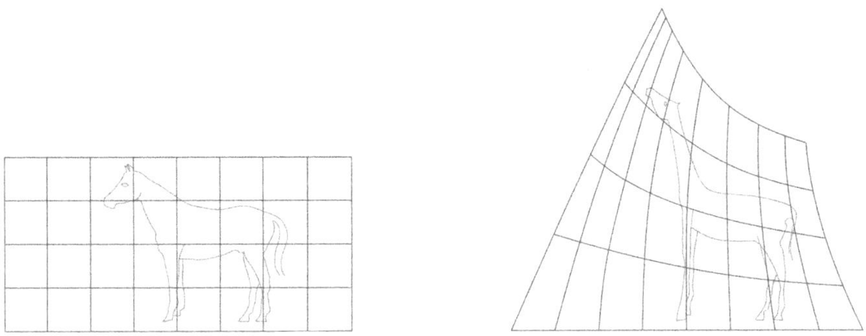

Figure 9.5 (A) Left: Horse/Giraffe Diagrams. CODA, 2012. Courtesy of Caroline O'Donnell. (B) Right: Horse/Giraffe Diagrams. CODA, 2012. Courtesy of Caroline O'Donnell.

The diagram shows a hypothetical Thompson-style horse to giraffe transformation (Figures 9.5A and B). First, the Cartesian grid is laid over a horse-like form. Second, the top-left corner is pulled upward, implying an invisible motivating force outside the grid at that corner. A third diagram includes the potential motivating context: the juicy leaves high up on the acacia tree, which no other competing animals can reach (Figure 9.6).

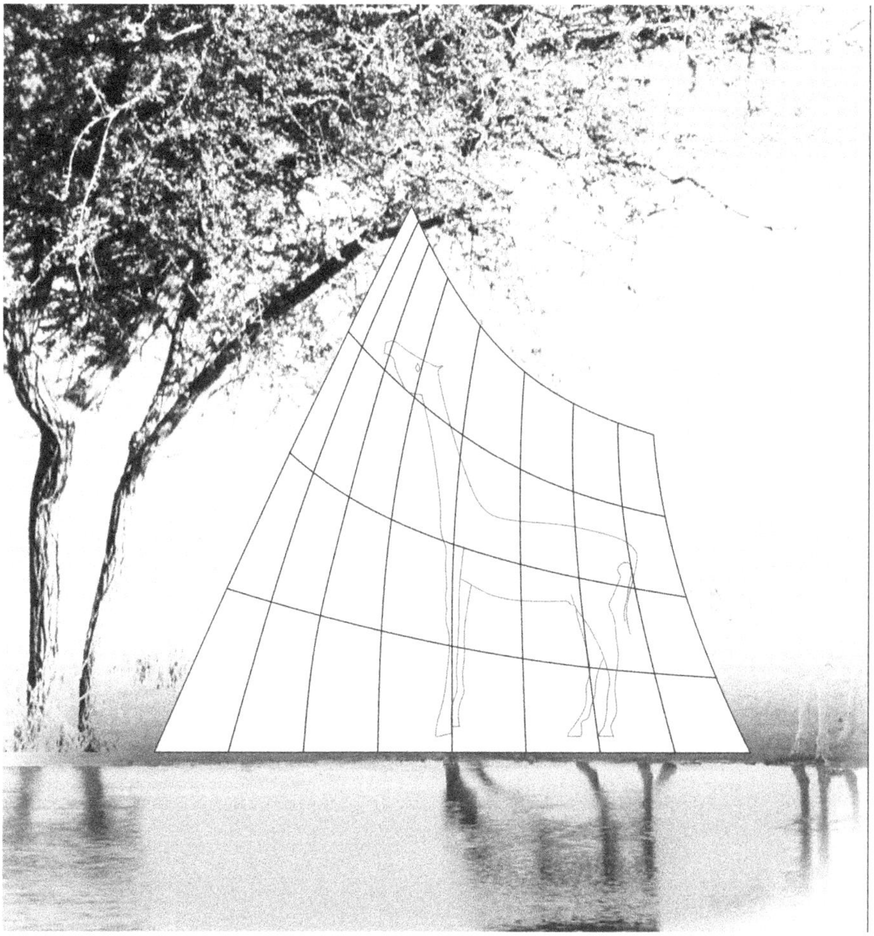

Figure 9.6 Horse/Giraffe Diagram with context. CODA, 2012. Courtesy of Caroline O'Donnell.

Of course, as mentioned with the finches, once we begin to look more closely, we find that the smooth transition of the grid would become much more complex. In the case of the giraffe, bulges would occur, for example, in the heart, which has had to become several times larger as a consequence of the pumping action, or the tongue, which has had to grow a second skin in response to the tree parasites, and so on (Figure 9.7).

A final diagram combining the bulging and the context forces us to think not only of the figure, but of the mechanical systems inside and the environment outside. If morphologist and architects alike thought to draw in these vectors and the phenomena causing them, we might be better equipped to think of our work as being between the two, rather than being the object in the grid or the grid itself. Instead of an architecture obsessed with formal variation for the sake of variation, the object itself may have lost its weight and instead we might think more about the relation of the object to the force.

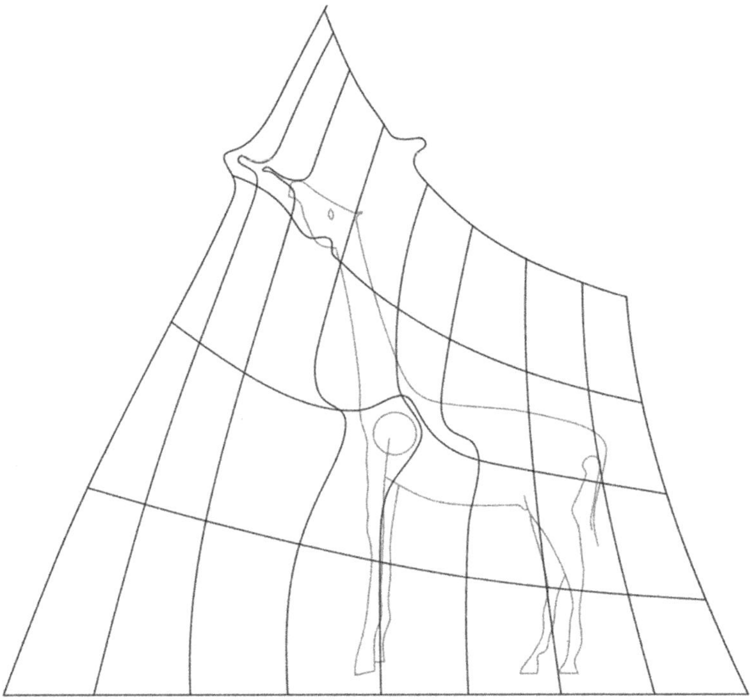

Figure 9.7 Horse/Giraffe Diagram with bulges. CODA, 2012. Courtesy of Caroline O'Donnell.

Even Lynn, who brought this thinking to the table and cited the external motivators, has said (and it is hinted at in his firm's name) that he is interested in "form." It is perhaps time for architects to move beyond the visible and grasp the invisible motives of form implied by Thompson's deformed grid, to draw in those motivators, and to follow the path to new forms of architecture.[47]

Notes

1 Ingraham, Catherine, *Architecture, Animal, Human: The Asymmetrical Condition* (New York: Routledge, 2006), 41.
2 J.-A. Meissonnier had already shown temples at the same scale in elevation view, from the same era.
3 Steadman, Philip, *The Evolution of Designs* (Cambridge: Cambridge University Press, 1979), 4.
4 Durand, Jean-Nicolas-Louis. *Precis Des Lecons D'Architecture Donnees A L'Ecole Polytechnique*, Hachette Livre BNF (Paris, 2013), 4.

5 Madrazo, Leandro, "Durand and the Science of Architecture," *Journal of Architectural Education*, Vol. 48, No. 1 (1994), 12.
6 Argan, Giulio, "On the Typology of Architecture" in *Theorizing a New Agenda for Architecture: An Anthology of Architectural Theory 1965–1995*, edited by Kate Nesbitt (New York: Princeton Architectural Press, 1996), 242–6.
7 Durand, Jean-Nicolas-Louis, *Precis des lefons d'architecture donnees a l'Ecole Royale Polytechnique*, Vol. 1, D3 (Paris, 1819), 1.
8 Darwin, Charles, *On the Origin of Species*, 6th edn. (Digireads Publishing, 2010), 48.
9 For an elaborated discussion of Thompson's critique, see Gould, Stephen J., "D'Arcy Thompson and the Science of Form," in *Topics in the Philosophy of Biology*, edited by Marjorie Grene and Everett Mendelsohn (Boston: D. Riedel Publishing Co., 1976), 249–51.
10 Thompson, D'Arcy, *On Growth and Form*, ed. John Tyler (Cambridge: Cambridge 1961) [originally 1917], 1026.
11 Gould, S. J., "D'Arcy Thompson and the Science of Form," *New Literary History*, Vol. 2, No. 2, Form and Its Alternatives (Winter, 1971), 246. [use earlier reference.]
12 Ibid., 246.
13 Thompson, D'Arcy, *On Growth and Form*, ed. John Tyler (Cambridge: Cambridge 1961) [originally 1917], 15.
14 Ibid., 1050.
15 Ibid., 889.
16 Ibid., 241.
17 Gould, Stephen Jay, "D'Arcy Thompson and the Science of Form," in *New Literary History*, Vol. 2, No. 2, Form and Its Alternatives (Winter, 1971), 245.
18 Gould, Stephen Jay, "D'Arcy Thompson and the Science of Form," in *Topics in the Philosophy of Biology*, edited by Marjorie Grene and Everett Mendelsohn (Boston: D. Riedel Publishing Co., 1976), 10.
19 D'Arcy Thompson, *On Growth and Form*, ed. John Tyler (Cambridge: Cambridge 1961) [originally 1917], 10.
20 A version of this diagram had appeared in 1944 in his essay, "Principles of Palladio's Architecture," *Journal of the Warburg and Courtauld Institutes*, Vol. 7 (1944), but contained only ten villas, and the pattern, excluding Villa Capra (Rotonda).
21 Rowe, Colin, *The Mathematics of the Ideal Villa, and Other Essays*. (Cambridge, MA: MIT Press, 1976).
22 Ibid.
23 Lynn, Greg, "New Variations in the Rowe Complex," *Folds, Bodies, and Blobs* (Brussels: La Lettre Volée, 1998), 202.
24 Ibid., 56.
25 Stiny, George Nicholas and Mitchell, William J. "The Palladian grammar," *Environment and Planning B: Planning and Design*. Vol. 5, No. 1 (1978), 5–18.
26 Stiny, George Nicholas, "Computing with Form and Meaning in Architecture" in *Journal of Architectural Education*, Vol. 39, No. 1 (1985), 9.
27 Hersey, George L. and Freedman, Richard, *Possible Palladian Villas: (Plus a Few Instructively Impossible Ones)* (Cambridge, MA: MIT Press, 1992). *All Possible Palladian Villas* was the original title of the book, until it was pointed out that, despite the thousands of variations, the options that had been produced did not represent a comprehensive set.
28 Ibid., 125.

29 Stiny, George and Gips, James, *Algorithmic Aesthetics: Computer Models for Criticism and Design in the Arts* (Berkeley, CA: University of California Press, 1978), 131.
30 Lynn, Greg, "Multiplicitous and Inorganic Bodies," *Folds, Bodies, and Blobs* (Brussels: La Lettre Volée, 1998), 34.
31 Ibid., 47.
32 Ibid., 212.
33 Ibid., 56.
34 Ibid., 56.
35 Ibid., 76.
36 Ibid., 39.
37 This thinking is described as having shifted from a focus on form to one "focused on the process of formation, to dynamic constitutive systems and ecologies, to techniques, building blocks, modules, evolution and diversity." Detlef Mertins, "Variability Variety and Evolution in Early 20th Century Bioconstructivisms," in *Research and Design: The Architecture of Variation* (New York: Thames and Hudson, 2009), 55.
38 Several digital practices today, however, are leading the way in engaging dynamic, biological, responsive systems, including, to name a few, Jenny Sabin Studio/Sabin Design Lab at Cornell AAP, Kokkugia, Smart Geometry Group, Philip Beesley, Ecologic Studio, MinimaForms, and Epi-phyte Lab. Such practices are drawing into the whiteness of the computer's background dynamic systems and forces to which their outputs must respond, as well as drawing in scientific models for architectural production.
39 The occurrence of Thompson's work in architectural publications is innumerable. See, for example, Philip Beesley and Sarah Bonnemaison's *On Growth and Form: Organic Architecture and Beyond* (Riverside Architectural Press, 2008) which includes essays relating Thompson's work with Architecture, as well as a bibliography.
40 D'Arcy Thompson, *On Growth and Form*, ed. John Tyler (Cambridge: Cambridge 1961) [originally 1917], 11.
41 Ibid., 1033.
42 Ibid., 1026.
43 Gould, S. J., "D'Arcy Thompson and the Science of Form," *New Literary History*, Vol. 2, No. 2, Form and Its Alternatives (Winter 1971), 233
44 Schluter, Dolph, *The Ecology of Adaptive Radiation* (Oxford: Oxford University Press, 2000), 10–11. Note: Another theory called "character displacement" proposes that two lineages will become different as a result of selection against competition for the same resources. This will lead to niche displacement.
45 The term "context" here refers to the surroundings of the animal. In evolutionary biology, the understanding of finch evolution depends critically on the phylogeny (history of splitting events that gave rise to current diversity) and thus in a way the "historical context" of the organism itself becomes relevant.
46 Thompson, D'Arcy, *On Growth and Form*, ed. John Tyler (Cambridge: Cambridge 1961) [originally 1917], 1033.
47 This thesis is elaborated in Caroline O'Donnell, *Niche Tactics: Generative Relationships between Architecture and Site* (New York, Routledge, 2015).

10

Diagrams of Entropic Forces: New Growth and Form

Philip Beesley

Instability and wandering, shifting boundaries without certain territory is risky. However, could those qualities possibly contribute to substantial architecture?

Building upon D'Arcy Wentworth Thompson's famous maxim "form is a diagram of forces," this discussion explores precarious *entropic* systems—the natural forces that create flux and instability—within new interactive architecture. Thompson's vision of the essential organizations and dynamics within natural life is positioned as a core for emerging collaborations within the author's work within the Living Architecture Systems Group. These experimental architectural systems are now at the early stages of integrating distributed kinetic structures and hybrid control systems that include machinic curiosity. Concepts guiding this work are derived from emerging conceptions of living systems offered by recent physics research. Ilya Prigogine's Nobel Prize-winning mid-twentieth-century analysis of dissipative form, Gavin Crooks's renewed definition of dissipative adaptation voiced a decade ago, and Jeremy England's recent writing on evolutionary dynamics provide strikingly innovative interpretations of how physical laws shape forms of life. These thinkers have contributed to a substantial reform of the meaning of entropy. A renewed definition that emphasizes maximum potency and potential is presented. This interpretation can contribute to new architecture.

In *On Growth and Form*, Thompson wrote:

> any portion of matter … may in all cases alike be described as due to the action of force. It is a diagram … of the forces which have been impressed upon it when its conformation was produced, together with those which enable it to retain its conformation … [It is] also the conformation of the organism itself, whose permanence or equilibrium is explained by the interaction or balance of forces.[1]

Thompson offered transformed understanding of the underlying interlinked forces that make up natural form while at the same time retracing classical understanding in his emphasis on permanence and equilibrium. This dual quality—transformation by integrated dynamic systems influenced by Thompson's insights, while at the same time assuming that life means opposition of the decaying flux of the world—has influenced numerous twentieth-century thinkers. In Ernst Schrodinger's famous mid-century treatise *What Is Life* (1944) the author asserted that the essential ingredient of life is "negentropy," the ordering force that opposes the decay and disorder of the world.[2] If negentropy is order, then entropy would be the antithesis of order. Resonating with Shakespeare's famous meditation that "we are destined to dust," invoking *entropy* can imply that the universe is a giant soulless machine slowly running out of steam and lapsing into disorder. A fundamental opposition of living systems to the decaying world is implicit within that conception. If we use conventional definitions that identify entropy with disorder, and if we follow the second law of thermodynamics which declares that entropy inexorably increases, we might agree that our ultimate fate is disease and death. The exquisitely ordered states of living organisms, repairing and renewing themselves until they die, might indeed seem the opposite of entropy.

The form-languages of architecture have been influenced in fundamental ways by this conception. Classical architectural design can be seen as dominated by the pursuit of low entropy and high order. The Roman writer Vitruvius described these qualities as combining into "firmitas," one of the key qualities of architecture, even the art's central core. Closed boundaries, reducing and sheltering interiors from the exterior, create fundamental kinds of architecture. Forts, paradise gardens, and palaces alike have secure, bounded walls. Those forms, together with their dense, concentrated building materials of stone and brick, create low entropy within their bounded territories. The constructions made by these materials create clear differences between interior and exterior, maintaining stable boundaries and sheltered and managed interiors. In turn, these closed and bounded forms impact their surrounds. High entropy tends to be created outside walled buildings. Rebounding in response to the dense bounded forms, turbulent forces to the exterior tend to be amplified. To the outside, wilderness may freely play, while to the inside a calm and stable sanctuary may be maintained.

While the human interpretations that underlie the term "entropy" are ancient, the word is modern. The root of entropy was coined amid searches for efficient work during the Industrial Revolution. Entropy was coined in the early eighteenth century as the amount of energy lying within physical materials that is not available for "work." Taken from "tropos," Greek for turning and transformation, this original interpretation denotes a fundamental cost, measured as the change in units of heat energy within a given material needed to change it from one state to another, measured against absolute temperatures. Used in this way entropy can be focused on the cost of work needed for the industry of transforming raw, disorganized ingredients into organized, useful material. The higher the entropy, the less work that can be easily generated from that material and the more work required in transforming that material into something useful. The fundamental laws of thermodynamics state that as time passes within closed environments where no material can enter or leave, entropy always increases. If the objective is yielding maximum work from a machine, entropy can be positioned as an emphatically negative word. Within the closed boundaries of machines and vessels, things intermix, decay into their lowest energy states possible, reach equilibrium, and die. There are many examples of highly organized raw materials from nature that contain low entropy: diamonds, glass, and stone. Those materials tend to require large amounts of energy to change their internal organization. They tend to resist internal change.

When concentrated and stable, low-entropy materials are used to make buildings: they can create quality of durability and permanence. Surely architectural firmitas seeks the secure and the eternal, opposites to decay. If entropy were used in considering classical architecture, it could similarly be negative, opposing, and disabling firmitas. It would be tempting to equate entropy with a host of conditions lying outside the stable foundations and walls of a city: lawlessness, wilderness, loss of civic order.

However, does entropy inevitably mean "disorder," and does that term inevitably lead to death and decay? New voices, drawing on physicist Ilya Prigogine's insights last century and applying those to open, living systems, state that entropy can be a profoundly positive quality. Recent voices even assert that entropy is a fundamental quality that helps to create and renew life. How could unstable, apparently disorderly, forces be compatible with life? If renewed definitions reveal the refreshing and renewing qualities of entropy, embedding those directly within the intermeshed forces and dynamic topologies that make up the forms of our natural world, might this contribute to new kinds of architectural design?

A strikingly new conception that seems to counter the assumption that life depends on resisting entropy was offered by Ilya Prigogine in his 1978 Nobel address.[3] In his research, he asserted that the closed systems of classical science and historical mechanical engineering are fundamentally different than the conditions of the natural world. Upending conceptions of the inevitable decay of forms within the world, he described wholes realms of *dissipative* forms: the barred, clumping textures of cumulus clouds; rolling standing waves in the ocean and in constantly shifting dunes, formed by constant energy shedding, dissipation-holding their forms, steady amid constantly fluxing exchanges of force. Dissipative forms lie around us. They can be seen in veils of smoke billowing at the outer reaches of a fire, the barred, braided fields of clouds; torrents of spiraling liquids; mineral felts efflorescing within an osmotic cell reaction.[4] Prigogine asserted that these kinds of forms were far from accidents or temporary phenomena. Instead, he demonstrated that they could be consistently found within constantly cycling systems, the natural result of the far-from-equilibrium forces that regularly move through the world. Dissipative forms might seem ephemeral, but they are in fact tenacious and durable, holding their organizations in dynamic balance while at the same time materials and energy perpetually cycle through them.

The UK physicist Gavin Crooks developed Prigogine's vision by offering renewed conceptions of entropy directly linked to dissipative forms.[5] Instead of "disorder," he suggested that dissipative forms demonstrated particular resonant nodes within the flux of natural forces, constantly shedding forces and helping to create entropy within their surrounds. Building from Crooks and Prigogine, the MIT-based physicist Jeremy England has recently proposed that life itself epitomizes maximum entropic production.[6] Constant cycling of materials in reactive fluxes are orchestrated in the metabolisms and perturbations fostered by life, yielding a maximum of entropy. Far from living things opposing the flux of the world, threatened by order, England asserts that life is directly rendered as an optimum that combines both maximum entropy production and resilient, durable organization. Instead of resistance to decay, this new approach offers "dissipative adaptation" of the vast forces within the world itself. Schrodinger's earlier assertion that life depends on "negentropy" appears to be fundamentally questioned and even contradicted by the research of Crooks and England.

This renewed interpretation is accompanied by a shift from the assumption that the world should be analyzed as a closed system. Classical physics and machinic control have tended to focus on closed, bounded worlds: the finite world in which action is always met by an equal and opposite reaction. In contrast, open environments are fundamental to Prigogine, Crooks, and England. Instead of "total" environments that have finite boundaries, their collective work is founded upon open systems: on the far-from-equilibrium state of the earth immersed in the heat bath of the sun, and the surface of the earth continually immersed in the shifting forces of tectonic plates, and the constant flux of air and ocean. From outside and inside, energy and materials continuously exchange. Time passes, and the actions that occur within time are only reversible in theory, not in reality. Time passes. The world is open, not closed. This vision invites substantial renewal seeking a maximum of potent interaction, creating fertility. With this, the word entropy needs redefinition. Instead of "disorder," entropy can be redefined as a distinct force that seeks a maximum of diverse freedom and a maximum of potential.

With such a fundamental shift in conception of energy and material exchange within passing time, it seems that assumptions guiding the boundaries of architecture might have substantial reasons to shift in turn. The vision of potency and fertility invoked by England and his predecessors invites translation into architecture. Within the design of classical architecture, free citizenship is made possible by the finite *urbs* where stable fortified enclosing walls of the city provide resistance and hardened boundaries. Might an alternative architecture construct dynamic, constantly shifting open thresholds, seeking constantly renewed thresholds of exchange?

The qualities evoked by these renewed conceptions can be seen within a series of new projects by the Living Architecture Systems Group.[7] This collaborative research and design group is pursuing precarious, reactive qualities designing systems that maintain durability and coherence while at the same time generating entropy. They use paradigms of dissipative structures and diffusion as guides for their design and their forms. Their organization is characterized by punctuated oscillation and quasiperiodic geometries, with shifting boundaries that fluctuate. Deliberate ambivalence is inherent to the approach, yielding qualities where things convulse and stutter in emerging vitality, seeking intense mutual relationships of exchange with surrounding environments.

Astrocyte (2017) (Color Plate 3) and *Noosphere* (2018) (Color Plate 4) are installations created by this collaborative group. Both are composed of an interwoven series of spherical shells containing artificial intelligence surrounded by a floating landscape of delicate skeletal forms. Astrocyte refers to the form of a nerve cell, containing radiating connections to a myriad of surrounding cells. Noosphere refers to life on Earth as a vast organism with a thinking skin.[8] In a next state of evolution the noosphere could transform the biosphere with collective consciousness, and even collective sympathy and empathy. Noosphere incorporates details from Astrocyte while shifting from acrylic to expanded metal constructions and adding distributed sound systems. These constructions continue a series that the Living Architecture Systems Group has pursued for the recent decade.

Individual building components have been developed following practical design strategies that have attempted highly efficient material use employing thin sheets of material, cut and extruded into skeletal interlinking structural components. The structures create resilient scaffolds that support meshes of electronic controls alternating with liquid vessels containing active chemical reactions. Microprocessors with interlinked connections operate in a similar way to neurons intermeshed in natural brains. Chemical skins within the liquid-containing cells bear combinations of oil, inorganic chemicals, and water-based solutions stimulated by LED light, suggesting new kinds of self-renewing skins for future building. Triangular arrays of thin filaments organize the structural component designs, creating resilient scaffolds that can handle widely varying forces within their shifting mesh works. Feather-like tiles in overlapping arrays clothe the scaffolds, creating porous membrane covers that offer widely varying dynamics that can alternately resist and attract their surroundings. When closed, these tile-works provide strong shelter. When open, the systems tend to attract interaction, actively responding and amplifying the movement and exploration of viewers and occupants in surrounding space.

The structural mesh of Astrocyte uses overlapping strands within hexagonal pyramid cells that possess strength in minimal material. The structure is made up of a series of lightweight meshwork spherical shells that arc high above occupants, making simple shelters. The shells use thermally formed acrylic, using digitally fabricated cutting patterns that permit stretching into deeply elongated hollow doubly curvet trumpet-shaped cells whose skeletal patterns have been optimized for production with minimal waste (Figure 10.1A). Each segment of the curved space truss system uses hundreds of these cells, each organized in overlapping strands of material that balance each, creating strength from thin laceworks of material (Figure 10.1B). The individual cells form arrays of close-packed rings with projecting spines that intersect with neighboring cells, much in the same way that natural bone structures employ multiple filaments spanning between outer and inner shells. The intermeshed stems create resiliency, able to support widely varying forces and shifting motions (Figure 10.1C).

(A)

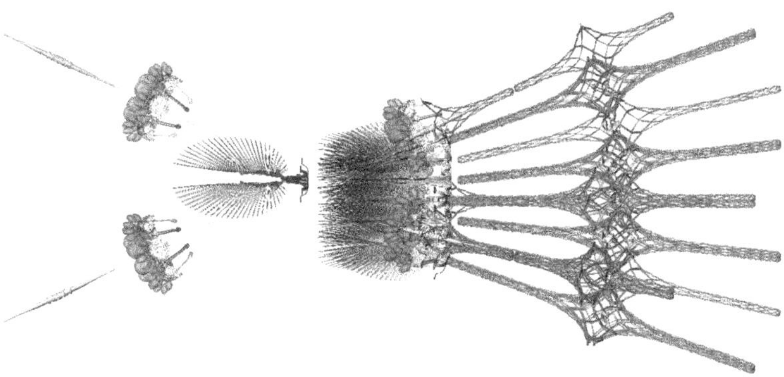

(B)

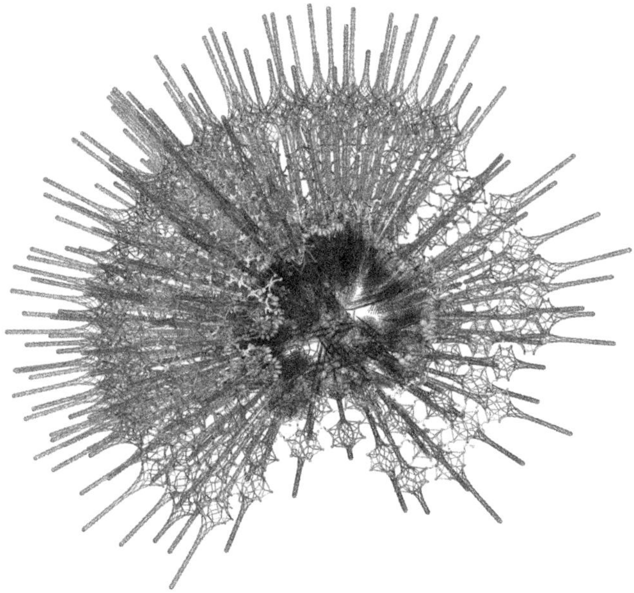

(C)

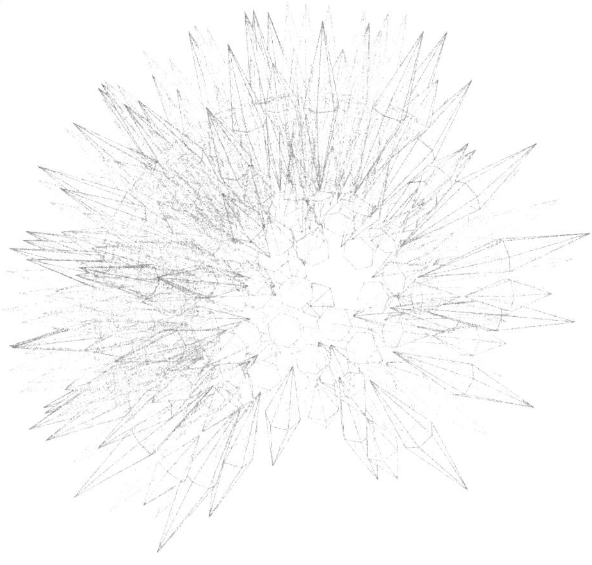

Figure 10.1 Philip Beesley, (A) Exploded View of Astrocyte, (B) View of Astrocyte Digital model, (C) Underlying Digital Model Organization, Astrocyte, by Living Architecture Systems Group/Philip Beesley Architect Inc./4DSOUND et al. (2017).

Covering the inner surfaces of these skeletal scaffolds are radial arrays of mechanisms and lighting instruments. The massed lining of each hemispherical shell has been designed to respond to feather-light variations of air movement in the surrounding environment, responding with shimmering cyclical motion. Deeply fissured fronds are fitted with spinning miniature motors carrying offset weights that impart vibrating movement to the assemblies. Tapered glass vessels inserted into 3-D printed housings carry high-powered LED lights, concentrated by narrow-beam reflectors that concentrate the light into axial beams shining through the central cores of the structural components. Cradling each light and surrounded by frond-clusters are cellular manifolds made from multiple glass vessels. A combination of oil, inorganic chemicals, and aqueous solutions create chemical skins within these prototype cells.

The work follows a conception of architecture that extends outward and inward from boundaries, encouraging crossing and passage. Outward, tendrils and plumes interweave with surrounding layers of air. Clusters of laser-cut translucent polymer fronds are arranged by grouping and bundling their angled geometries around close-fitting inner sheath structures. Cantilevered resilient stays are inserted, and arrays of individual impact-resistant acrylic chevron links supporting these bundles are chained together with elastic joints to form a diagrid of corrugated mesh with diffusive, viscous performance, permitting outward extensions. Halo-like clouds of hovering material ripple and vibrate in response to shifting forces of the environment.

Component designs within these scaffolds tend to be skeletal in form, created by orienting vertices within each component cell where elements need to intersect with neighboring cells, carrying internal component details through drawing vectors between those points. Material is laid down along those axial paths by using automated fabrication, including fused deposition and digital cutting. The thickness and shape of each of those paths are refined in cycles, seeking balances in which all potential forces can be dynamically and shared with neighboring components. Rigid assemblies tend to reflect and pass on forces without absorbing them, and that force-shedding can increase the potential for damage in neighboring assemblies. Because of this, where assemblies of components need to be rigid—for example, where they are required for protection of delicate electronics or liquid chemical solutions—secondary assemblies tend to be needed, capable of blending and cushioning the accumulated forces. Flexure and elasticity are retained by voiding out large surfaces and volumes, guarding against the accumulation of major differences in adjacent assemblies.

In contrast to rigid frames, the assembled scaffolds that result from this kind of design tend to resemble natural tensegrity-based systems. Mesh works of relatively long tensile filaments are embedded with small compressive struts. Chains of individual compressive details are linked together into overlapping diagonal arrays containing flexible joints. These resilient structures can carry widely differing loads and accommodate composite draping and flexure while at the same time large organizations are preserved. Unlike the resistance and protection against intervention of classical wall-building, these structures tend to function by introjection, drawing external forces inward, passing and sharing them amid deeply interwoven arrays of

(A)

(B)

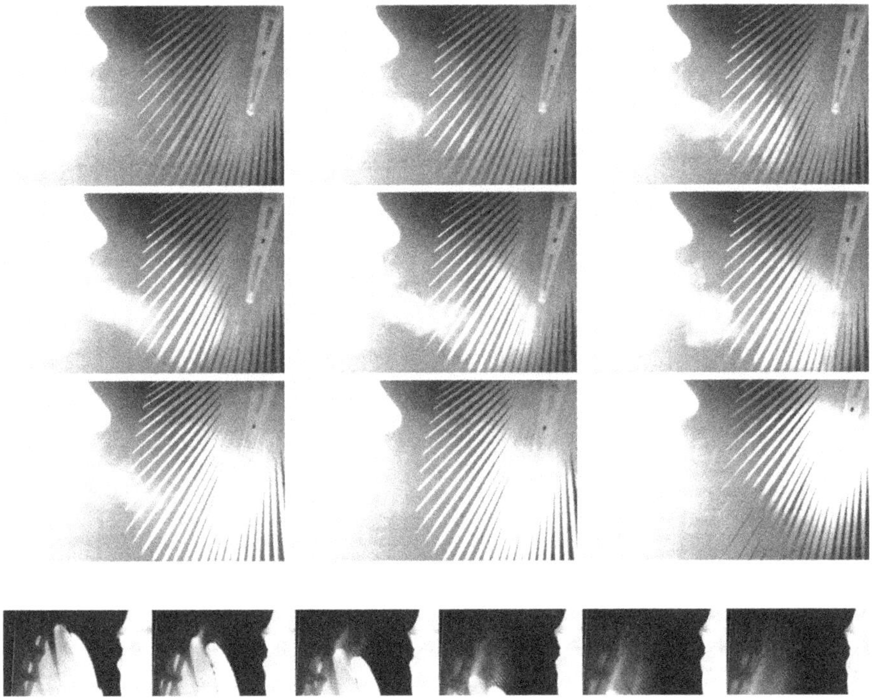

Figure 10.2 Philip Beesley, (A) Thermal photograph of Frond Filter assembly, series 2 design analysis, Philip Beesley Architect Inc., 2013, (B) Thermal photograph of Frond Filter assembly, design analysis, Philip Beesley Architect Inc., 2013.

individual cells. Similarly, dynamics that come from within these porous enclosures tend to emanate outward, shedding their forces in rippling oscillations of exchange.

Tile-shaped components used for covering these structures and for stirring and propelling air movements are designed for "precarious" highly reactive behaviors. Thin flexible sheets of durable polymer material are shaped into comb-like rows of tapered filaments, creating frond-shaped filters. Rows of filaments are tested in cycles, finding the maximum possible lengths that can maintain their positions without collapse while at the same time maintaining maximum flexibility for reacting to external influence. Design methods include testing and visualizing how these components behave within surrounding flows of air. Capacity for heat exchange serves as a practical measure of this capacity for reaction.

The illustration included here demonstrates a design method that uses thermal photography for testing prototype designs of these elements (Figure 10.2A). A frond filter is positioned to hang vertically as a conductive thermal screen. It partly masks the glowing face looking straight toward the camera. In the spaces between the narrow, tapering tines of the frond, the facial structure is etched as a map that betrays the knotted bundles and fibrous clusters lying below the smooth profiles of the jaw and throat. The cool tip of the nose is partly covered by the converging stem that gathers the frond's tines at its center. The stem follows the shadow, fading to dark, while above at the pineal gland and bridge of the nose it glows white-hot. Lingering in the surface of the frond is a shifting map of facial energy (Figure 10.2B).[9]

If "notch" filters applied to thermal detectors are tuned to precisely address gaseous material, minute air currents and concentrations of carbon dioxide and oxygen can be seen concentrating and dispersing around our bodies, extending in laminar flows along inner and outer surfaces of building envelopes. There is a tangible sense of living and inorganic exchange in these images. After the warming breath ceases, the heated center moves upward into the stem, while the outer tips of the frond quickly reverse their flow and darken, pulling cold from the surrounds.

The intimate dimensions implied by these experimental works imply form-languages for designing buildings. Instead of valuing resistance and closure, new form-languages for architecture could foster mutual relationships and maximum interaction. The densely layered forms of a jungle are made of diffusive, deeply interwoven materials that expand and interact with their surroundings. A new city designed to easily handle unstable conditions of cooling and heating in rapid cycles of shedding heat and warming and collecting heat again might well look like a hybrid forest where each building is made from dense layers of ivy-like filters and multiple overlapping layers of porous openings. Similarly, the kind of diffusive forms seen in reticulated snowflakes and the microscopic manifolds of mitochondria have a common form-language of intense reticulation. Their increased surface areas make their reaction-surfaces potent. By drawing on physical examples of coupled interactions and interconnections, a renewed kind of unapologetically fragile architecture can be found.

Might architecture foster that productive instability and help to create a maximum of entropy within its internal systems and their larger environments while at the same time maintaining homeostasis for its inhabitants? Form, as Thompson wrote, is a diagram of forces. Seen through the lens of recent physics, the profound interconnectedness of Thompson's maxim can be renewed: dissipative architectural forms are a diagram of entropic forces.

Notes

1 Thompson, D. W., *On Growth and Form* (Cambridge, England: Cambridge University Press, 1942 [1917]).
2 Schrödinger, Erwin, *What Is Life?: The Physical Aspect of the Living Cell; with Mind and Matter and Autobiographical Sketches* (Cambridge: Cambridge University Press, 1992).
3 Prigogine, Ilya, "The 1977 Nobel Prize in Chemistry," Nobelprize.org. Nobel Media AB 2014, http://www.nobelprize.org/nobel_prizes/chemistry/laureates/1977/press.html
4 Paragraphs adapted and revised from Philip Beesley, "Precarious Living Reactions: New Paradigms for Robotic Architecture," in *Towards a Robotic Architecture. Framework and Processes,* edited by Mahesh Daas and Andrew Wit (ORO Editions, 2018).
5 Crooks, Gavin E., "Entropy Production Fluctuation Theorem and the Nonequilibrium Work Relation for Free Energy Differences," *Phys. Rev.* E 60, 2721, September 1999.

6 England, Jeremy, "Why Trees Don't Ungrow," *Aeon*, 2018, https://aeon.co/essays/does-the-flow-of-heat-help-us-understand-the-origin-of-life
7 Early Living Architecture Systems Group projects are illustrated within Philip Beesley et al. *Hylozoic Ground: Liminal Responsive Architecture*, Riverside Architectural Press, Toronto, 2010, and Philip Beesley et al. (eds.), *Sentient Chamber*, Riverside Architectural Press, Toronto, 2016.
8 The sculpture and its title *Noosphere* follow the conceptions of Teilhard de Chardin, in his *The Phenomenon of Man*, Harper and Row, New York, 1956.
9 Adapted from Beesley, Philip, "New Thermal Architecture," in *High Definition: Design for Zero Tolerance*, edited by Bob Sheil (London: Architectural Design, Academy Editions, 2013), 90–9. ISBN 978-1-118451-85-4.

11

Tracing Threads of the Living Organism

Ellen K. Levy

Why does D'Arcy Wentworth Thompson's masterwork, *On Growth and Form*, continue to stimulate new expressions in the fine arts? Beyond the attention commanded by the images and language and the model Thompson, himself, presents as a polymath and environmentalist is the unstated promise that artists might learn to create a world the way nature does—from forces to forms. One can perhaps view much of the text and accompanying figures in *On Growth and Form* as a kind of instruction manual for generative techniques. However, no single quality sums up this tome; *On Growth and Form* evokes multiple associations. Its morphology is pluripotent as is its influence on developments that have impacted contemporary art. These developments include complex systems, classification systems, and origins of life theories in which artists, using a wide range of media, address issues such as morphogenesis, taxonomy, and structure. In this chapter, I trace some of these threads.

Largely bypassing explanations of natural selection, adaptation, and evolutionary fitness, Thompson identified problems of form and growth as largely mathematical and physical. History is an aspect of his work that paleontologist Stephen J. Gould believed was essential but overlooked. The following passage from *On Growth and Form* underscores this:[1]

> Immediate use and old inheritance are blended in Nature's handiwork as in our own. In the marble columns and architraves of a Greek temple we still trace the timbers of its wooden prototype, and see beyond these the tree-trunks of a primeval sacred grove; roof and eaves of a pagoda recall the sagging mats which roofed an earlier edifice; Anglo-Saxon land-tenure influences the planning of our streets and the cliff-dwelling and cave-dwelling linger on in the construction of our homes! So we see enduring traces of the past in the living organism—landmarks which have lasted on through altered function and altered needs.
>
> D'Arcy Wentworth Thompson[2]

The quotation suggests Thompson's belief that the patterns found in the physical universe cannot be reduced to information alone. Many artists discussed in this text share Thompson's emphasis on materiality and history.

Complex Systems and Morphogenesis

Thompson contributed to the development of complex systems, which inform much scientific research and artistic production. A brief background includes work of biophysicist Alfred Lotka who, in 1910, observed phenomena now associated with complexity, among them, chemical oscillations, predator–prey relationships, and autocatalysis in response to feedback systems.[3] At a later date, mathematician Alan Turing showed that molecular-level interactions can lead to morphogenesis and cell differentiation. He cited Thompson in his own major work, "The Chemical Basis of Morphogenesis."[4] Turing recognized that Thompson's application of mathematical modeling to visualizing form development established a basis for understanding morphogenesis.[5] For example, reaction-diffusion systems provided developmental biology a working model for the growth of patterns on animal pelts and shells. Tracing the emergence of complexity, media artists Christa Sommerer and Laurent Mignonneau used the term "rational morphologists" to identify those involved early on with its principles, including Thompson, Conrad Hal Waddington, Richard Owen, and Hans Dreisch.[6] These dynamic principles were readily adapted by artists.

Artist Philippe Parreno generated his 2014 artwork, "With a Rhythmic Instinction to Be Able to Travel beyond Existing Forces of Life (Green, Rule #1)," by implementing successive states of an algorithm with a cellular automaton. It establishes dynamical transitions from one state to another based on mathematician John Horton Conway's program "The Game of Life" (Figure 11.1). Cellular automata feature some of the central characteristics of self-organized complex systems in that they proceed by applying local, simple rules.[7] Cells are typically blackened according to whether surrounding cells are black or white. Emergent features may result, and the outcome relies on many factors interacting at different phases over time. Parreno prepared a context for the automata consisting of numerous drawings of fireflies in various developmental stages.[8] The emergence, presence, and disappearance of lit cells determine the sequence of images and the intensity of sound accompanying his animation. Each sequence of drawings of a depicted firefly from various life stages enacts a life cycle that concludes with its "death" that lingers on the monitor for several seconds. Soon after, another life cycle ensues. It is Parreno's emphasis on dynamically simulating life cycles in multiple ways that differentiates his work from other contemporary artists who also incorporate algorithmic productions of text, sounds, and images and imbues his work with evolutionary-like processes. His choice of tools, ranging from graphite to new media, reinforces the viewer's sense of watching a life form through a progression of developmental stages.

Sculptor Janet Echelman explores complexity and morphogenesis, creating environments and unexpected configurations with reinforced fishing nets. Her works respond dynamically to the forces of wind, water, and light. In *Sculpture* magazine this polymath elaborates, "I began with the history of the site, a centuries-old fishing village that became an industrial zone in the last few decades. There are references to smokestacks and their red-and-white striped patterns, the angled masts and cables of

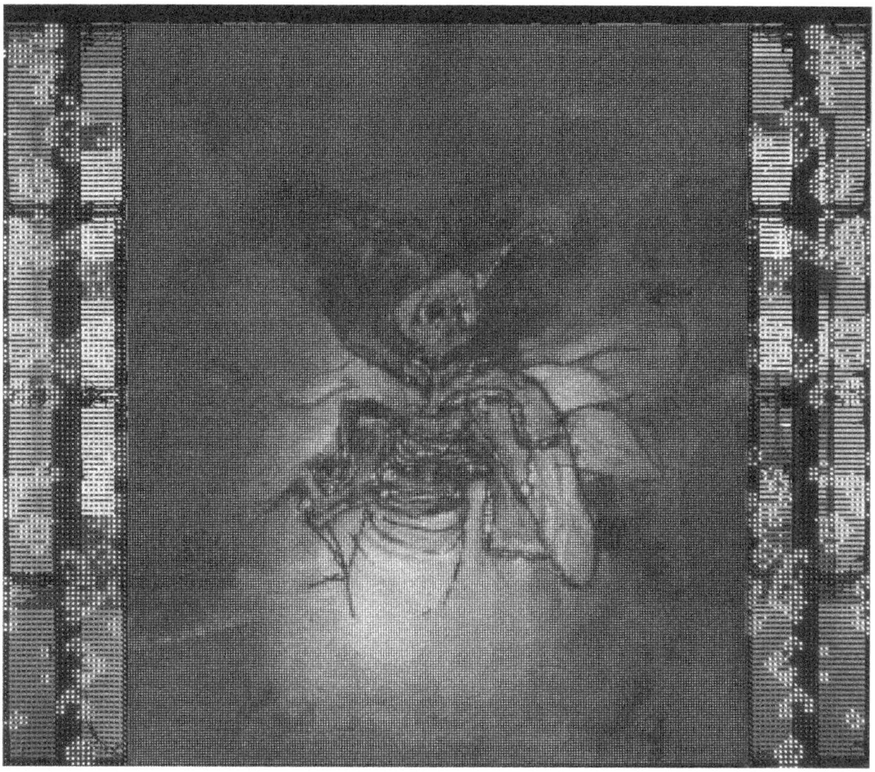

Figure 11.1 Philippe Parreno, "With a Rhythmic Instinction to Be Able to Travel beyond Existing Forces of Life (Green, Rule #1)," 2014. Eight Martin Professional EC-20 LED panels, ten Martin Professional EC-10 LED panels, Mac mini, speakers and amplifiers, dimensions variable. Courtesy of Pilar Corrias. Photo by Andrea Rossetti.

Portuguese ships, the patterns and forms of fishing nets and Portuguese lace."[9] In my conversation with the artist, she noted that the works can resemble sea anemonies and undergo shape-shifting in real time when acted upon by wind. Echelman's netting initiates a range of analogies. As it folds and billows, the netting readily suggests phases of evolutionary development such as cell and organ differentiation. The artist forges a vivid comparison; one might think of the rope-like protein that forms a protective netting and surrounds the DNA at the core of the nucleus.[10] An early work of Echelman's appears to have undergone invagination, the process of being folded back on itself to form a cavity. Echelman's distorted net suggests an unforced relationship to the deformed grid in Thompson's famed topological transformations. She notes that her works may conjure pre-Cambrian life forms before the advent of multicellular life.[11] Her reference is clearly to Gould's book *Wonderful Life: The Burgess Shale and the Nature of History* (1990), which considers questions of contingency regarding the great diversity of fossils from the Burgess Shale. Gould asks whether and to what extent the same life forms might result if the process were to recur, finding it unlikely.[12] Echelman similarly courts chance in the morphing of her forms; the effects of weather in interaction with her structures are unpredictable.

Echelman's work effortlessly spans the three categories discussed in this anthology—fine arts, design, and architecture—while embodying Thompsonian principles of form generation. She works with a range of professionals and materials, utilizing high-tech fibers in addition to netting. In collaboration with media artist Aaron Koblin, an open source programming language called Golang was used to project sound and color onto Echelman's work *Skies Painted with Unnumbered Sparks*, installed in Vancouver, BC[13] (Cover). She invited interaction; spectators were empowered to use their mobile devices to draw on the sculpture with light. The colors of biomorphic forms were altered in digital animations projected onto her work; the graphics were rendered in real time. The ropes are made of a fiber that Echelman states is fifteen times stronger than steel.[14] Her monumental sculpture spans 745 feet between the 24-story Fairmont Waterfront and the Vancouver Convention Center.

Echelman places her sculptures in a variety of public environments, and her works and titling often have ecological overtones. In describing *1.8* (2016) sited in London (Color Plate 5), she explains that the title refers to the length of time in microseconds that the earth's day was shortened because of an earthquake that emanated from Japan and redistributed the earth's mass.[15] Her stated aim is to show "the interdependency that we all have with the larger cycles of the planet."[16]

Thompson's pluripotent influence is palpable also in Oliver Laric's 2018 video *Betweenness* and sculptures. They were part of his installation "Year of the Dog," which explored both morphogenesis and allometry. His animation narrates the dynamic growth patterns of different parts of an organism. It features the unpredictable morphing of a black line drawn on a white ground that shifts continuously in scale and proportioning and can evoke processes of cell mitosis and conjure an array of animals, both real and mythological. For example, one still from *Betweenness* dynamically portrays an image of a bee as if snared between two lines, at right angles to each other. For me, the image evokes positioning between two different evolutionary lineages, vertical and horizontal. Viewers following the evolving line may also become aware of the critical points where a form can be identified as an entity or as one of several entities.

In related 3-D-printed and painted polyurethane sculptures that the artist titles *Hundemensch* (2018), the silhouettes of a human ear, a frog, and a crab appear in relief below the surface of each representation of a genetically modified creature's back (Figure 11.2). Laric states

> I am not focusing just on metamorphosis, but also on the moments in between moments … Along with the video I am showing three resin sculptures cast from the same mold of a half-human, half-dog animal holding another smaller dog. The three are like distant cousins, both similar and different, and, in a way, I'm uncovering the process of how that difference is made.[17]

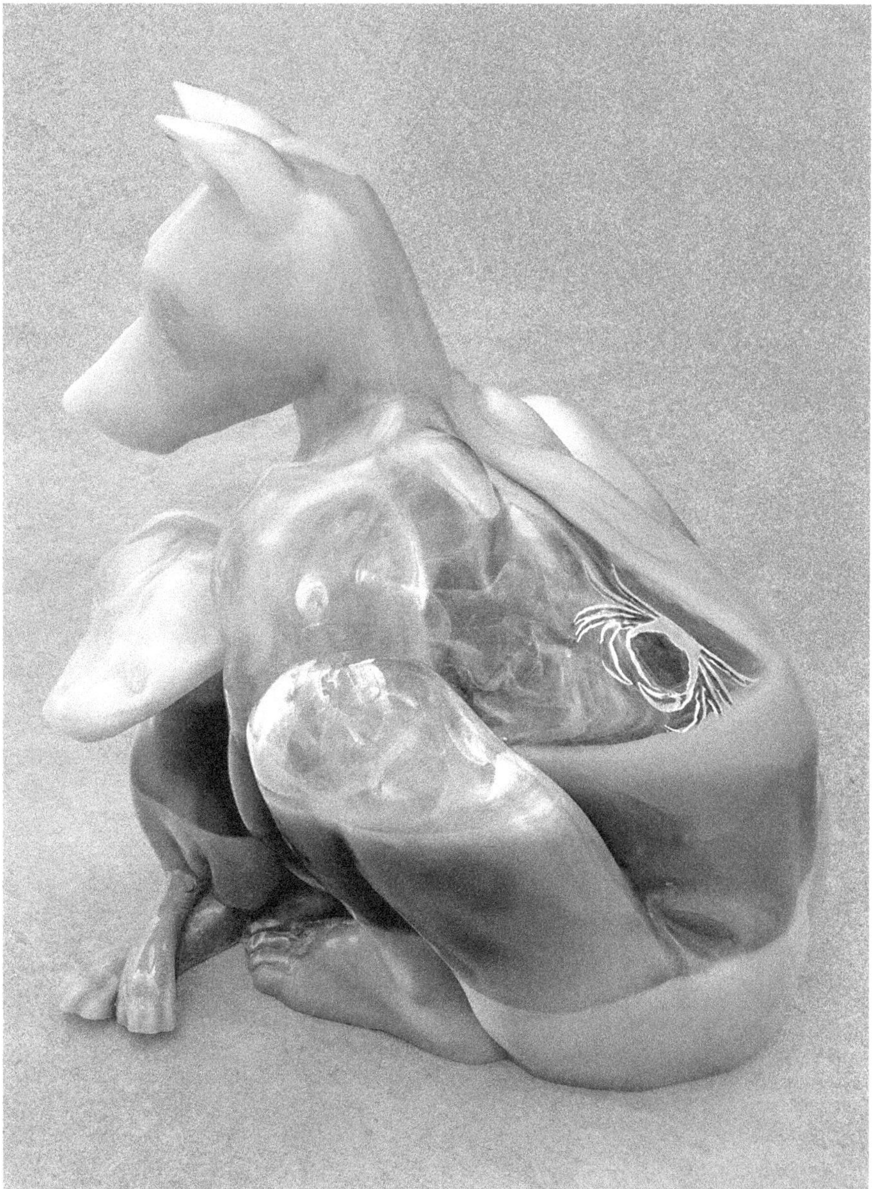

Figure 11.2 Oliver Laric, *Hundemensch*, 2018. Polyurethane, pigment, 53 × 52 × 58 cm, 20 3/4 × 20 1/2 × 22 3/4 in, Unique (LARIC-2018-0168). Courtesy of the artist and Tanya Leighton, Berlin. Photo by Gunter Lepkowski.

Classification

An important thread of Thompson's influence can be seen in classification issues. Thompson was interested in using form to create lineages, but his caveat was that the causes of the shapes of the structures must also be identified. Typical taxonomies assign the classification of organisms in accordance with principles of their adaptation to changing environments. The founder of phylogenetic systematics, Willi Hennig (1913–1976), articulated a method by which phylogeny can be reconstructed and resolved as a hierarchical classification.[18] The significance of physico-chemical processes and constraints to organismal development, Thompson's bailiwick, resurfaced as complexities of inheritance became apparent. One such development, horizontal gene transfer (HGT) also known as lateral gene transfer (LGT), literally and metaphorically uprooted the belief that Darwinian mutation and recombination are the sole reasons for evolutionary change. Unlike vertical gene transfer in which reproduction is from parent to offspring, in HGT the transfer of hereditary material is sideways, through transformation (direct uptake of DNA), transduction (transfer of DNA from one bacterium into another via bacteriophages), or conjugation (vis donors and recipients). HGT wreaked havoc with classification systems, fundamentally challenging the reconstruction of phylogenetic systematics (the scientific study of the evolutionary relationship of organisms) and questioning the concept of species.[19]

As early as 1918, Frederick Griffith discovered HGT when determining that live pneumococci acquired traits from other pneumococci, but little was done with the information.[20] Microorganisms were allotted no place in the neo-Darwinian evolutionary synthesis of the 1930s and 1940s; only the two kingdoms of plants and animals counted. In 1967, bacteriologist Lynn Margulis revived interest in the earlier endosymbiont hypothesis of organelle origins.[21] Her theory of endosymbiosis proposed that eukaryotes (life forms with nuclei) may have emerged when one prokaryote (life forms minus nuclei) absorbed and ingested another smaller one, which became the nucleus.[22] Margulis hypothesized that they then formed a single new being.[23]

Carl Woese's work in molecular phylogenetics, the study of the genetic relationships among microorganisms, caused enormous upheavals in classification. He completely changed our understanding of evolution and effected taxonomic revisions that presented entirely new principles of organismic descent. Woese provided evidence by devising ingenious comparative sequencing techniques of ribosomal RNA (rRNA). In the 1960s he compared amino acid sequences; he later extended his research to encompass RNA mechanisms and then genomes in the 1990s. Woese showed eukaryotes were organized in a more complex manner than eubacteria and archaebacteria because they did not result from Darwinian mechanisms but reflected composite origins as a "symbiotic collection".[24] By 1977, Woese and George E. Fox discovered Archaea as a separate group of prokaryotes and proved that the original eukaryotic and prokaryotic division was wrong. In opposition to the prokaryote–eukaryote dichotomy and on the basis of ribosomal ribonucleic acid (rRNA) evidence involving 16S rRNA, Woese proposed three fundamental domains of life: the *Archaea*, the *Bacteria*, and the *Eucarya*.

Ongoing debates emphasize how critical issues of classification had become. Along with Theodosius Dobzhansky, Ernst Mayr, a formidable evolutionary biologist and founding neo-Darwinist, helped forge evolutionary biology and genetics into a single, coherent science known as the modern synthesis. Mayr's main concerns as documented in his 1942 book, *Systematics and the Origin of Species*, were species as taxonomic entities, populations, and units of evolution. Initially, HGT was peripheral to Mayr's interests in species but one that he later acknowledged as important after its prevalence in nature became established.

As reported to me by a former biology student at Harvard during the 1950s, Harvard's Museum of Comparative Zoology Library (MCZ) drew bemused responses from students by classifying its vast holdings of books, monographs, and journals as if they, themselves, were biological specimens assigned to taxonomic groupings. On a visit in 2018 to check out this reputed idiosyncrasy, I learned that, indeed, some of the collections are still classified in the old MCZ system categories. They are divided into large subject areas such as A (Aves), Cr (Crustacea), Po (Porifera), and E (Echinodermata) and await Library of Congress nomenclature. The retention of such unique taxonomic groupings in an institution renamed The Ernst Mayr Library ironically echoes some of the historical controversies in classification among evolutionary biologists. As recently as 1998, Mayr suggested that the now accepted classification of life into three primary domains, Archaea, Bacteria, and Eucarya—originally proposed by Woese and Fox—be abandoned in favor of the earlier Prokaryote–Eukaryote classification.[25] Woese's response was impassioned: "Dr. Mayr's biology reflects the last billion years of evolution; mine, the first three billion. His biology is centered on multicellular organisms and their evolutions; mine on the universal ancestor and its immediate descendants."[26]

For many evolutionary biologists, the reassessment of taxonomy due to considerations of HGT was vital to understanding the metabolic and phenotypic diversity encountered in the microbial world. Decades after the publication of *On Growth and Form*, it was recognized that the prevalent "tree of life" configuration to represent lineage was inadequate.[27] The partial merging of ancestral lineages through HGT more accurately describes a reticulated evolution similar to a rhizome.

At times artists weigh in on these controversies, and a remarkably thoughtful exhibition held at Grinnell in 2017 testifies to that fact.[28] Today, some artists are very knowledgeable of the issues and create atypical taxonomies to bring attention to pressing ecological issues. For example, artist Christy Rupp applied art historical practices of organization, reproduction, collection, and arrangement to address why our impact on nature has often been negative. Rupp reshuffled categories of organic specimens with categories of art, thereby conflating classification systems of art and biology. Rupp's exhibition "Catastrophozoic" (2017) was dominated by an installation of sixteen birds taken from centuries of art history and portrayed them ensnared by discarded plastic bags that constitute a large part of our waste products.[29] Rupp constructed her bird taxonomy from references to works by a variety of other artists, including Constantin Brancusi, Frida Kahlo, Lee Bonticou, Louis Bourgeois, Maurits Cornelis Escher, and Juan Miro. Rupp's sculpted goldfinch was modeled after seventeenth-century Dutch painter Carel Fabritius's goldfinch and was fashioned from plastic netting, with discarded pieces of packaging layered to simulate iridescent plumage.

Rupp's works, often created from chicken bones collected from fast-food restaurants, are intended to promote awareness of human patterns of consumption. She critiques our destruction of the environment through making sculptures of fossils, linking them with species extinction and damage caused by fossil fuels. Like Thompson's collection of specimens and drawings of bones, her work draws our attention to external skeletons and the patterns of hexagonal mesh netting packed together. Her taxonomy is replete with portrayals of toxic molecules and genetically modified insects. The installation, *Extinct Birds Previously Consumed by Humans (From the Brink of Extinction to the Supermarket)* (2008), depicts a Great Auk, a bird in the penguin family (Figure 11.3). It recalls Thompson's King Penguin taxidermy specimen, a highlight of Thompson's museum in Dundee. It is also a reminder of Thompson's environmental activism concerning wildlife, particularly seals. The label at Dundee states that the emperor penguin was brought back from the Antarctic and presented to Thompson by Sir Ernest Shackleton after his 1907–1909 Nimrod expedition. Over-hunting the auk for food and prized down feathers led to its extinction.

Another significant way to visualize a lineage in the plant and animal worlds involves tracing the hybridization (both natural and human-made) that has occurred. Plant-breeder Wilhelm Olbers Focke's 1881 influential book on the hybridization of flowers examined the artificial, manmade process of crossing two genetically different organisms to result in a third organism with a different, often preferred, set of traits. Hybridization is also recognized as a frequent natural occurrence that offers a source of variation and an evolutionary process that may enhance speciation. In plants, the common process of hybridization presents an alternative to the hereditary transfer of genetic material. Hybridization systems challenge the definition of species boundaries.

Hybridization is also common in art and involves the crossing of different elements, be they aesthetic, physical, or conceptual. Artist Gemma Anderson's hybrid forms are unchartered by standard scientific classification systems, but they have an aesthetic logic and precision stemming from her background in scientific illustration. She designates her approach "isomorphology," consisting of a way to relate species that would otherwise be considered unconnected. She states, "Isomorphology appropriates a base from the Linnaean system and the museum system, which, although imperfect, provides an organized natural world … Isomorphology is therefore an innovative and complementary approach, and one that intends to blur normative animal, vegetable and mineral boundaries."[30] The 2012 copper etching that she identifies as a "knot like nematode, drawn from specimens at the Natural History Museum" is a representative example of her invented system (Color Plate 6). Anderson constructs a generative grammar of organic forms based on observational sketching and focuses on resemblances between animal, vegetable, and mineral specimens. She draws productively on philosopher of science John Dupré's *The Disorder of Things: Metaphysical Foundations of the Disunity of Science* (1993). In view of the evidence from HGT that "evolutionary trees" do not describe prokaryotic evolution over evolutionary time, Dupré proposed an alternate method of taxonomy calling for cross-classification.[31] He encouraged taxonomic pluralism for entities that can be sorted according to multiple classification schemes; his method involves testing the validity of each classification. Anderson does not use a "type" method in her own classification

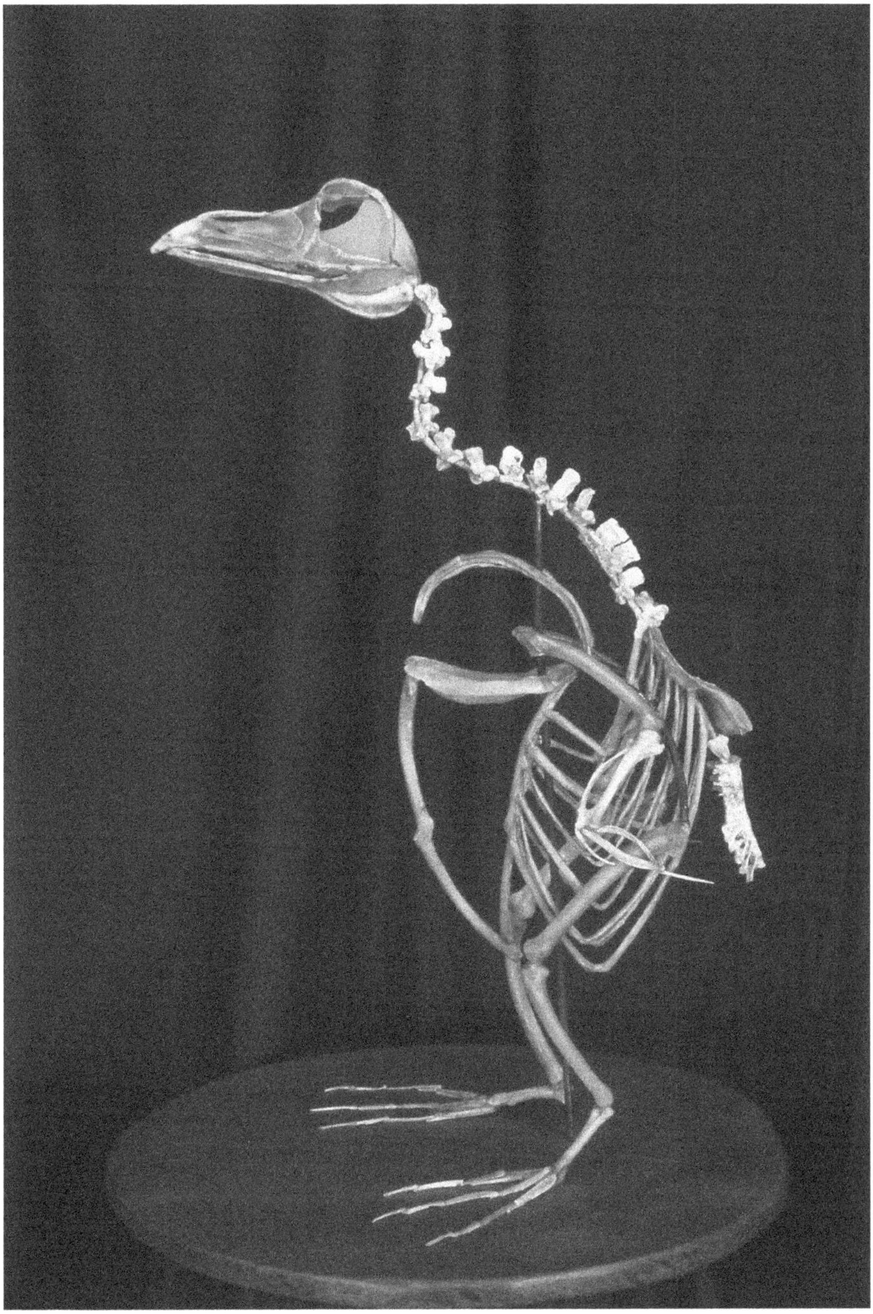

Figure 11.3 Christy Rupp, Great Auk from *Extinct Birds Previously Consumed by Humans (From the Brink of Extinction to the Supermarket)*, 2008, 30 × 17 × 22 inches, chicken bones and mixed media. Courtesy of Christy Rupp.

system but constructs images of increasing degrees of symmetry, and she summarizes her method as an effort to navigate morphological diversity. Steeped in Thompson's insights, she explores the natural world and creates a flexible set of forms that is both empirical and abstract. In addition to Dupré, the Swiss-German modernist Paul Klee's influence is present; Anderson captures his sense of playful invention.

Anderson's work in scientific illustration has added to the richness of her art practice and the reverse. She decries the loss of the hand in zoological taxonomy; she believes that combining older techniques (the camera lucida) with the newer technologies of DNA analysis and the scanning electron microscope can provide a better comprehension of the specimens under investigation.[32] Anderson points out that one of her team of scientific illustrators, George Edgecombe, is a systematist studying the evolutionary relationships among arthropods. She lauds his use of a combination of morphological and molecular data, integrating fossils with information literally drawn from living examples.[33] Edgecombe's work is a reminder that several scientists have relied on observation to produce notable visual amalgams of art and science. They include Pieter Harting, who initiated the field of biomimetic inorganic chemistry, and Ramón y Cajal, who produced exquisite drawings of neurons. Both created delicate hand drawings rendered while looking through a microscope.

In contrast with Anderson's approach, artist Mark Dion classifies classification systems, themselves. By documenting the history of research expeditions in elaborate installations, he queries the values of various classification systems devised in different periods. For example, he utilized an existing archive of images that portrayed changing conceptions of nature along with the spread of colonialism by recreating historical scenes and nature illustrations from the archives of the Department of Tropical Research Field Expeditions.[34] William Beebe, a popularizer of ecological thinking and biological science, included women as lead scientists and field artists, conducting investigations in jungle field stations and floating laboratories. Around 1900, the funders of these expeditions sought to control natural resources in the Caribbean and South America, and they exploited indentured workers from India and descendants of slaves from Africa and Native Americans. In the catalog produced about these expeditions we learn that Beebe's underwater investigations utilized a bathysphere, a spherical submersible used in the 1930s, with a telephone. A researcher would dictate what he saw underwater. Artists would then create illustrations based on the secondhand information if specimens were unavailable.[35] Since the artists rarely saw the objects they drew, the result was invention rather than strict realism. Numerous images in the archive depict the working conditions in the floating labs of the research vessels, and Dion recreated several modeled on Beebe's original bathysphere (Figure 11.4A and 11.4B). Dion's installation, *The Department of Tropical Research: Aquatic and jungle field stations in 2 parts*, held at The Drawing Center during 2017 is reproduced in this text.

Complexities of classification raised by synthetic biology are explored by some contemporary artists with access to public laboratories and the use of gene-editing tools. The access to tools has enabled several to create, catalogue, and, at times, chart new organisms.[36] The origin of this technology is a bacterial immune system known as CRISPR-Cas9 (Clustered Regularly Interspaced Short Palindromic Repeats Crispr Associated Systems). It is a controversial resource for bioartists who may work with live tissues, bacteria, living organisms, and life processes.[37]

(A)

(B)

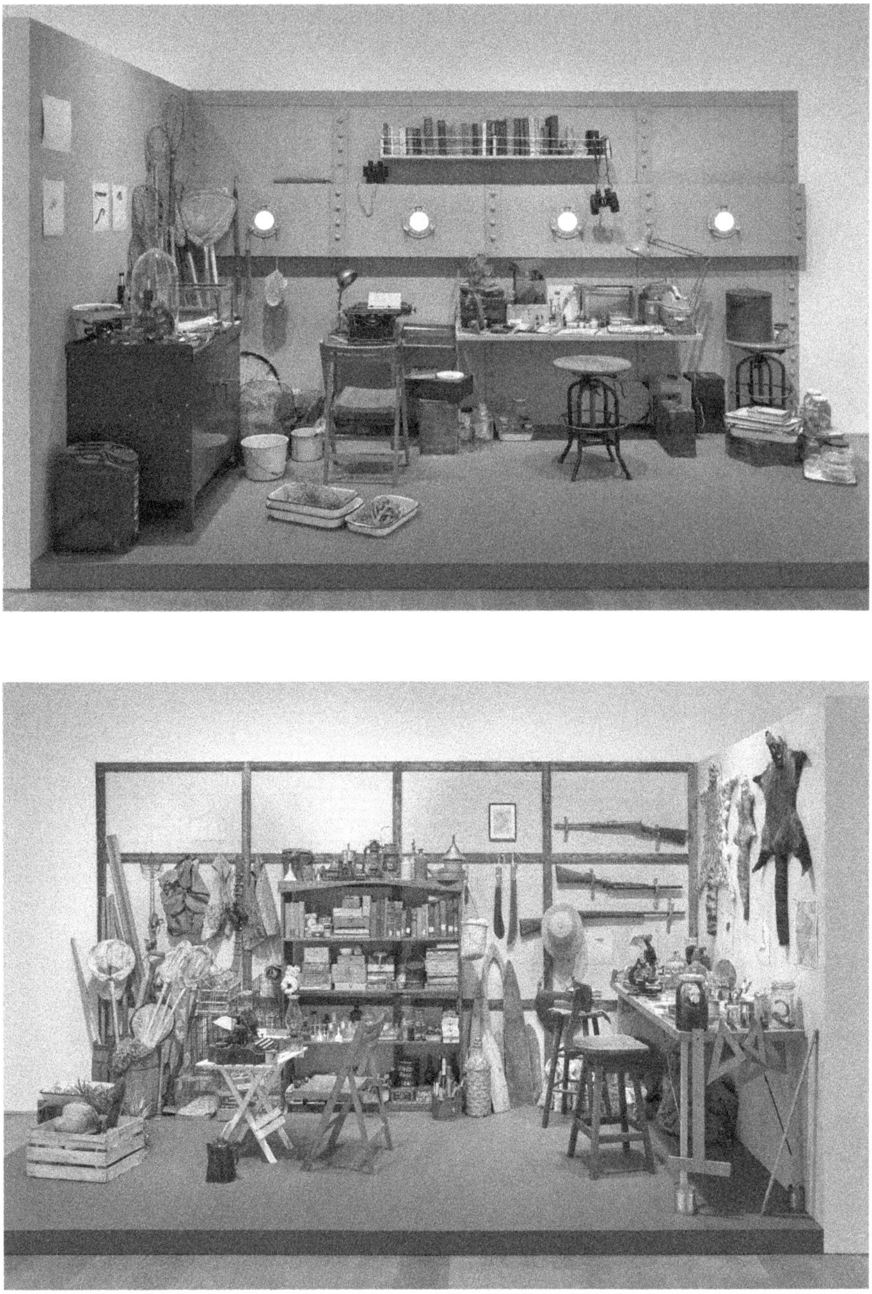

Figure 11.4 Mark Dion, *The Department of Tropical Research: Field stations in 2 parts, 2017.* (A) Top, Aquatic and (B) Bottom, Jungle. Mixed media, 97 × 168 × 84 inches; 4 × 426.7 × 213.4 cm. Installation view, Mark Dion: Exploratory Works: Drawings from the Department of Tropical Research Field Expeditions, The Drawing Center, New York, 2017. Courtesy of the artist and Tanya Bonakdar Gallery, New York/Los Angeles. Photo by Martin Parsekian.

Marta de Menezes is currently the artistic director of Ectopia, an experimental art laboratory in Lisbon that promotes projects at the intersection of art and science. She is also Director of Cultivamos Cultura in the South of Portugal. Collaborative art and science research is conducted at Instituto Gulbekian de Ciência in Oeiras (Portugal), where artists have access to biomedical research. De Menezes's bioart often takes the form of a thought experiment in which no actual physical alteration occurs. To carry out her work, De Menezes locates laboratories that are already working with the mutations desired by the artist. She addresses issues of life's origins, the natural vs. the unnatural, and hybridity, emphasizing the philosophic implications. She conceived *Truly Natural* (2017) as a carefully researched thought experiment visualized through drawings and models. She did not use CRISPR-Cas9 technology, often abbreviated as CRISPR. She instead relied on utilizing research data from a given laboratory's own CRISPR technology that had edited the genome of a spontaneously mutated mouse, identified as a balb/c Nude Mouse. The mouse had mutations of the forkhead box (FOX) NI genes, indicating that its immune system was compromised. (FOXN1 is mainly expressed in the thymus and skin epithelial cells.) De Menezes's choice of an animal model revolved around the knowledge that the balb/c Nude Mouse's genes are critical to differentiation and survival and comprise a diverse group of "winged helix" transcription factors involved in development, metabolism, cancer, and aging.[38] De Menezes's project involved creating a document of the removal of mutations selected by domestication. She theoretically proposed returning the mouse to a state where no genes had been subject to man-made manipulation.

In *The Origin of Species—Post Evolution—Maiz* (2017–), she collaborated with philosopher Maria Antonia Gonzalez Valerio. The proposal involved identifying the mutations with bioinformatics tools that permit comparison between the genome of a current organism and the feral counterpart from which it was selected. CRISPR-Cas9 was used to silence a transgene with the aim of theoretically re-creating an organism closer to its early state that would reverse the course of biohistory.[39] She pointed out in an accompanying brochure that she selected corn because she considers it is a bio artefact in many ways. She further explains that corn has undergone domestication for thousands of years by Mesoamerican cultures, but, importantly, its growth and spread are linked to its cultural meaning. The figure shown in this chapter is her taxonomic chart (Color Plate 7).

CRISPR is promising, although some scientists believe it is not yet sufficiently understood to be applied therapeutically in vivo.[40] One concern is that the technology might inadvertently alter other genes in unpredictable ways.[41] It is worth pointing out that de Menezes's is primarily theoretical, questioning, and philosophical—in ways most scientists are unlikely to pursue.

Origins of Life

Decisions involving ancestry inherently raise questions about life's origins. Several years before the 1917 publication of *On Growth and Form*, Thompson organized a talk on life's origins in 1912 for a conference presided by Sir Edward Sharpey-Schäfer, then president of the British Association of Science.[42] A main question at that time was whether the earliest forms of life on earth were autotrophs (organisms that produce complex organic compounds from simple substances present in their surroundings) or heterotrophs (organisms depending on both plants and animals for nutrition). Morphology during Thompson's period could raise causal and developmental questions but was unequipped to answer those involving time and life's origins.

Woese became increasingly intrigued by the question of life's origins. During his research, Woese had theorized a construct he called a progenote that would have preceded the cell during evolution.[43] He assumed its genome was RNA rather than DNA, regarding the genetic code as universal with an ancestry among the progenotes. HGT had greatly modified the terms of a discussion about ancestry. As Woese pointed out, although the communal nature of HGT invalidates the idea of a single common ancestor, this does not mean there is no universal common ancestry.[44] Furthermore, Woese concluded that the idea of a common ancestry emphatically suggests an evolutionary mechanism is at work apart from one of a Darwinian nature. To be able to trace life's origins, Woese realized the importance of aligning the evolution of the genetic code with the evolution of the cell. By 2006 he conceived the genetic code, itself, in fully evolutionary terms. Like the challenge Thompson posed to Darwinian adaptation, Woese made it clear that the ancestral organisms of the code are not bacteria that have adapted due to environmental stresses. The molecular phylogenetic reconstructions that Woese brought to public recognition issued from non-Darwinian, non-adaptive changes in a molecule (a conclusion that initially inspired taunts and disbelief).

Artworks, too, can raise questions about life's origins. The idiosyncratic photographic archive that Rachel Sussman created in *Oldest Living Things* (2014) groups the world's oldest organisms.[45] For her collection, Sussman assembled an index of millennia-old organisms. Her subjects are portrayed in their environmental niches and intimate how they came to have their shapes.[46] Science writer Philip Ball's description of lichen colonies spreading over rocks via principles of self-organization is apposite to Sussman's photographs of lichen.[47] Thompsonian principles similarly apply to brain coral, which resembles brains because, for different reasons, both need to increase the proportion of surface-layer to total mass in order to provide more surface area. The buckling that occurs during growth to provide such surface area is the hallmark of altered function and needs. Sussman's photographs of 2000-year-old whitish brain coral intimate the Thompsonian-like reasons for its appearance.

The interplay of organism and environment is integral to Sussman's work as seen in her photographs of 500,000-year-old actinobacteria that live in the Siberian permafrost. The extreme environment of the permafrost is characterized by an abundance of microorganisms, including bacteria, archaea, phototrophic cyanobacteria, along with green algae, fungi, and protozoa. These ancient organisms are survivors; they have withstood permanently frozen, oligotrophic (nutrient-poor) conditions, including complete darkness, constant gamma radiation, and extremely low rates of metabolite transfer.

Sussman's investigations chart a search for origins of life in the cosmic arena. A more recent project is a handwritten timeline of the universe in her exhibition, "(Selected)

History of the Spacetime Continuum" (2016). The timeline starts before the Big Bang and extends billions of years into the future.

Artist Tauba Auerbach probes the origin of life through rigorously engaging issues of structure, particularly symmetry. Research of the origins of life on earth is generally linked with the chirality (handedness) of life's molecules. Findings published in *Nature* (1991) suggested that, in order for life to emerge, something first had to break the symmetry between left-handed and right-handed molecules.[48] Life on earth is made almost exclusively of left-handed amino acids and right-handed sugars, and only right-handed versions of DNA and RNA are found in living organisms. Molecular chirality has ramifications for form; it can cause a change in morphology from spherical to helical.[49]

Thompson's interest in the origins of life and handedness long preceded Auerbach's. Spiral structures fascinated Thompson, especially the rare appearance of chirality in the narwhal's horn, in actuality a tooth, that projects beyond the head of the whale. In those few cases where more than one horn exists, the grooves are not mirror images, but both spiral in the same direction.[50] Thompson speculated that its cause might be asymmetrical propulsion from the tail.[51]

Auerbach concentrated several bodies of work on chirality, producing forms in metal and borosilicate glass while creating interlocking and threaded three-dimensional structures, cut from plywood and aluminum. Inspired by Martin Gardner, her exhibition, "The New Ambidextrous Universe," was held at the Institute of Contemporary Arts in London (2014). *Square Helix (Z)* (2014), a long, thin sculpture mounted on a plinth, consists of two metal rods, one orange and one blue, suggesting to me the mysterious unwinding of the double helix structure of DNA. *Knit Stitch* (2014) may recall the hypnotic repetitions involved in binding fabric. In *S Helix* (2014), Auerbach bent a glass rod and set it on a colored plinth. The refraction of chameleon paint set off orange, gold, and yellow hues.[52] *Chiral Fret (Meander)/Extrusion/Ghost* (2015) is formed with woven canvas on wooden stretcher and provides an example of the subtlety of her work, with roots in drawing and the materiality of canvas (Color Plate 8).

Auerbach held a later exhibition at Paula Cooper Gallery, "Projective Instrument" (2016), that emphasized mathematical principles and drew on theosophical speculations on "hyperspace" by Claude Bragdon, the American architect active at the turn of the last century. In *Ornament as Instrument* (2016), Auerbach rendered projective geometry as an active structural principle. She wound borosilicate helices through one another to create *A Flexible Fabric of Inflexible Parts*. Documentation stated that each element began with the structure of a helix and underwent numerous iterations; rotationally symmetric patterns were interlaced, twisted into a third dimension and then extruded.

Not coincidentally, Auerbach included an extensive library in an exhibition that gave prominence to Theodore Cook's *The Curves of Life* (1914) along with many other historically notable geometers. Although I did not see Thompson's book displayed, Auerbach's immersion in symmetry, mathematical proportions, and correspondences between the morphology of art and nature thoroughly resonates with Thompson's sensibility.

Implications of New Developments in Science for Art

Another thread of Thompson's influence is now seen in recent molecular and cell studies that reinforce Thompson's message that forms resulting from adaptation are to a large extent determined by mathematical and physical principles applied to matter. Genes were not discussed in the first (1917) edition of *On Growth and Form*, but some of the principles Thompson espoused were later shown to function within the genes themselves. Many scientists now view epigenetics as operative factors in inheritance. Epigenetic components are additional inheritable changes caused by modification of gene expressions rather than directly by the genes (e.g., methylation). Increasing numbers of scientists confirm that developmental constraints and epigenetic modifications affect not only morphogenesis but also stem cell regulation and tissue transitions.[53] Biophysicist Evelyn Fox Keller points out that epigenetics and the role of the cell have not been sufficiently explored in evolutionary development.[54]

Recognition of HGT, bacteria, material forces, and epigenetic factors influences the ways scientists visualize patterns of evolution, growth, and development; evolution and development are considered to be intertwined entities.[55] Artists involved with biology are aware of some of these developments but will often intuitively follow their aesthetic preferences. If we tend to associate principles of emergence, complexity, pattern making, and dense local connectivity in art with Thompson, we may assume the resulting art might look fundamentally different from art that is influenced by Darwin. For example, media artists applying natural selection algorithms typically assign values to adaptation, selection, and fitness. However, the visual differences between art rooted in Thompson or Darwin are likely hard to parse. A fully contemporary view of natural selection would theoretically allow for the functioning of complex systems and for a degree of epigenetic modification.[56] One anticipates that artists will integrate some of these factors, although it may be difficult to partition environmental and genetic effects through computation and physical modeling.

Conclusions

The artistic descendants of Thompson are often involved with aspects of complexity, classification systems, and life's origins, while exploring a wide range of media, both computational and biological. Their works lead to provocative questions, if not always answers. Some of the pluripotent Thompsonian threads that artists have appropriated from *On Growth and Form* include materials that can self-organize (Parreno), mathematical proportioning (Auerbach), and the contingency of unforeseen events (Echelman). A generative metamorphosis lends tangibility to expressions of form (Laric).

Evolutionary biologist Jan Sapp pointed out that the horizontal transfer of concepts, techniques, and individuals among a variety of fields was essential to successfully exploring major aspects of evolutionary biology.[57] The interdisciplinary exchanges he cited were all within the sciences, including the fields of microbiology, molecular biology, natural history, and ecology. By contrast, this chapter has examined ramifications of artists appropriating some of the content and methods of these scientific fields or working as art/science collaborators within labs. Neither artists nor scientists evolve in isolation; they coevolve. However, consilience with science is not really the point of the art discussed. In fact, most of the art we considered here directs our attention in ways that are very different from science. The scientists tend to ask "how" questions, involving causal mechanisms, while the artists ask "why" questions involving significance. For example, for scientists, the organisms photographed by Sussman might primarily offer scientists a challenge "to identify gaps in knowledge regarding early development, the growth rate/size curve, mortality, regeneration, competitive effects, colonization, and succession on rock surfaces."[58] By contrast, Sussman emphasizes the fragility of life and questions about why certain species endure. The classification systems devised by artists (Anderson, Rupp, Dion, and de Menezes) are rarely intended to document actual genealogical descent. Unlike Hennig, they aim to affect culture and consciousness.

Thompson insisted that both proximate and ultimate causes of form be identified: "All the while, like warp and woof, mechanism and teleology are interwoven together, and we must not cleave to the one nor despise the other; for their union is rooted in the very nature of totality."[59] To trace Thompson's continuing influence among artists is to recognize the varied ways artists bring together physical forces, stuff, and organic form. In addition, some of the artists have engaged environmental issues that Thompson, the polymath and environmentalist, also considered. Within the context of making forms, artists raise critical questions about our origins, our poor environmental stewarding, the siloing of our professional fields, and our racial and gender prejudices. *On Growth and Form* is rooted in living systems; the book's impact rocks back and forth in time, makes new demands of observers, and continues to result in novel aesthetic forms.

Notes

1 Gould, S. J., "Evolution and the Triumph of Homology or Why History Matters," *American Scientist*, Vol. 74 (1986), 66.

2 Thompson, D. W., *On Growth and Form*. 2nd edn. (Cambridge: Cambridge University Press, 1942), 1020–1.

3 Lotka, A. J., "Contribution to the Theory of Periodic Reaction," *J. Phys. Chem*, Vol. 14 (1910), 271–4; Lotke, A. J., "Analytical Note on Certain Rhythmic Relations in Organic Systems," *J. Science Progress, Proc. Amer. Acad*, Vol. 14 (1920), 406–17.

4 Turing, A., "The Chemical Basis of Morphogenesis," *Phil. Trans. Roy. Soc. B*, Vol. 237, No. 641 (1952), 237.

5 Thompson, D. W., "Morphology and Mathematics," *Transactions of the Royal Society of Edinburgh* 50, Vol. 4, No. 27 (1915), 857–95; Ball, P., "Forging Patterns and Making Waves from Biology to Geology: A Commentary on Turing," *Phil. Trans. R. Soc. B*, Vol. 370, No. 1666 (1952), 370.

6 Sommerer, C. and Mignonneau, L., "Modeling the Emergence of Complexity: Complex Systems, the Origin of Life and Interactive On-Line Art," *Leonardo*, Vol. 35, No. 2 (2002), 162.

7 Margolus, N. and Toffoli, T., *Cellular Automata Machines: A New Environment for Modeling* (Cambridge, MA: MIT Press, 1987).
8 Corrias, P., "Philippe Pareno," www.pilarcorrias.com/artists/philippe-parreno/ (accessed July 30, 2018).
9 Morgan, R. C., "Janet Echelman's Porto Princess," *Sculpture*, Vol. 24, No. 6 (July/August 2005), 51.
10 https://www.medindia.net/news/protein-that-protects-nucleus-also-regulates-stem-cell-differentiation-penn-study-reveals-124398-1.htm (accessed June 23, 2020).
11 Frank, R., "Sculpting Urban Airspace," *Sculpture*, Vol. 30, No. 7 (September 2011), 23.
12 Gould, S. J., *Wonderful Life* (New York: Norton, 1987), 283.
13 Morse, T., "Working the Crowd," *Art News*, September 2014, 81.
14 https://www.youtube.com/watch?v=npjTmG-TBHQ&feature=emb_title (accessed June 23, 2020).
15 Lay, T. and Kanamori, H., "Insights from the Great Japan Earthquake," *Physics Today*, Vol. 64, No. 12 (2011), 33. https://physicstoday.scitation.org/doi/10.1063/PT.3.1361 (accessed September 30, 2018).
16 Janet Echelman website, http://www.echelman.com/project/1-78-dubai/ (accessed July 30, 2018).
17 Madsen, K. V., "Interview with Oliver Laric," *Artforum* (February 26, 2018). https://www.artforum.com/interviews/oliver-laric-talks-about-his-show-at-metro-pictures-in-new-york-74445 (accessed July 30, 2018).
18 Hennig, W., *Phylogenetic Systematics* (Urbana, IL: University Illinois Press, 1966).
19 Boto, L., "Horizontal Gene Transfer in Evolution: Facts and Challenges," *Proc Roy Soc B*, Vol. 28, No. 277 (October 2010), 819–27. http://rspb.royalsocietypublishing.org/content/277/1683/819 (accessed July 30, 2018).
20 Griffith, F., "The Significance of Pneumococcal Types," *Journal of Hygiene*, Vol. 27, No. 2 (January 1918), 113–59.
21 Sagan, L., "On the Origin of Mitosing Cells," *Journal of Theoretical Biology*, Vol. 14 (1967), 225–74.
22 Margulis, L., *Origin of Eukaryotic Cells: Evidence and Research Implications for a Theory of Origin and Evolution of Microbial, Plant, and Animal Cells on the Precambrian Earth* (New Haven, CT: Yale University Press, 1970); Margulis, L. and Sagan, D., *Microcosmos: Four Billion Years of Evolution from Our Microbial Ancestors* (New York: Summit Books, 1986), 86.
23 Quammen, D., *The Tangled Tree: A Radical New History of Life*. (New York, London: Simon & Schuster, 2018), 114, 115.
24 Woese, C. and Fox, G. E., "Phylogenetic Structure of the Prokaryotic Domain: the Primary Kingdoms," *PNAS*, Vol. 74 (1977), 5088–90, 5088; Saap, J., *The New Foundations of Evolution: On the Tree of Life* (Oxford: Oxford University Press. 2009), 176.
25 Mayr, E., "Two Empires or Three?" *PNAS USA*, Vol. 95 (1998), 9720–3.
26 Woese, C. R., "Default Taxonomy: Ernst Mayr's View of the Microbial World," *PNAS*, Vol. 95, No. 19 (1998), 11043–6. http://www.pnas.org/content/95/19/11043 (accessed October 30, 2020).
27 Archibald, D., *Aristotle's Ladder, Darwin's Tree: The Evolution of Visual Metaphors for Biological Order* (New York: Columbia University Press, 2014).
28 Brown, J. and Wright, L., *Making Life Visible: Art, Biology, and Visualization*, Grinnell College (cat), 2017.
29 Cross Contemporary Art, Saugerties, NY, exh. September 29–October 29, 2017.

30 Anderson, G., *Drawing as a Way of Knowing in Art and Science* (London: Intellect Press, 2017), 80.
31 Dupré, J., *The Disorder of Things: Metaphysical Foundations of the Disunity of Science* (Massachusetts, London: Harvard University Press, 1993), 16–60.
32 Anderson, G., "Endangered: A Study of Morphological Drawing in Zoological Taxonomy," *Leonardo*, Vol. 47, No. 3 (2015), 232–40.
33 Anderson, 235.
34 Exploratory Works: Drawings from the Department of Tropical Research Expeditions, Drawing Center (April 14–July 16, 2017), *Curated by Mark Dion, Katherine McLeod, and Madeleine Thompson.*
35 Dion, M., McLeod, K. and Thompson, M., "Exploratory Works: Drawings from the Department of Tropical Research Expeditions" (New York, The Drawing Center's Drawing Papers, 2017), exh. cat.
36 In the United States, gene-editing tools are more generally available to the public, with and without varying levels of oversight and in the absence of a public policy and central governmental regulation.
37 The controversies involve issues of government regulation and ethics.
38 Pignata, C. 1., Fusco, A. and Amorosi, S. "Human Clinical Phenotype Associated with FOXN1 Mutations," *Adv Exp Med Biol*, Vol. 665 (2009), 195–206.
39 De Menezes acknowledges a debt for the concept to George Gessert.
40 Shim, G. et al., "Therapeutic Gene Editing: Delivery and Regulatory Perspectives," *Acta Pharmacol Sin*, Vol. 38 (June 2017), 6.
41 Flavell, L. CRISPR/Cas9 and Cancer in *Immuno-Oncology News*. https://immuno-oncologynews.com/crisprcas9-and-cancer/ (accessed July 30, 2018).
42 Report of the British Association for the Advancement of Science. https://archive.org/stream/reportofbritisha20adva/reportofbritisha20adva_djvu.txt (accessed July 30, 2018).
43 Quammen, D. *The Tangled Tree: A Radical New History of Life* (New York, London: Simon & Schuster, 2018), 205, 206.
44 Vestigian, K., Woese, C. and Goldenfield, N., "Collective Evolution and the Genetic Code," *Proc. Natl Acad. Sci. USA*, Vol. 103 (2006), 10696–701; Doolittle, W. Ford, "The Practice of Classification and the Theory of Evolution, and What the Demise of Charles Darwin's Tree of Life Hypothesis Means for Both of Them," *Philos TransR SocLond B Biol Sci*, Vol. 364, No. 1527 (August 12, 2009), 221–8. https://www.ncbi.nlm.nih.gov/pmc/articles/PMC2873000/ (accessed July 30, 2018).
45 http://www.rachelsussman.com/nautliusspruce/ (accessed July 30, 2018).
46 Sussman, R., *The Oldest Living Things in the World* (Chicago, IL: University of Chicago Press, 2014).
47 Ball, P., *Branches: Nature's Patterns: A Tapestry in Three Parts* (Oxford: Oxford University Press, 2009), 51.
48 Goldanskii, V. and Kuzmin, V. V., "Chirality and Cold Origin of Life," *Nature*, Vol. 352 (July 11, 1991), 114.
49 Ajayaghosh, A., Parayalil, C. and Varghese, R., "Self-Assembly of Tripodal Squaraines: Cation-Assisted Expression of Molecular Chirality and Change from Spherical to Helical Morphology," *GDCh* (December 12, 2006), DOI: 10.1002/anie.200603611 (accessed July 30, 2018).
50 Thompson, D. W., *On Growth and Form*, 2nd edn. (Cambridge: Cambridge University Press, 1942), 1097.

51 Thompson, D. W. (1939), "The 'Horn' of the Narwhal, Volume in Honour of the Scientific Activity of N. M. Knipovich (1885–1939)" *Institute for Marine Fisheries and Oceanography, VNIRO*. Moscow, U.S.S.R., 347–35.
52 White, C. (April 29, 2014), "*Tauba Auerbach*: The New Ambidextrous Universe," *Studio International*, 16.
53 Hallgrimsson, B. and Hall, B. K. (eds.), *Epigenetics: Linking Genotype and Phenotype in Development and Evolution* (Berkeley, CA: University of California Press, 2011), 472.
54 Keller, E. F., "Physics in Biology—Has D'Arcy Thompson Been Vindicated?," *The Mathematical Intelligencer*, Springer, Vol. 40, No. 4 (December 2018), 33–8.
55 Omaya, S., *Evolution's Eye: A System's View of the Biology-Culture Divide* (Durham and London: Duke University Press, 2000), 115–27.
56 Depew, D. J. and Weber, B. H., *Darwinism Evolving: Systems Dynamics and the Genealogy of Natural Selection* (Chicago, IL: University of Chicago, 1988).
57 Sapp, J., *The New Foundations of Evolution: On the Tree of Life* (Oxford: Oxford University Press, 2009), 317.
58 Armstrong, R. A. and Annaler, G., "Lichenometric Dating (Lichenometry) and the Biology of the Lichen Genus Rhizocarpon: Challenges and Future Directions," Geografiska Annaler: Series A, *Physical Geography*, Vol. 98, No. 3 (2016), 183-206.
59 Thompson, 7.

12

The Growth and Form of ArtNano Innovations: Inspirations from D'Arcy Wentworth Thompson's *On Growth and Form*

Todd Siler

ArtNano Innovations, the enterprise I founded with Geoffrey Ozin in 2011, has as its aim the use of multimedia artworks and aesthetic experiences to explore the beneficial possibilities of nature-inspired innovations in nanoscience and nanotechnology. It explores the production and possibilities of nanomaterials: the smallest human-made functional structures billionths of a meter; specifically, 1 to 100 nm. The immense versatility of nanomaterials helps us meet global challenges in aerospace, agriculture, biotechnology, climate control, communications systems, defense, electronics, nanomedicines for treating life-threatening infections and diseases, renewable energy systems, security systems, and many other applications that shape our future. Ozin has acknowledged the debt nanotechnology owes to the polymath D'Arcy Thompson, stating, "Although tremendous strides have been made in the fields of biomineralization and biomimetics since the work of D'Arcy Thompson, his basic paradigm for the growth and form of silica and calcareous microskeletons, through spatially directed mineralization of inorganic material in a living organism, has essentially withstood the test of time."[1]

Creating nanomaterials takes the same kind of ingenuity and transdisciplinary thinking that Thompson showed when researching the structures of nature's lifeforms and patterns of growth. In fact, Thompson's explorations in mathematical biology come to mind when reading *Nanochemistry: A Chemical Approach to Nanomaterials* (2009).[2] The approach to integrating art-science-technology and welcoming cross-disciplinary collaborations has resonance with Thompson, given his participation in such diverse activities as founder of the St Andrews Preservation Trust and membership in the Edinburgh Mathematical Society. Not surprisingly, Ozin and his colleagues are quick to acknowledge the original insights and research of the nineteenth-century German biologist and artist Ernst Haeckel and Thompson's work on morphogenesis, along with Dutch biologist and naturalist Pieter Harting on synthetic morphology (2009).[3] As Ozin et al. (2009) have written, "One can argue that materials self-assembly had its beginnings with the observations of Harting that organics are somehow involved in the organization of inorganic in biomineral formation."[4] Ozin and his team describe six nanoconcepts (size, shape, surface, self-assembly, degree of perfection/imperfection, and utility), which are currently used to produce four basic building blocks of nanomaterials—namely, nanocrystals, nanowires, nanosheets, nanotubes.[5] Myriad combinations of these nanoconcepts form the nanomaterials that are remedying our most urgent global problems: curbing and recycling CO_2 emissions from both atmospheric and industrial sources. Ozin et al. have developed technology that coverts greenhouse gases to fuels for renewable energy systems as well as value-added chemicals.[6] Value-added chemicals are also called "specialty" chemicals, "effect" chemicals, "performance" chemicals, and "formulation" chemicals, which are used in a wide range of industries to provide key effects, such as converting CO_2 to methanol (CH_3OH); methanol to hydrocarbons, and methanol to gasoline.

Thompson's Theory of Transformations[7] has inspired the imaginative work of multimedia artists, such as Richard Hamilton's Growth and Form (1951, 2014)[8] reconstructed installation in 2014 by Museo Nacional Centro de Arte Reina Sofía, Madrid, and Tate Modern, London, as well as contemporary nanochemists and materials scientists who have created unique forms of bioart-like structures that resemble the molecular and cellular asymmetrical shapes of tissues, cell-aggregates, hexagonal skeletons, and other biomorphic forms illustrated in Chapters IV and V of *On Growth and Form*.[9] Compared to more conventional biomorphic art forms, the nanoartists and nanoscientists use novel synthetic creations as "a vehicle for communicating scientific advances to a broader audience ... [this] miniaturized art enables the direct engagement of sensory aspects such as sight and touch for materials and structures that are otherwise invisible to the eye," write Yetisen et al. (2008).[10] Their artistic media include "semiconductors, microfluidics, and nanomaterials ... including micromachining, focused ion beam milling, two-photon polymerization, and bottom-up nanostructure growth."[11]

Many of the nanoart creations are purposefully designed to illuminate the hidden facets and properties of these minuscule forms of matter that are directly related to the delicate, densely packed cells and alveoli we see in Haeckel's siliceous diatoms and radiolarians; Thompson described the formation of these structures in connection with the biosilicification process.[12] The new nanosculpted objects offer more than engrossing aesthetic experiences[13] and new possibilities for ArtScience explorations.[14] They contribute to the production of versatile microtemplates for "morphosynthesis," a term coined by Charles T. Kresge, who invented the supramolecular templating technique for supramolecular materials.

Continuing my forty years of immersion in ArtScience practice, during the past seven years I've created metaphorical drawings, paintings, sculptures, and art installations that explore what nature makes and what we make of nature (Color Plate 9). Inspired by the empirical research of nanoscientists and materials scientists, this exploratory work considers the interactions of physical forces that influence all forms of growth and processes of growing forms that really matter to a future confronting climate change.

Thompson and Holism

Thompson recognized how all-encompassing and relevant numbers are to everything in nature. Natural numbers, integers, rational numbers, real numbers, and complex numbers (Figure 12.1) all have a presence in his mind and sense of matter. Myriad manifestations of math were self-evident to him, whether he was investigating the surface-tensions between cell-aggregates[15] or pondering the angles of the partitions connecting/separating the cells in a dragonfly's wing that exert inward-and-outward pressure on one another[16]. Thompson translated his observations into a theory of transformations that "make forces of nature visible,"[17] an aim later realized by the sculptor and inventor Kenneth Snelson in his gravity-defying, three-dimensional structures, such as Needle Tower (1968) and Dragon (2000–2003), that embody tensegrity design principles.[18] The term "tensegrity," coined by R. Buckminster Fuller, refers to "the characteristic property of structures under tension that are contiguous and structures under compression that are not."[19] Fuller's Geodisc Domes are exemplary embodiments of tensegrity; an equal distribution of structural stresses occurs between the triangular elements that form the hemispherical polyhedrons.[20]

It takes little stretch of curiosity to connect tensegrity principles with the spherical skeletal configurations of radiating spicules, such as the *Aulastrum triceros* (Hkl), in which Thompson observed: "the skeletal matter tends to be developed normally to the surface of the sphere, that is to say along the radial edges where the external vesicles (now compressed into hexagonal prisms) meet one another three by three."[21] These observations and comparative studies suggest how Thompson used numbers as "rulers" for enumerating forms of growth and the growth of forms.

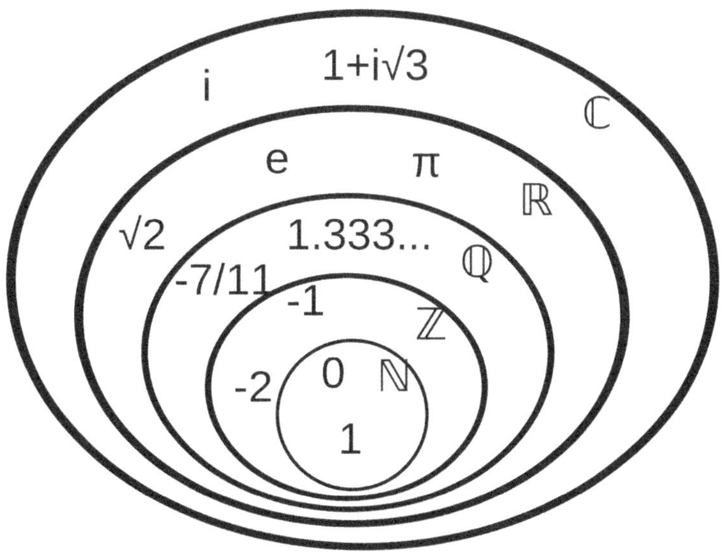

Figure 12.1 The diagram, "Subsets of the complex numbers," was created by fr:Utilisateur:HB on August 25, 2014. *Source:* Wikimedia Commons.

To borrow Ozin's phrase for materials synthesis, "art, design, and architecture embody a language of shape."[22] This language can consist of inorganic materials with curved structures that resemble the geometrical forms visible in scanning electron micrograph images of morphogenesis.[23] Crystallographer Alan L. Mackay describes morphogenesis in a way that bridges inorganic and organic materials:

> The apparent conflicts of our time between molecular biology (often labelled "reductionism") and the study of whole organisms, societies, etc. (often labelled "holism" and now "organicism") are really artificial. The most fundamental physics (work by Aspect, Bell, Bohm, etc.) has confirmed that all parts of the universe are indeed connected to each other, often in ways which are counter-intuitive and only to be discovered by subtle experiments. Since we simply cannot consider everything simultaneously we have to try to select what is important, (bearing in mind that at critical junctures very small influences may be decisive)—the eigenvectors corresponding to the largest eigenvalues of the all-embracing interaction matrix, although this metaphor implies a linear approximation and nature is, in general, definitely non-linear.[24]

Many of the impressionistic works of ArtNano Innovations express nanomaterials' natural and synthetic complexity. Their complex forms and process of formation have qualities that resist quantitative measure. Even though Thompson described the functional structures of life forms with precision, it seems he sensed that nature cannot be completely represented mathematically, yet he resisted vitalism and its beliefs. This dichotomy creates a fundamental conflict. The symbolic nature of language is all-encompassing compared to the *mathematical* language of nature. Our symbolic languages encompass infinite forms and functions of human expression that enable us to communicate information in multiple ways and means.[25] Indeed, the history of ancient, modern, and postmodern art confirms this fact. Nature isn't only about "numbers" or only numbers; nature is much more than we can measure and transcends our limited measure-meant systems.

Precision vs Accuracy vs Approximation

By comparison to today's computer-aided visualizations of organic and inorganic forms, Thompson's mathematical renderings approximate the change and growth of dynamic forms that are now directly visible through tools such as four-dimensional electron microscopy.[26] Regardless, Thompson's drawings are nonetheless accurate renditions that are generalizable to a world of functional forms. Unaided by scanning electron micrograph (SEM) imaging tools and transmission electron micrograph (TEM) images, which nanoscientists and nanotechnologists rely on for seeing, studying, visualizing, building, and applying nanomaterials, Thompson relied on his talent for rendering details. His elegant hand-drawn images enabled him to understand the relationships and "parameters" between various forms of matter—offering fresh interpretations of life's forms, functionality, and utility, such as the often-surprising deformations he observed in aquatic lifeforms. Thompson has written:

> Among the fishes we discover a great variety of deformations, some of them of a very simple kind, while others are more striking and unexpected. A comparatively simple case, involving a simple shear, is illustrated by Figs. 146 and 147. The one represents, within Cartesian co-ordinates, a certain oceanic fish known as Argyropelecus offers. The other represents precisely the same outline, transferred to a system of oblique co-ordinates whose axes are inclined at an angle of 700; but this is now (as far as can be seen on the scale of the drawing) a very good figure of an allied fish, assigned to a different genus, under the name of Sternoptyx diaphana. The deformation illustrated by this case of Argyropelecus is precisely analogous [italics mine] to the simplest and commonest kind of deformation to which fossils are subject as the result of shearing stresses in the solid rock.[27]

Thompson intuitively explored the concept of "parametrization," which mathematician Shai Haran summarizes in describing parametrization as a circumstance where "every point on a line projects to a point on a circle or sphere, or form."[28] The parametrization of forms is as implicit as it is prevalent in Thompson's descriptions of forms. The same holds true for all forms of nanomaterials (Figure 12.2).

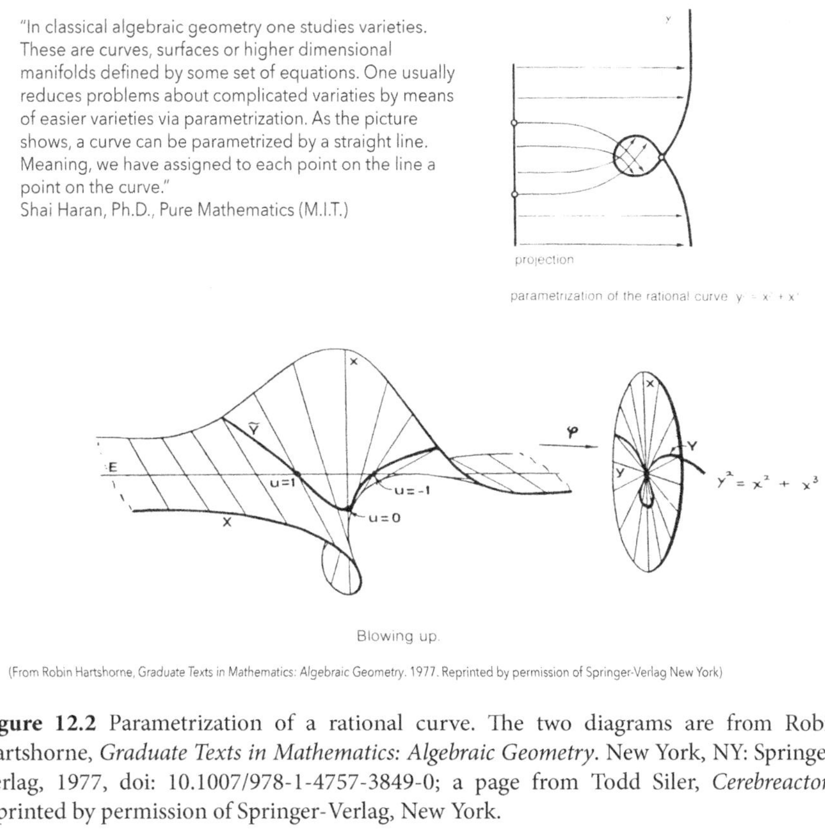

Figure 12.2 Parametrization of a rational curve. The two diagrams are from Robin Hartshorne, *Graduate Texts in Mathematics: Algebraic Geometry*. New York, NY: Springer-Verlag, 1977, doi: 10.1007/978-1-4757-3849-0; a page from Todd Siler, *Cerebreactors*, reprinted by permission of Springer-Verlag, New York.

Thompson's Approach to Knowledge

Thompson's exploratory work in mathematical biology anticipated many subsequent approaches in transdisciplinary intersections. We can perceive aspects of Thompson's interpretive works in contemporary ArtScientists whose integrative studies of Nature's structures and forces are realized in cross-disciplinary collaborations. I coined the term "ArtScience" in 1975 to describe how art and science are naturally interconnected ways of knowing the world and representing what we know.[29] My approach to the ArtScience process integrates arts-and-science-based experiential learning methods and creative inquiry that involve four basic iterative steps: Connect-Discover-Invent/Innovate-Apply. This process combines the tools, symbolic languages, best practices, and approaches to innovative thinking.[30] ArtScience enables the application of tacit and explicit knowledge, generating new ideas based on our knowledge of nature in novel and imaginative ways. To this end it fosters both openness and skepticism: two essential good habits for growing and utilizing human knowledge as we see fit. Acts of creative learning, discovering, and innovative thinking in the context of this chapter are self-evident in the evolution of chemistry, for example, which has grown from molecular chemistry to supramolecular chemistry to constitutional dynamic chemistry to adaptive chemistry.[31]

ArtNano Innovations: Envisioning Possibilities of NanoWorld

Inspired by the empirical research of Ozin and his team of nanochemists and materials scientists, my exploratory art considers the interactions of physical forces that influence all forms and processes of growth. I'm especially drawn to envisioning how nanomaterials can be shaped, sculpted, or engineered to control climate change.

ArtNano Innovations addresses how potentially infinite nanomaterials—fashioned from a Periodic Table of Nanomaterials—will rapidly change some of the core technologies we rely on to meet urgent global challenges on an ongoing basis. Nanomaterials are synthesized from the bottom up (i.e., from atom to atom clusters to nanomaterials to materials)[32] and are fabricated from the top down. (Figure 12.3).

I seek fresh insights into how intermediate forms of matter are shaped by various organizational principles (e.g., self-assembly, co-assembly, directed (templated) assembly, hierarchical assembly).[33] Nanostructures work to improve a world of technological innovations, such as solar cells, batteries, energy devices and climate control systems, computers, and advanced medicines and mechanisms for maintaining our personal health and managing the wellness of global environments. My art interprets the potential of nanomaterials. I suggest how nanomaterials may, ultimately, affect nearly every thing human-made, including aeronautics, aviation, automobile industries waste management, and renewable energies. As physical scales change, so change the properties, phenomena, and potential purposes of the materials created from these nanomaterials. The process of making nanomaterials is complicated, in so

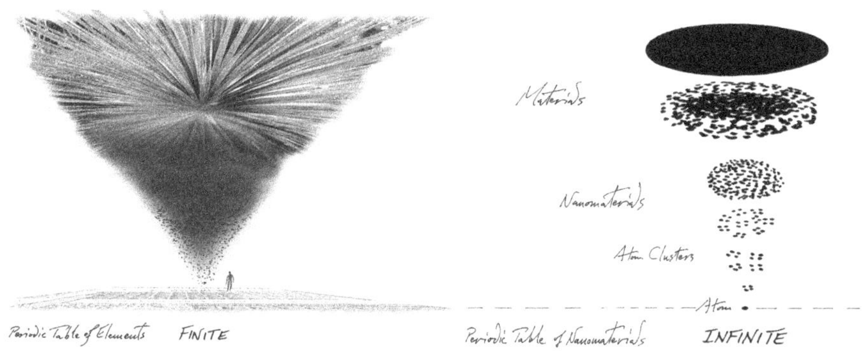

Figure 12.3 *The Periodic Table of Elements* (FINITE) vs *Periodic Table of Nanomaterials* (INFINITE) (2012–2013). Courtesy of Todd Siler.

far as it drives the evolution of the concepts and theories that shape this nascent field of knowledge. Describing the creative process can be equally complicated since it is essential that the public be well informed about this field in order to make informed decisions about the uses and benefits of nanomaterials.

When Function Determines Form, Nature as We Know It Changes as Well

Thompson's principles resonated with architect Louis Sullivan, who coined the insightful phrase "form follows function." From nanoscience's perspective, function determines form. The utility of a nanomaterial determines its size, shape, surface, self-assembly, and degree of perfection/imperfection. "Size is essential." Ozin points out that nanomaterials change their behavior as they change their size, even as little as by a single atom:

> It seems ironical to realize there are no new nanomaterials. Rather, they are just reconstructed forms of known materials sculpted at the nanoscale. All the atomic compositions and atomic arrangements of the materials are known. But it is their physical size and shape and accessible surface properties plus their self-assembly into purposeful higher tier "panomaterials" with structural features formed over multiple length scales, from nanometers to millimeters to centimeters to meters and beyond that creates, for example, the NanoAdvantage as intimated by this work of ArtNano Innovations.[34]

Nanoscientists often start with the desired end-use or functionality and then work their way back to the fundamental building blocks they'll need to accomplish their goal. Ludovico Cademartiri and Ozin state:

> There are always (at least) two ways to "see" a concept or a problem: from the outside in and from the inside out," The first takes the nanoscale problem and imagines it at a macroscopic length scale; nanocrystals packing into microchannels can be "seen" as golf balls packing into holes; nanowire meshes can be "seen" as spider webs; hexagonal close-packed arrays of nanochannels resemble honeycomb in a beehive ... This approach allows you to draw conclusions based on your sensory experience with the macroscopic world, without getting lost in the crowd of unfamiliar microscopic details ... It allows you to visualize a problem or a phenomenon in your mind by putting yourself into it; it removes the length-scale barrier which makes intuition harder.[35]

The Value of Metaphorms and Metaphorming

Searching Thompson's investigative approach to understanding the complex layers of relationships connecting the functional structures of biological and physical matter, I find the following reflection by Thompson reveals how he used analogies to empirically probe physical reality:

> When we find ourselves investigating the forms assumed by chemical compounds under the peculiar circumstances of association with a living body, and when we find these forms to be characteristic or recognizable, and somehow different from the same substance is wont to assume under other circumstances, an *analogy* [italics mine], captivating though perhaps remote, presents itself to our minds between this subject of ours and certain synthetic problems of the organic chemist. There is doubtless an essential difference, as well as a difference of scale, between the visible form of a spicule or concretion and the hypothetical form of an individual molecule.[36]

Thompson's use of analogical thinking is hardly unique to his scientific discovery process. It's invaluable to the discovery process more broadly that propels and promotes both self-learning and collaborative learning.[37] The Nobel laureate chemist and poet Roald Hoffmann uses metaphorical catalysts for stimulating breakthrough thinking and the discovery process. Hoffmann has written: "The images that scientists have as they do science are metaphorical. The imaginative faculty is set in motion by mental metaphor. Metaphor shifts the discourse, not gradually, but with a vengeance. You see what no one had seen before (Color Plate 9)."[38] More than a "mental metaphor," a "metaphorm" combines all forms of connection-making tools (e.g., metaphors, analogies, signs, symbols, numbers, stories, figures of speech, euphemisms, premises, models, and other things) to express, represent, and communicate our thoughts, feelings, ideas, opinions, knowledge, and experiences.[39] I refer to the creative process of using metaphorms as "metaphorming (Color Plate 10)."[40] This process underlies virtually everything that unites Ozin's work with that of his colleagues, as well as our ArtScience collaboration: in which we aim to transform the forms and functions of new and existing materials in new ways.

In many respects, our collaborative search for evidence-based truths about the natural world parallels the bold path cut by the quintessential ArtScience explorer Leonardo da Vinci (1452–1519), whose creative inquiries into the forms and mechanics of life relied heavily on visual metaphors and physical analogies to communicate essences of complex everyday phenomena. In this regard, his ground-breaking work bridges the type of investigative studies that lead to the scientific method and which is intimately connected to Thompson's exploratory work and approach to new-knowledge creation and sharing.[41]

I learned there are many ways to see these interconnected nanoconcepts (size-shape-surface-defects-self-assembly-function-utility) through experiential learning and metaphorming.[42] I discovered there's a deep connection between the scientific process of building the tiniest human-made structures billionths of an inch and my artistic process of creating massive monotype prints and paintings, some measuring 12ft. x 200ft, using various forms and forces of Nature. As shown here, I used an assortment of hand-made natural templates to paint and print varying textural surfaces. My approach embraces unpredictability and uncertainty, which contrasts with the calculated process nanochemists use, such as centrifugation- or evaporation-induced self-assembly, to organize colloids into nanocrystals, nanowires, nanosheets, and nanotubes.

In effect, what I do on the macroscopic scale, nanochemists do on the nanoscale with their "template" system. This system is defined as "a nanostructure (solid, liquid, or gaseous), which can be used to direct the assembly of a nanostructure. A droplet can be used as a template. A glass slide on which you deposit a film of nanocrystals can be considered a template. A bubble in a liquid can be used as a template."[43] I've learned how nanochemists employ centrifugation to self-organize colloids, for example, in silica. They use this technique in a precisely controlled manner to create nanoscale integrated circuits for computers or nanomaterials for optimizing solar cells. Ozin points out, "Amazingly, even the addition or subtraction of a single atom can influence the chemical, physical and biological properties of nanoscale materials."[44] I understand this centrifugal process, since I use a related, yet different, printing/painting process that produces novel forms and objects. One significant difference is the way in which I use nature, as a reference for my paintings. My process is intentionally open-ended and more unpredictable than the process deployed by nanoscientists. Moreover, it's an individual effort of artistic and scientific experimentation that's less collaborative and more personally contemplative.

Picturing NanoWorld (1 to 100 nanometers): A Collaboration

In depicting the things nature makes and what we make of these things, I find myself drawn to the same gravity and beauty of the creative process that nanoscientists experience while synthesizing and forming nanomaterials (Figure 12.4).

Collaborations with Geoffrey Ozin, whom I will call Geoff when discussing our collaborations, have been part of my artistic practice, for over seven years.[45] His

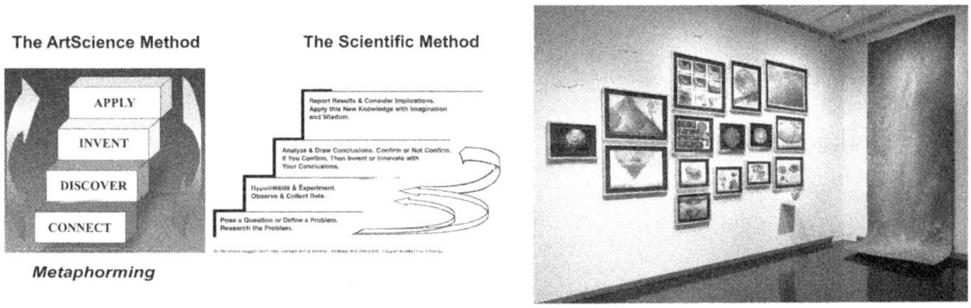

Figure 12.4 Comparing The ArtScience Method to The Scientific Method, and showing how both relate to the four-step Metaphorming process used in ArtNano Innovations. Art installation, "Metaphorming Nature: Connecting Human/Nature's Creative Potential," mixed media on synthetic canvas. Courtesy of the CU Art Museum, University of Colorado, Boulder and www.ArtNanoInnovations.com.

thoughts reflect a deep understanding of the complementarity of Art & Science, and not simply their general partnership. He recognizes why and how all art is interpretation. And interpretation requires creativity. Thompson understood this interconnectivity of living systems in ways that anticipated how nanoscientists explore systems of nature, and his understanding helped advance the field of integrative studies. These studies aim to grasp the interconnectedness of the natural world—or "the nature of nature," to borrow a phrase the German-Swiss artist Paul Klee. I share this focus in my work. Everything we've created since the beginning of time is a product of the human brain and its creative process. We intuitively grasp that every thing human-made reflects the brain's handiwork. In the painting or metaphorm titled *Envisioning Minds & Nature Forming Nanomaterials That Form All Materials* (2011–2014), an imaginary "Book of Nature" takes centerstage (Color Plate 11).[46] One innovation, in particular, that excites me and drives my collaboration involves the development of a Periodic Table of Nanomaterials and Nanomaterials Genome.[47] Eventually, this will become a useful way to create a potentially infinite number of nanomaterials, each with multiple applications.[48]

Geoff and I imagined the possibilities and limitations of combining the work on the Human Genome Project with emergent Materials Genome Project, which was initiated in 2011 and instigated in 2010 by the Nanoinformatics Roadmap 2020. Central to our visualizations was a Nanomaterials Genome (NMG) database, comprising the most advanced relational data-mining tools equipped with state-of-the-art visualization techniques. The NMG would leverage a network of inference engines designed to connect and interpret patterns of nanomaterials information. His project generally consists of a multidimensional Periodic Table of Nanomaterials, but it can also take the form of a telescopic kaleidoscope in which an infinite number of nanomaterials can be created by the endless combinations and permutations of nanomaterials' elemental compositions. These metaphorms intend to envision "forms and the forces that give rise to it," to quote Thompson (Figure 12.5).

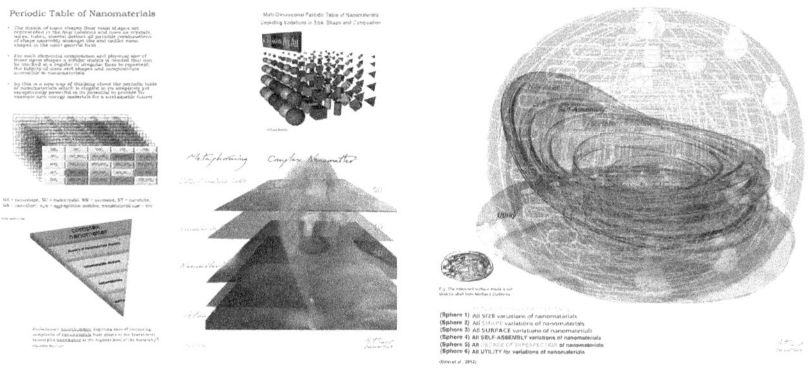

Figure 12.5 Todd Siler, *Evolution of the Periodic Table of Nanomaterials* (2012–2013) monotypes on 140 lbs. Watercolor paper, 25 × 51 inches. Courtesy of ArtNano Innovations.

Geoff Commented on My Contribution to the Project

> To amplify on Todd Siler's innovative NanoWorld, at the core of nanotechnology, is a periodic table of about a hundred elements. They have been combined in myriad ways through chemistry and made into countless nanosystems built from individual nanocomponents with nanometer-scale dimensions. These versatile nanocomponents can be fashioned in a spectrum of shapes resembling spheres, wires, rods, sheets, and tubes. And these shapes, in turn, can be assembled into hierarchical architectures intentionally designed to have form and function frequently akin to those found in the natural world.[49]

The wall of ArtNano Innovations presents at-a-glance some key concepts, practices, innovations, and applications of nanoscience. It highlights one application, in particular: addressing the "#1 Global Challenge facing humanity: Climate change."[50] The drawings, paintings, monotype prints, and sculpture interpret a series of peer-reviewed articles by Ozin in *Materials View* encircling the topic "Transitioning from a Fossil & Nuclear Energy Economy to a Clean & Green Energy Economy."[51] To Geoff's point, "Grasping Our Growing Gigatonne CO_2 Challenge"[52] remains everyone's problem just as creating a sustainable future does for humankind. In order for the NanoLeaf concept (Figure 12.6) to become a core part of our clean renewable energy solutions, it must compete with the production of methane and methanol. To this end, nanochemists have considered developing "Ultra-Black Ru nanoparticles/Silicon Nanowire (SiNW) Catalysts" designed to tap "the potential of visible and near-infrared photons."[53] My metaphorical painting and sculpture explore the process of designing and synthesizing nanocrystals, nanowires, nanosheets, and nanotubes that can be used to make "photoactive nanostructure materials for converting gaseous CO_2 to a solar fuel using light-assisted, heterogeneous catalytic converters powered by the sun"[54] (Figure 12.7) My art has been informed by

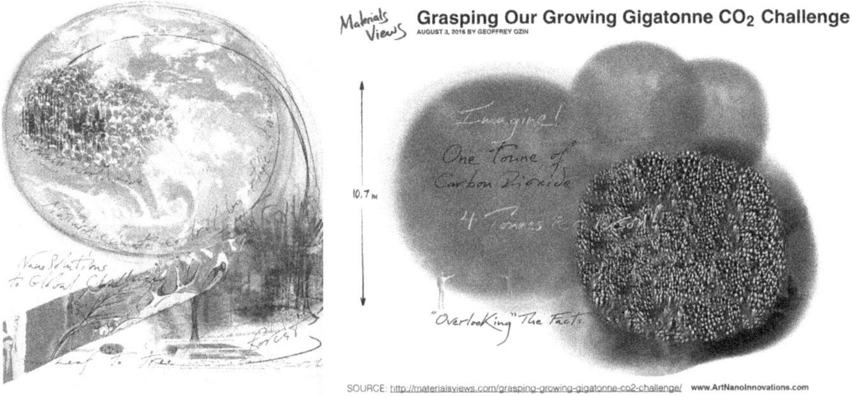

Figure 12.6 Todd Siler, *NanoSolutions to Global Climate Change: Cutting-To-The-Chase Using the "NanoAdvantage"* 2012 monotype digital print accompanied by a sculpture, titled "The Trillion Dollar NanoSolution to Global Artificial Photosynthesis: Why & How CO_2 Is a Friend Not a Foe to Our Sustainable Future," 2014. Courtesy of ArtNano Innovations.

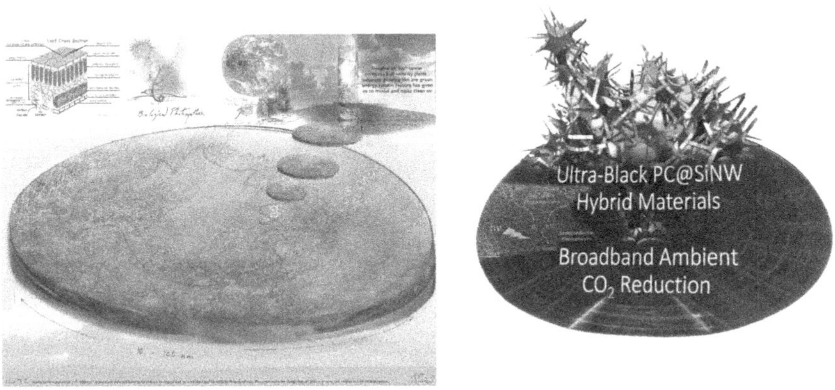

Figure 12.7 Todd Siler, *Metaphorming NanoCrystals (1 to 100 nm) for NanoLeafs used in Artificial Photosynthesis*, 2013, monotype digital prints. Courtesy of ArtNano Innovations.

numerous Skype chats with Geoff over the years. As simply as possible, Geoff described what his team does, elaborating on some critical design decisions that underscore their empirical work. Our conversations are always freewheeling:

Geoff: Imagine replicating what ordinary plants, algae and bacteria do daily. Every second, plants are utilizing the light and heat from the sun to convert carbon dioxide and oxidize water into a solar fuel.

Todd: Like these images imply? I ask, pointing to Figure 12.7; my mixed-media drawings highlight the evolution of an ArtNano innovation that performs this common photosynthetic cycle endlessly.

Geoff: Correct. As I understand your visual suppositions, they intimate how a leaf produces the oxygen and water we need to fuel our future from this most abundant fuel source: carbon dioxide.

Todd: He pauses for a moment, pondering the sequential steps they take actualizing this seemingly simple innovation, adding: "Of course, the process of building and implementing this technological innovation require a worldwide community of nanoscientists and technologists who must leverage their empirical work on heterogeneous catalysis solutions to light-assisted, gas-phase CO_2 reduction catalysis utilizing both the light and heat from the sun."

Geoff searches his desk and holds up a copy of this progress report he and his colleagues have just completed, titled "Solar Powered CO_2-to-Fuel." "Now percolate on this report I just sent you. See if it inspires any new metaphorms!"[55]

I always appreciate the creative freedom Geoff entrusts me with. He's always given me permission to question anything-and-everything his field has discovered or created. No matter how basic or abstract my ideas or premises are, he first welcomes them with an open mind before critiquing them. Geoff relayed, "It is deep down in the nanometer scale world of materials where my scientific career began more than four decades ago and it is a truism that much of my inspiration working in this field has derived from the aesthetics of the shapes and colors that pervade this small world, and their orchestration into purposeful technologies."[56]

Summary

One wonders what more Thompson may have discovered examining the natural structures that fascinated him. Imagine if this polymath had used SEM and TEM imaging tools which nanoscientists and nanotechnologists rely on today for creating and studying these astonishingly tiny functional structures and systems serving practical innovations for myriad industries. As Ozin and his ArtScience collaborators continue to advance nanoscience theories and technological innovations, they simultaneously reaffirm how many breakthroughs grow out of "aesthetic experiences" and accidental discoveries. On an equally important and related note, Cadamartiri and Ozin have written: "When all approaches to a problem seem to fail, start asking 'why?' Our mind limits the range of our thoughts to make us quicker in taking decisions: it does so by using assumptions, that is, notions we assume to be true without a definitive proof. But science is not about believing: it is about moving towards the truth."[57]

ArtNano Innovations offer several viable solutions that connect and build on both the explicit and implicit knowledge evident in Thompson's elegant visualizations of forms, which he accompanies with eloquent descriptions of their functions. Like Thompson's exploratory work, ArtNano Innovations aims to explore the patterns of growth of human knowledge of nature and how we utilize that knowledge to shape supramolecular materials in pure or applied inorganic chemistry[58] and to make "new materials discovery using machine-enhanced human creativity."[59] This nascent area of research can either confound or inspire us to dig deeper and wider—seeking truths that transcend the limitations of one body of knowledge over another; one perspective above all others. As Eric Schmidt, Executive Chairman of Google, reminds us all: "'None of us is as smart as all of us.' One of the hallmarks of nanoscience is its interdisciplinary nature—its practice requires chemists, physicists, materials scientists, engineers and biologists to work together in close-knit teams."[60] Creative collaborations will enable us "to transition from a Fossil and Nuclear Energy Economy to a clean and Green Energy Economy powered by an efficiently 'electrified' world."[61] Nanoscience and nanotechnology are currently "touted as the engine of innovation that will drive the next industrial revolution." That's no hyperbole. I imagine that the spirit of Thompson would concur.

Notes

1 Ozin, G. A., Arsenault, A. C. and Cademartiri, L., *Nanochemistry: A Chemical Approach to Nanomaterials*. Forward by Chad A. Mirkin (Cambridge, UK: Royal Society of Chemistry (RSC) Publishing, 2009), 19.

2 Ibid. (2009), 18–21.

3 Ibid. (2009), 17–19.

4 Ibid. (2009), 16. Also, read Ball, Philip, *The Self-Made Tapestry: Pattern Formation in Nature* (Oxford: Oxford Press, 2001); Ozin et al. (2009) have written: "Ball's obsession with materials and patterns can be considered to represent a 21st century revival of D'Arcy Thompson's classic text *On Growth and Form*" (p. 26).

5 Ozin, G. A., Arsenault, A. C. and Cademartiri, L., *Nanochemistry: A Chemical Approach to Nanomaterials* (2009), 13. Also, read Cademartiri, Ludovico and Ozin, Geoffrey A. Foreword by Lehn, Jean-Marie Lehn. *Concepts of Nanochemistry* (Weinhelm, Germany: VCH-Wiley, Verlag GmbH & Co. KGaA, 2009), 19–41.

6 http://www.chem.utoronto.ca/edistillations/2018/03/greenhouse-gases-to-fuels.html; www.materialsviews.com/category/opinion/; www.nanowizardry.info, www.solarfuels.utoronto.ca.

7 Bonner, John T. (ed.), Chapter IX: "On the Theory of Transformations, or The Comparison of Related Forms" [268–325] Abridged Edition of D'Arcy Thompson, *On Growth and Form* (Cambridge, UK: Cambridge University Press, 1961), 175.

8 GROWTH AND FORM, Richard Hamilton 1951 (2014), Installation, Reconstruction. Diverse materials, MACBA Collection. Gift of Rita Hamilton. Installation reconstructed in 2014 by Museo Nacional Centro de Arte Reina Sofía, Madrid, and Tate Modern, London, on the occasion of the exhibition "Richard Hamilton."

9 Bonner, John T. (ed.), Abridged Edition of D'Arcy Thompson, *On Growth and Form*. (Cambridge University Press, 1961); see the "artificial tissue" formed from drops of sodium chloride (Fig. 41, 105), the "network of cell-aggregates in the 'Reticulum plasmatique' (Fig. 55) and 'Aulonia hexagona'" (Fig. 56), 156.

10 Yetisen, A. K., Coskun, A. F., England, G., Cho, S., Butt, H., Hurwitz, J., Kolle, M., Khademhosseini, A., Hart, A. J., Folch, A., and Yun, S. H., "Art on the Nanoscale and Beyond," *Adv Mater*, Vol. 28, No. 9 (March 2, 2016), 1724–42. http://www.ncbi.nlm.nih.gov/pubmed/26671704 (accessed January 3, 2019).

11 Ibid. (2015), 2, 10, 11.

12 Ozin et al., Nanochemistry: A Chemical Approach to Nanomaterials (2009), 19.
13 Yetisen et al., "Art on the Nanoscale and Beyond," 1724–42.
14 Ibid., 23; Read Stephen Wilson, *Art and Science Now: How Scientific Research and Technological Innovation Are Becoming Key to 21st-century Aesthetics* (London, England: Thames and Hudson, 2012).
15 Bonner, John T. (ed.) (1961), 88.
16 Ibid., 94–8.
17 Snelson, Kenneth, *Forces Made Visible. Essay by Eleanor Heartney*. Additional text by Kenneth Snelson (Lenox, MA, Manchester, VT and New York: Hard Press Editions in association with Hudson Hills Press, 2009).
18 Snelson, Kenneth, *The Binary World: Tensegrity, Weaving and the Binary World. Newton's Third Law and the Duality of Forces*. http://kennethsnelson.net/Tensegrity_and_Weaving.pdf (accessed January 3, 2019).
19 Paraphrased from this source: https://en.wikipedia.org/wiki/Tensegrity.
20 "The term was coined … as a portmanteau of 'tensional integrity'." The other denomination of tensegrity, *floating compression*, was used mainly by Kenneth_ Snelson. U.S. Patent 3,063,521, "Tensile-Integrity Structures," November 13, 1962, Buckminster Fuller. U.S. Patent 3,169,611, "Continuous Tension, Discontinuous Compression Structure," February 16, Kenneth Snelson. https://en.wikipedia.org/wiki/Tensegrity.
21 Bonner, John T. (ed.) (1961), 159, 160.
22 Cademartiri, Ludovico and Ozin, Geoffrey A., *Concepts of Nanochemistry* (Weinhelm, Germany: VCH-Wiley, Verlag GmbH & Co. KGaA, 2009).
23 Ibid.
24 Mackay, Alan L., Introduction of "Crystal Souls—Studies of Inorganic Life," *Forma*, Vo. 14, No. 1, 2 (1999), 9–11.
25 Siler, Todd, Truizms: Seek & See Truths to Inspire Innovation (Denver, CO: ArtScience Publications, 2012; Apple's iBook Store for iPad and iPhone: goo.gl/k85hP). This collection of drawings with picture-statements and statement-pictures interprets personal and universal truths that shape our lives and future. Also, read Siler, Todd and Ozin, Geoffrey A., "Cultivating ArtScience Collaborations That Generate Innovations for Improving the State of the World," SEAD: White Papers for an NSF-funded initiative of the SEAD Network (2014); https://seadnetwork.wordpress.com/white-paper-abstracts/final-white-papers/cultivating-artscience-collaborations-that-generate-innovations-for-improving-the-state-of-the-world/
26 "Scientific Accomplishments: Instrumentation Research, Methodology, and Standards for Nanotechnology (PCA 4). 4D Electron Microscope for Directly Visualizing Atomic-scale Motion." https://www.nano.gov/sites/default/files 4delectronmicroscope_nsf_0.pdf (accessed January 3, 2019); Barwick, B., Park, H. S., Kwon, O.-H, Baskin, J. S. and Zewail, A. H., "4D Imaging of Transient Structures and Morphologies in Ultrafast Electron Microscopy," *Science*, Vol. 322 (2008), 1227.
27 Bonner (1961), 298.
28 This explanation on parametrization by Shai Haran, PhD, Pure Mathematics (MIT) is an excerpt from Todd Siler *Cerebreactors* (New York: Ronald Feldman Fine Arts Inc, 1980), 14; this ArtScience publication was produced for this solo exhibition, "Top Secret: Inquiries into the Biomirror," Ronald Feldman Fine Arts, New York, NY, 1981. On the subject of parametrization, see https://en.wikipedia.org/wiki/Parametrization; https://mathinsight.org/line_parametrization. Also, read Hughes-Hallet, Deborah; McCallum, William G.; Gleason, Andrew M. (January 1, 2012). *Calculus: Single and Multivariable* (John Wiley, 2012), 780. ISBN 9780470888612. OCLC 828768012.

29 Siler, Todd, "The ArtScience Program for Realizing Human Potential," *Leonardo Journal/International Society for the Arts, Sciences, Technology*, Vol. 44, No. 5, 417–24.

30 Brown, Adam, Snelson, Kenneth, Root-Bernstein, Robert and Siler, Todd, "ArtScience: Integrative Collaboration to Create a Sustainable Future." 2011. http://www.leonardo.info/; http://www.mitpressjournals.org/doi/pdf/10.1162/LEON_e_00161

31 Lehn, Jean-Marie, "Towards Complex Matter: Supramolecular Chemistry and Self-organization," *European Review*, Vol. 17, No. 2 (2009), 263–80; Lehn Jean-Marie, *Supramolecular Chemistry: Concepts and Perspectives* (New York: VCH, 1995), 89–138.

32 Cademartiri, Ludovico and Ozin, Geoffrey A., *Concepts of Nanochemistry* (2009), 21, 22.

33 Ibid. (2009), 27–9.

34 Ozin, Geoffrey and Siler, Todd (2011), "ArtNano: Small Science Big Artwork." On Forming ArtNano Innovations. The co-founding of www.ArtNanoInnovations.com to explore a wide range of innovations for creating a sustainable future. Document for the World Cultural Council.

35 Cademartiri, Ludovico and Ozin, Geoffrey A., *Concepts of Nanochemistry* (2009), 23.

36 Bonner, John T. (ed.), Chapter IX: "On Spicules and Spiculat Skeletons," [136, 137] Abridged Edition of D'Arcy Thompson, *On Growth and Form*. Cambridge University Press, 1961.

37 Seifter, Harvey, Principal Investigator and Director of the "The Art of Science Learning," Phases 1&2, 2017, National Science Foundation-funded innovation initiative; www.artofsciencelearning.org.; Also, read Root-Bernstein, Robert, STEMM Education Should Get "HACD," *Science* (July 2018) Col. 361. 6 (6397), 22–3. National Academies of Sciences, Engineering, and Medicine, The Integration of the Humanities and Arts with Sciences, Engineering, and Medicine in Higher Education: "Branches from the Same Tree" (National Academies Press, Washington, DC, 2018).

38 Roald Hoffmann's vision of "The Metaphor, Unchained." *American Scientist* (2006) (94), 407.

39 Siler, Todd *Think Like a Genius: Use Your Creativity in Ways That Will Enrich Your Life* (New York: Bantam Books, 1997; Ballantine Bantam Dell Publishers eBook, 2017), http://www.penguinrandomhouse.com/books/167113/think-like-a-genius-by-todd-siler; *Metaphorming Worlds*. The Taipei Fine Arts Museum (Taiwan, Republic of China, 1995). Introductions by Dr. Robert Root Bernstein and Marilynne S. Mason; *Cerebralism: Creating a New Millennium of Minds, Bodies and Civilizations* (New York: Ronald Feldman Fine Arts, 1993); Metaphorming Minds: Envisioning the Possibilities of Nature (Israel: Old Jaffa Press, 1991); *Breaking the Mind Barrier: The Artscience of Neurocosmology* (New York: Simon and Schuster, 1990); *Metaphorms: Forms of Metaphor* (New York: The New York Academy of Sciences, 1988).

40 Siler, Todd, "The ArtScience Program for Realizing Human Potential," *Leonardo Journal/International Society for the Arts, Sciences, Technology*, Vol. 44, No. 5 (2011), 417–24.

41 Read Carlo Zammattio's essay, "The Mechanics of Water and Stone," in Reti et al., *The Unknown Leonardo* (New York: Harry N. Abrams, 1990), 198.

42 Root-Bernstein, Robert and Root-Bernstein, Michele. *Sparks of Genius: The 13 Thinking Tools of the World's Most Creative People* (Boston: Houghton Mifflin Company, 1999), 156–8; Seifter, H. et al., "The Art of Science Learning," Phases 1&2, National Science Foundation-funded innovation initiative, 2017. http://www.artofsciencelearning.org/metaphorming/ (accessed January 3, 2019).

43 Cademartiri, Ludovico and Ozin, Geoffrey A. (2009), 61.
44 Ozin, Geoffrey and Siler, Todd (2011), "ArtNano: Small Science Big Artwork." On Forming ArtNano Innovations. The co-founding of www.ArtNanoInnovations.com to explore a wide range of innovations for creating a sustainable future. Document for the World Cultural Council.
45 "50 Years of Materials Research and Still Searching," Vignettes of Research High lights, Geoffrey A. Ozin, Materials Chemistry and Nanochemistry Research Group, Solar Fuels Cluster, Center for Inorganic and Polymeric Nanomaterials, Chemistry Department, University of Toronto.
46 Siler, Todd, "Neuroart: Picturing the Neuroscience of Intentional Actions in Art and Science," *Frontiers in Human Neuroscience* (July 2015) 9 (410); www.frontiersin.org; "Neuro-Impressions: Interpreting the Nature of Human Creativity," in *Frontiers in Human Neuroscience* (October 2012), 6 (282). (http://www.ncbi.nlm.nih.gov/pmc/articles/PMC3469876/)
47 Qian, Chenxi, Siler, Todd and Ozin, Geoffrey A., "Exploring the Possibilities and Limitations of a Nanomaterials Genome," *Small*, Vol. 11, No. 1 (2015), 64–9.
48 Ibid.
49 Ozin, Geoffrey A., "Todd Siler's NanoWorld: Think Billionths of a Meter," art review for The inaugural ArtNano works, "NanoWorld" sponsored by Ronald Feldman Fine Arts New York (www.feldmangallery.com) at The Armory Show, Pier 94, New York, on March 5–9, 2014.
50 Marien, M. (ed.), Future Survey. The Millennium Project: "15 Global Challenges facing humanity." 2012. http://www.millennium-project.org/millennium/challenges.html (accessed January 3, 2019).
51 Geoffrey A. Ozin's op-ed, titled "A Clean & Green Energy System powered by an efficiently 'electrified' or 'solarfield' world," considers this reality, and I quote: "It's Your Choice! An Unsustainable Fossil & Nuclear Energy Economy? Or, a Clean & Green Energy Economy?" In Materials Views—Opinion Edition (accessed January 3, 2019).
52 Ozin, G.A., "Grasping Our Growing Gigatonne CO_2 Challenge," *Advanced Science News*. 2016. http://www.materialsviews.com/grasping-growing-gigatonne-co2-challenge/ (accessed January 3, 2019).
53 Ozin, Geoffrey A., *Nanochemistry Views* (Materials Chemistry and Nanochemistry Research Group, Center for Inorganic and Polymeric Nanomaterials, Solar Fuels Cluster, University of Toronto, 2015); http://nanowizard.info/wp-content/media/monograph-V1-II.pdf.
54 Ibid. (2015).
55 Email and Skype conversations and reflections on ArtNano Innovations, 2015.
56 Ibid. (2015) These conversations and reflections on ArtNano Innovations include visual notes, articles, blogs, and public presentations between Geoffrey Ozin and Todd Siler.
57 Cademartiri, Ludovico and Ozin, Geoffrey A., *Concepts of Nanochemistry* (2009), 19.
58 Ozin, Geoffrey A., "1999 Pure or Applied Inorganic Chemistry Award Lecture Curves in Chemistry: Supramolecular Materials Taking Shape," *Can. J. Chem*, Vol. 77 (1999), 2001–2014.
59 Ozin, Geoffrey A. and Siler, Todd, "Catalyst: New Materials Discovery: Machine-Enhanced Human Creativity," *Cell*, Vol. 4, No. 6 (June 2018), 1183–9.
60 Ozin et al., *Nanochemistry: A Chemical Approach to Nanomaterials* (2009), xviii.
61 Ozin, Geoffrey A., "A Clean & Green Energy System powered by an efficiently 'electrified' or 'solarfield' world," in *Materials Views—Opinion Edition*.

13

On Growth and Form and Lightweight Structures

Sarah Bonnemaison

Many people feel a special sense of happiness when dipping a ring in soapy solution and blowing through it to create a cloud of bubbles. Time comes to a standstill. For Peter Sloterdijk, "There is a solidarity between the soap bubble and its blower that excludes the rest of the world."[1] This intense exclusion is where his philosophical reflection begins:

> The child stands enraptured on the balcony, holding its new present and watching the soap bubbles float into the sky as it blows them out of the little loop in front of his mouth. Now a swarm of bubbles erupts upwards, as chaotically vivacious as a throw of shimmering blue marbles. Then, at a subsequent attempt, a large oval balloon, filled with timid life, quivers off the loop and floats down to the street, carried along by the breeze. It is followed by the hopes of the delighted child, floating out into space in its own magic bubble as if, for a few seconds, its fate depended on that of the nervous entity. When the bubble finally bursts after a trembling, drawn-out flight, the soap bubble artist on the balcony emits a sound that is at once a sigh and a cheer. For the duration of the bubble's life the blower was outside himself, as if the little orb's survival depended on remaining encased in an attention that floated out with it.[2]

Sloterdijk reflects on the limit between interior and exterior of spherical bubbles and meditates on our changing conception of the Cosmos as our collective bubble. As he says, "Who first conceived of the idea that the world is nothing but the soap bubble of a globalizing breath? What being-out-of-itself would then be all that is actually?"[3] The philosopher takes the reader over three volumes, starting with one bubble, the earth, the cosmos; then two joined bubbles and ultimately foam, a multitude of bubbles. These light, ephemeral, translucent self-contained objects caught the imagination not only of children, but also of the engineer Robert Le Ricolais, the architect Frei Otto, and, of course, the biologist D'Arcy Wentworth Thompson.

Organic Wholeness

Sloterdijk reflects on the fundamental link between bubble and dwelling.

> The sphere is the roundness endowed with an interior, exploited and shared. Because dwelling always means forming spheres, small and large, humans are the creatures that establish circular worlds and look outwards towards the horizon. To live in spheres means to produce the dimension in which humans can be contained. Spheres are creations of spaces with an immuno-systemic effect for ecstatic creatures worked from outside."[4]

Behind this notion of dwelling as "forming spheres" and "looking outwards towards the horizon" lurks a conception of the organic by the German Romantic polymath August Wilhelm Schlegel. In his treatise on *Art and Literature* from 1808, he wrote, "Organic form is innate; it unfolds itself from within and reaches its determination simultaneously with the fullest development of the seed."[5]

Let me draw a parallel between the notion of being "inside the sphere looking out" and the notion of "unfolding itself from within." The organic metaphor of the seed entails an idea of a self-regulating system: a little totality that carries the reasons for its own change and development within itself. This figure of a self-regulating system "unfolding from within" was influential on generations of architects working with the seed as an organic metaphor—such as Frank Lloyd Wright or Buckminster Fuller. In a similar vein, Eileen Gray in the 1920s writes about a house as a shell. Comparing human need to retreat to an animal in its shell, she introduces nature as a model, arguing against a mechanistic view of architecture.[6] She says:

> The house is not a machine to live in. It is a shell for human beings, their extension, their retreat, their spiritual emanation. Not only its visual harmony, but its entire organization, all the aims of my work, come together to render it human in the most profound sense of the word.[7]

Such metaphors, "the seed," "the shell," are just small indications of the enormous debt architects owe to nature, as a model, a source of inspiration and *justification*. Thompson's influence on architecture and particularly his book *On Growth and Form* is part of an enduring theme of drawing from nature in architectural theory and innovations.

Over the years, *On Growth and Form* has been a major reference and even a guide for architects. It was required reading when I was an architecture student in New York in the 1980s; then it was a reference in Bodo Rasch and Frei Otto's office when I worked there as a young architect. In my design practice of tensile structures, I returned to it repeatedly for inspiration—from the surface tensions of frog eggs to the growth patterns in shells. The centenary celebration of its publication in 1917 is an opportunity to reflect on what Thompson offers architects. *On Growth and Form* has been a guide on how to observe nature, draw out patterns of behavior, and apply such patterns to the design of buildings—especially in terms of supporting structures and materials. In other words, architects rediscovered "organicism" as an approach to design, but with an added concern of treading lightly on earth—hence the bubble as a recurring leitmotif.[8]

Organicism as a Recurring Theme in Architectural Education

Many architecture schools from the 1960s to 1980s explored the relationship between architecture and nature in a way that was indebted to Thompson. I will look at a few schools here, of which I have had firsthand experience: Pratt Institute (and its link to University of Pennsylvania) and the Massachusetts Institute of Technology.

In the 1980s, the architectural curriculum at Pratt Institute introduced first-year design students to geometries of the natural world, and *On Growth and Form* was required reading.[9] In a course taught by Haresh Lalvani, students learned that all regular volumes could be ordered with numerical characteristics, like a periodic table, an approach very much influenced by Buckminster Fuller. For example, a particular type of polyhedron undergoes geometric transformations according to mathematical rules. These transformations are organized in a table where each row and column depicts a particular transformative action on the polyhedron. The resulting tables are "families of form." Students also learned that the world was full of regular polyhedrons, recognizable to the trained eye, so we were always on the lookout for geometrical patterns in the world around us—from ornaments on brownstones to pine cones. In seminars, we classified and elaborated the relationships between "families of form," pursuing these in design studios inspired by Archigram, Fuller, and the Japanese metabolists Kenzo Tange and Kisho Kurokawa.

At our studio desks, we were surrounded by honeycombs, seashells, photographs of spider webs, and chicken bones. Thompson's book was a major reference for understanding the deeper logic of bone structure for designs that clearly separated elements working in compression (like bones) from those working in tension (like organs). This allowed us to apply the precision of transformative geometries to create insect-like structures to inhabit our cities. The bones became columns and trusses, and the organs became soft envelopes for habitation nestled in space structures.

As *bricoleurs*, we would collect all sorts of industrial parts for their form, color, and material to build our architectural models. Like collecting shells and rocks on the beach, these objects (from glass radio tubes to metal springs) remained unaltered but acquired meaning through placement and association. Our proposals were often crystalline mega-structures inserted in the old city fabric, like coral reefs along rocky crevices.

At Pratt Institute, the curriculum was enriched by thinkers such as Anne Tyng (who was based at the University of Pennsylvania, where she taught morphology). Working with Louis Kahn, for thirty years she developed designs for towers and skyscrapers that explored Jungian archetypes and packing geometries. This combination of Jungian psychology and light-filled spaces supported by frames of platonic solids tapped into the Renaissance belief that organic geometries carry with them powerful imbedded meanings. As a professor at Penn, she also exposed Pratt students to the daring structural explorations of engineer Robert Le Ricolais, whose research laboratory focused on lightweight structures for bridges and towers. To learn how structures are loaded, deform, and change over time, Le Ricolais and his students test-crushed structures to collapse; developing stronger and lighter design variations from their findings, analyses of failure, morphologies of collapse, and new stabilities that result.

The third example of Thompson's influence on architecture schools that I experienced was at the Massachusetts Institute of Technology. There, György Kepes had a significant influence on the curriculum—identifying a pedagogical opportunity in linking the scientific observation of natural phenomena to art and design education. His Centre of Advanced Visual Studies (CAVS) and his influential six-volume series *Education of Vision* are benchmarks in education linking creative experience and scientific observation. Kepes's insight into nature and design was echoed by his colleagues the engineer Waclaw Zalewski and the architectural theorist Maurice Smith.

Like many of my colleagues at school, we were politically committed to create an architecture that would be inspired by nomadic traditions and would go against the grain of land speculation and the constraints of building codes. Ephemeral lightweight structures, from inflatables to tensile structures, offered a building technology and materials that allowed great freedom to experiment.

The Institute for Lightweight Structures

Much like Le Ricolais, Otto headed the Institute for Lightweight Structures at the University of Stuttgart, called the IL for short. His research too was, in many ways, imbedded in the philosophy of Thompson. As Georg Vrachliotis recalls, "publications such as D'Arcy Thompson's *On Growth and Form* first published in 1917, were taken from the bookshelf and read anew."[10]

Clearly linked to the University of Stuttgart's engineering and architecture schools, the IL developed a method of observing natural phenomena to conceptualize and develop architectural form and structures. A number of engineers in the faculty used this method of observation and physical testing. One recurring field of inquiry focused on the behavior of liquids. In *Thinking by Modelling*, his edited volume on Frei Otto, Vrachliotis reported that:

> Not only Otto, but other engineers such as Le Ricolais and Tomaszewski all shared an interest in the physical properties of liquids characterized by their consistent responsiveness to external forces. Seen from this perspective, the experiments conducted with soap bubbles and soap film—already carried out more a hundred years earlier by Belgian physicist Joseph Antoine Ferdinand Plateau (1801–1883)—appeared as a brand new technique that was useful for architecture.[11]

Thompson also drew from Plateau's experiments to show that cells influence each other's shapes with a triangle of forces and compared the symmetries of bubbles meeting at points and edges. He showed that uniform growth can lead to unequal cell sizes but their shared surfaces are always at an angle of 120 degrees, even if the cells are of different sizes. Ultimately, he came to the significant conclusion that the shape of dividing cells is driven by the geometry of their structure. Hence geometry is the driving force behind the development of the frog eggs or the form of soap bubbles.

Otto and his team at the IL explored organic forms and phenomena while working within a scientific paradigm. They studied nature entirely objectively; that is, they believed that they could observe without having any effects on what they were observing. The IL took photographs to observe phenomena, and they observed in order to measure. They were uninterested in the unpredictable, the invisible, or the intangible. The idea, clearly inherited from the German tradition of organicism, was to reveal forces and to give them form in the most elegant manner possible. This entailed looking very carefully at the structural principles of natural lightweight structures from soap bubbles to spider webs, and also at inorganic heavy materials under stress to find patterns of behavior—such as cracks in a clay tablet as it dries or falling bricks spinning table to study earthquake forces.

Architects and engineers from across the world came to the IL to research topics set by Otto. These included tensile, adaptive, pneumatics, branching structures, bamboo construction, and so on. IL publications on these topics ordered the findings according to types, arranged like a grid that could be extended ad infinitum in the future as research continued. Lightness and material efficiency, it seemed, came from nature's ability to adapt. Hence Otto's architectural experiments embodied:

> the ideal that adaptable architecture was also striving at. On the one hand, they reacted to external influences in a highly sensitive way, in the process of change and transformation, on the other hand, they disclosed a momentum that followed an inner logic, thus demonstrating a certain obstinacy in the model and the material of which it was made.[12]

In this context, physical modeling was a theory, a way of thinking, and a mode of invention.

Otto's great contribution to architecture was to express the natural flow of forces—gravity, wind, tensions, and compression—with the minimum amount of material. For example, the aim was to balance the forces coming from within and from without to create an organic unity. Instead of engineering the structure to resist wind forces, the organicist approach conceives of building material as part and parcel of the forces that allow it to stand erect and be one with wind pressures and other natural phenomena. This design approach is called "form finding."

Form Finding and Form Making

"Form finding" was an approach to discover the optimum form for structures working in tension (such as tents) by modeling forces with soap film, but also for structures working in compression (such as vaults) by modeling forces with hanging chains that naturally adopted catenary forms. At the IL, "form finding" was clearly set in opposition to "form making," viewed as the imposition of a preconceived idea on material without incorporating knowledge drawn from the study of nature. By contrast, "form finding" implied the ability of soap film or chains to settle into its least stressful position according to its own physical natural laws. The organic qualities of the form found through this process were generated from the geometry of the flow of forces in the material (Figure 13.1).

The term was first used by the great architect and theoretician Karl Friedrich Schinkel. For him, *Formerfindung* (form finding) was central to a tectonic approach to architecture. As Caroline van Eck says, "In this orientation towards *Formerfindung*, Schinkel's tectonic approach to architecture clearly announces the science of tectonics as it was developed from the 1840s onwards by his student Karl Bötticher, in which organicism plays a [...] prominent role."[13] Established in this tradition, the research done at the IL was conducted with the upmost scientific rigor with the aim to apply and test new assemblies, tectonics, and material in architectural projects. "In the 1950s, various construction technologies developed, that made it possible to produce both concave and convex surface forms."[14] Reinforced concrete was the material of choice for architects such as Felix Candela in Mexico or Oscar Niemeyer in Brazil. For Otto working in Germany, dry assembly with wood, steel, and fabric were his materials of choice. In fact, the IL was financially supported by steel industries that desperately needed new outlets after the war and saw a good partnership in what came to be called "high tech" in British architecture. In retrospect, members of the IL say:

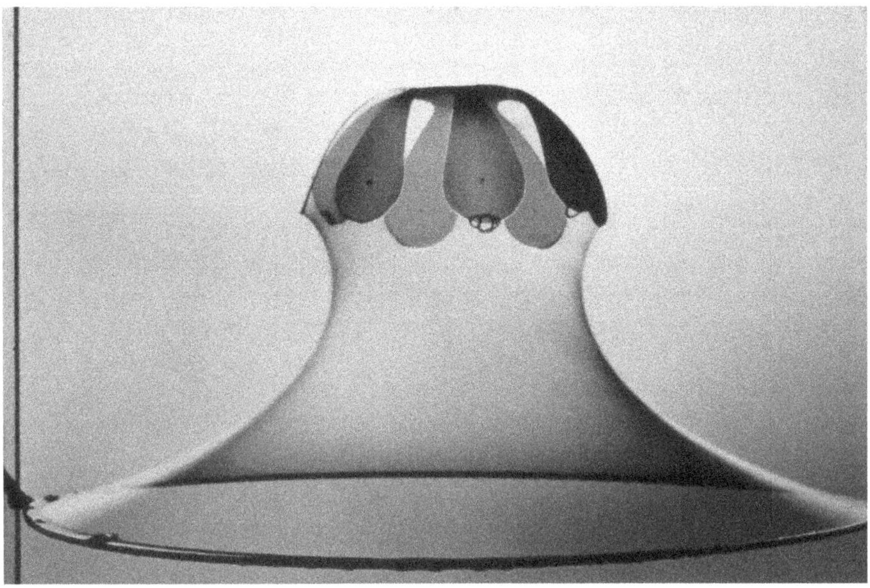

Figure 13.1 Extreme bulges produced on surfactant films by using springy rosettes. Courtesy of the Institute of Lightweight Structures and Conceptual Design, University of Stuttgart, Pictures Archives.

> Architects look at these soap-film experiments with great interest even today, which might be due to the fact that they retrospectively read them as a form of generative design, especially when seen from the contemporary perspective of digital design instruments. They are regarded as a means of partly automated 'form finding' based on mathematical rules. But in a historical context too, we can see how soap-film experiments attained fundamental significance as a three-dimensional modeling technique for non-orthogonal complex curved structures.[15]

In *Thinking by Modelling*, the authors collectively reflect on Otto's modeling techniques as a way of working through aesthetic and technical issues, arguing that these experiments were aimed not to find the perfect form, like the perfect solution to a problem, but rather to explore the relationship between material and forces in an open dynamic manner, providing a number of solutions and opening further venues for exploration. It was an open system where chance had a role to play:

> On the conceptual level the soap film experiments did not direct the architect's attention toward a single optimized form but rather to a whole set of similar forms—and not least to the transformation process of the fluid form itself. Frei Otto therefore used this property to test modifiable models. Manipulating the initial properties of one single trial setting, he created a whole series of different forms that were nevertheless related in their basic structure.[16]

The photographs of the soap film models can be misleading in that regard. They present "a solution" when upon reflection, "such soap-film experiments were particularly useful as a means to explore the changeability of forms. Thompson had already noticed the extreme inconsistency of soap film, describing it a highly fluid form that instantly reacts to external forces."[17]

A desire to bring lightweight structures into my design practice brought me to apprentice at FTL, the only architectural office in New York that specialized in tensile structures. There, I learned my craft and could see firsthand the results in a number of built projects. But like a salmon swimming up stream, it is at the office of Bodo Rasch in Stuttgart (an offshoot of Otto's office) that I really learned how to create optimized forms. The office team designed large retractable umbrellas to shade the pilgrims at the great mosque of Mecca. The integration of robotic arms, solar panels, and cutting-edge fabric was a great challenge to the built heritage of the mosque. But the client, the royal family of Saudi Arabia, were eloquent in guiding the office to create a beautiful hybrid of ancient and modern tensile architecture through the use of ornament.

When I started my own design office Filum Ltd with Christine Macy, I applied what I had learned to design and build tensile structures. Whether a small project like the portable and collapsible booth for Biopac company in California (1990) or a larger one like the Fuji Pavilion for Montreal's Botanical Gardens (1997), the result were well-designed, long-lasting structures.

Research-Creation Projects

Moving into academia gave me the opportunity to deepen my understanding of designing with lightweight structures—not only how, but why, and how to expand the possibilities for this kind of work and thinking. With the support of the Social Sciences and Humanities Research Council of Canada, I began to extend my work in tensile and lightweight structures as "research-creation" projects.

This started with an exploration into design method, moving from the objective, scientific, and positivist world-frame into the intersubjective, experienced, and perceived one. At that time a number of architects were beginning to look at "modernism" from a slightly different angle and—as many different readings came into play, and the notion of a dominant interpretation receded—I, too, wanted to use intertextuality and to make sense of lightweight structures in contemporary culture. While Roland Barthes's *The Death of The Author* argues for the distance between author and the text, Eeva Pelkonen says, "the organic form takes the maker out, to a certain extent (...) Besides the individual creator's will, there is some power elsewhere that is able to create architecture."[18] I began to see organic forms "offered as a set of inherited traits rather than individually authored by the architect."[19]

To effect this in my own work, I returned to the form-finding process with soap film models for their ephemeral qualities (they last only a few seconds) and for their ability to reconfigure. For example, a student of Le Ricolais reported on the dynamics of soap bubbles, which constantly reconfigured themselves each time the film burst.[20] Such spontaneous reconfiguration and its potential for revealing the unexpected are reminiscent of Marcel Duchamp's use of chance as a tool for artistic creation. André Breton saw in this an extraordinary potential. "The use of chance is effective only when it is stripped of any superstitious connotation, and therefore of any subjective projection. It is also what makes it 'extraordinary' in the eyes of Breton."[21] Hence the importance of "letting it happen," without projecting a need or a want.

On that note, let us now turn our attention to the first "research creation" project. Called *Gestures*, this project was built in the courtyard of the Art Gallery of Nova Scotia on the occasion of the year 2000 celebrations in Halifax. The intent was to explore the notion of chance as a design strategy for lightweight structures. I created a series of "scribbled" wire frame models, then dipped them in a soap solution. As the soap film surfaces appeared and then vanished, I watched the forms unfold from within. There was something magical about the process—forms were created by chance. But the soap film also brought great precision and the inexorable rules of physics to the randomness of the wires. The resulting form was a hybrid of chance and mathematics. Like Duchamp, I used chance in form finding not as an end but as a process that can create art.

Once built, the installation was a 15-meter-high gesture of colorful nets stretched between large wooded hoops. Over the summer, the structure shaded people at a series of events planned for the millennial celebrations. A telling anecdote of the power of this design approach is when an elder from the Eskasoni first Nation, Murdena Marshall, offered to name it. She gave it a Mi'kmaq name that can refer to both a butterfly and a hummingbird, thereby making an association between the organic forms of the installation and the fluttering wings of butterflies and small birds (Color Plate 13).

Life and Movement

One of the most persistent ways movement can be expressed is through organic forms. In fact, the idea of organic form in architecture is like a wave on the surface of the ocean—a side effect of a powerful undercurrent that has long sought to ally science and art in the discipline.

We can draw a thread connecting architecture that expresses ideas of movement at points when art and science crossed paths. It would connect Baroque architecture to Art Nouveau and proceed to lightweight structures and today's explorations with living architecture. In the Baroque era, we find perhaps the greatest examples of movement—generated by the intersection of science and art—in the pilgrimage churches of southern Germany. There, we encounter the work of architects Johann Balthasar Neumann and the Zimmerman brothers, who developed the curved volumes of these masterworks, shortly after Gottfried Wilhelm Leibniz published his studies of calculus in the late seventeenth century. The dramatic displays of movement in these churches—their tension, exuberance, grandeur, and complex curves—employed Leibniz's calculus to create curved geometries of the domes that attempted to approach the perfection of God's creation of the natural world.

With the influence of motion studies, movement analysis became a vector for formal and spatial innovation in early twentieth-century modern architecture. Examples include the kitchen designed by Lily Reich for the German Exhibition in 1931; Gray's "choreographic architecture" and furniture that is entirely conceived in relation to the body in movement, the moving partitions of Gerrit Rietveld's Schroeder House, Endless House of Frederick Kiesler, and Slow Moving House by Diller and Scofidio.

On the biology side, one can find confirmation that organic forms reveal the passage of time. Thompson says:

> Spirally arranged florets of the sunflower [...] are similar to one another in form, differ in age; and they differ in magnitude in a strict ratio according to their age. Somehow or other, in the logarithmic spiral the time-element always enters in; and this important fact, [is] full of curious biological as well as mathematical significance.[22]

Art historian Georges Didi Huberman sees an essential connection between life and movement:

> The movement is the most apparent of the characters of life; it manifests itself in all functions; movement is the very essence of many of them, and in every physical notion of force, on the other hand, tend to be reduced into one—that which engenders movement.[23]

If we could freeze the traces of human movement, might we envision an inhabited space such as a shell or a nest surrounding the body? With this image in mind, I embarked on an exploration to create "circular worlds" based on the traces of movement that would shape an individual's dwelling, like a human shell.

My next "research and creation" exploration aimed to turn movement into architecture, much in the same way that for Paul Klee, "pictorial art springs from movement, is in itself interrupted motion and is conceived as motion."[24] *Gestures* (2006), an installation at the Maritime Museum of the Atlantic, drew from disciplines outside of architecture: boat building, textile, motion capture, video, and dance.[25] Our research began by videotaping activities on the Halifax waterfront, human gestures that expressed the maritime culture, such as a man curling a rope around a cleat of a mast, a flag waving in the wind, a person pointing at a feature. Dancer Maria Osende choreographed a series of phrases that translated these everyday gestures into movements of the entire body. We then recorded her choreographies using motion-capture, tracing the movement in three-dimensional digital space. The movement paths curl down into small spirals, open up into large arcs, twist into curves, and fold back upon themselves (Figure 13.2).

Since "time measures motion," as Huberman remarks, "it becomes understandable that physical science, being physiological, has attempted the joined geometry of time and motion, through a graphic representation."[26] In this case, the technology of motion capture gave a graphic representation to join the geometry of time and human motion. Thirteen of these points, or "birds," located on the dancer's body, were tracked in virtual space according to their x-y-z coordinates. In the world of computer animation, such data is used to design avatars, but in this project we manipulated the trails of the "birds" to draw the major paths of the dancer's movements.

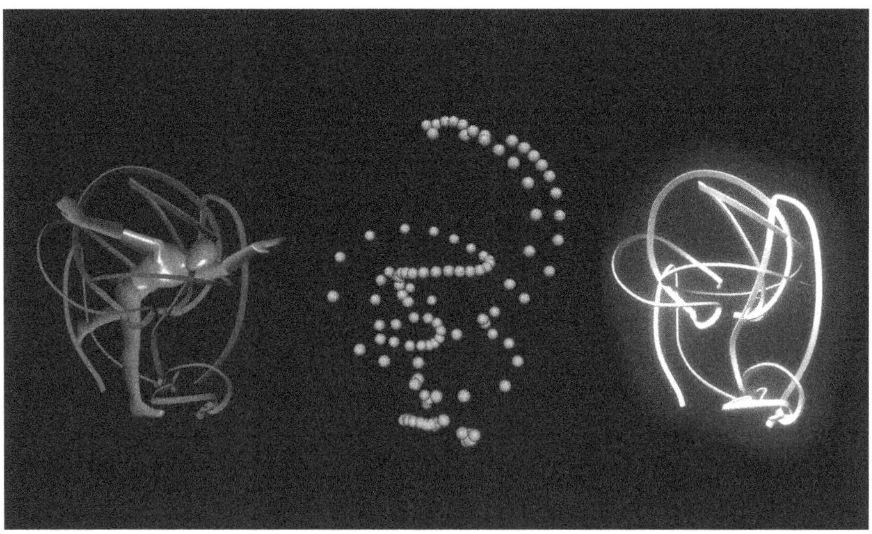

Figure 13.2 Sarah Bonnemaison, *From Traces to Form*, 2006. Courtesy of the Maritime Museum of the Atlantic, Halifax.

By tracing these movements, recording them first with lines and later with stereotomic models, I was able to give the dancer's movement phrase a three-dimensional graphic representation. The aim was to transcribe the dancer's phrases into expressive pavilions. But my intention was not to create a one-to-one relationship between the motion-trace and the form. Such a form would be realistic but would not express the idea of a dwelling like a shell. I removed many of the traces crisscrossing inside the sphere and carefully chose the ones that would become the supporting structure of the pavilions. Those had to obey the laws of physics in order to stand erect outdoors and withstand the loads of people climbing on them. So I reworked these trails to give them substance.

Then came the question of what materials would be used to build these human shells. The tradition of wooden boat building of the Canadian Atlantic coast offered an array of building techniques well adapted to complex curves. So through a number of steps, we further translated these structures into large curved wooden members. With the aid of craftsmen of the Maritime Museum, we learned techniques of steaming wood to curve it and riveting layers of curved wood into place. We then assembled them in the courtyard of the museum under the gaze of curious tourists.[27] The forms of the curves and sizes of the wood changed according to the speed of the movement, its scale (a small step or a large jump), and its quality (introverted or extroverted). These variations in the size of the wood members give an organic feel to the lines of the movement paths. They are pulsing and free-flowing lines, with the aim of creating a dynamic intricate space.

The final step to create the pavilions was the introduction of the tensile surfaces within the wood structures. This was necessary to create a sense of enclosure. If we think back to Leibniz, we know that the mathematical description of a curved line can be either derived or integrated. If it is derived, we obtain its simpler straight version; if the line is integrated, we obtain the surface under the curve. So to generate an enclosure or dwelling we integrated the curved lines of the wooden framework. As a result, the surfaces under the curves became surfaces built as stretched nets. Very much like the growth process of the mollusk as it connects one rib to the next with a thinner material slowly creating an enclosure. Similarly, the nets connect one rib to the next, thereby creating an enclosure.

As described earlier, the form of these stretched nets was designed through the form-finding process and employed soap-film models to generate an enclosure. Once the structures were built, each tensile surface contributed to distribute uniform tension throughout the integrated surface onto the wood members. When all the tensile surfaces were attached, the forces could flow throughout evenly and the whole structure settled in its own unique equilibrium.

Conclusions

The shell, like the bubble, is our home, our retreat. But it is also a place for "spiritual emanation," as Gray says. To build on this idea let us return to Sloterdijk, who wonders, "who conceived of the idea that the world is nothing but the soap bubble of a globalizing breath?"[28] As our mind expands we imagine the dome of the cosmos surrounding our planet earth. We can appreciate the feeling of security this dome has given us for so long. But the fine net created by today's networks of telecommunication around the earth has become our bubble. As Sloterdijk concludes:

> Now networks and insurance policies are meant to replace the celestial domes; telecommunication has to reenact the all-encompassing. The body of humanity seeks to create a new immune constitution in an electronic medial skin. Because the all-encompassing and contains structure, the heavenly firmament, is irretrievably lost, that which is no longer encompassed and no longer contained, the former contentum, must now create its own satisfaction on artificial continents under artificial skies and domes.[29]

Notes

1 Sloterdijk, Peter, *Spheres, vol. I: Bubbles Microspherology* (Cambridge, MA: MIT Press, 2011), 18.
2 Ibid., 17.
3 Ibid., 19.
4 Ibid., 121.
5 Cited in Hvattum, Mari, "'Unfolding from within' Modern Architecture and the Dream of Organic Totality," *The Journal of Architecture*, Vol. 11, No. 4 (2006), 497.
6 Gray, Eileen et Badovici, Jean, *E1027, une maison en bord de mer*, originally published in *Architecture Vivante* in 1929, Reprint, Marseille: Imbernon, 2015, 101.
7 Gray, 101.
8 Although Caroline Van Eck does not address the work of Le Ricolais or Otto in her book, it is a major reference for understanding the roots and development of organicism. Van Eck, Caroline, *Organicism in Nineteenth-Century Architecture: An Inquiry into its Theoretical and Philosophical Background* (Amsterdam: Architecture et Natura Press, 1994).
9 See the papers published for conference on the influence of *On Growth and Form* on architecture: *On Growth and Form*, eds. Phillip Beesley and Sarah Bonnemaison (Halifax, NS: Tuns Press, 2008).
10 Vrachliotis, Georg, Kleinmanns, Joachim, Kunz, Martin and Kurz, Philip (eds.), *Frei Otto, Thinking by Modeling* (Leipzig: Spector Books, 2017), 54.
11 Ibid., 54.
12 Ibid., 57.
13 Van Eck *Organicism*, 162.
14 Vrachliotis et al., *Frei Otto*, 57.
15 Ibid., 54.
16 Ibid., 57.
17 Vrachliotis et al., *Frei Otto*, 57.
18 Pelkonen, Eva, Gans, Deborah, and Kuz, Zehra (eds.), *The Organic Approach to Architecture* (Chichester: Willey-Academy, 2003), 61.
19 Pelkonen et al., *The Organic Approach*, 61.
20 Vrachliotis et al., *Frei Otto*, 57.

21 Troche, Sarah, "Marcel Duchamp: trois méthodes pour mettre le hazard en conserve," *Cahiers Philosophiques*, Vol. 4, No. 131 (2012). Blog de la Revue, DOI 10.3917/caph.131.0018, Reseau Canopé, 136.
22 Thompson, 496.
23 Didi-Huberman, Georges et Mannoni, Laurent, *Mouvements de l'air, Etienne-Jules Marey, photographe des fluids* (Paris: Gallimard, 2004), 185.
24 Paul Klee, quoted in Sigfried Giedeon, *Mechanization Takes Command* (New York: Oxford University Press, 1948), 109.
25 Credits for the installation *Gestures*: Architecture: Sarah Bonnemaison and Christine Macy; Dance: Maria Osende: Video: Ariella Palhke; Motion capture: NS Community College at the Truro campus; Textile dying: Robin Muller.
26 Didi-Huberman, *Mouvements,* 191.
27 The construction was done by students of Dalhousie University under my supervision in a summer term design-build course. See Macy, Christine, *Free Lab: Design-Build Projects from the School of Architecture, Dalhousie University, Canada, 1991–2006* (Halifax, NS: Tuns Press, 2008).
28 Sloterdijk, *Spheres*, 25.
29 Ibid., 80.

14

D'Arcy Thompson Going Forward: Seven Views

Between Chance and Necessity

Philip Ball

I feel a little sheepish in admitting that the statement of D'Arcy Thompson's that resonates most with me seems to be possibly an apocryphal one. No one seems able to say where—or if—Thompson said that "everything is the way it is because it got that way," although it is often quoted without qualification. I guess we're inclined to accept it as genuine because it seems so plausibly the sort of thing he would have said.

Daniel Dennett has suggested that the phrase can be applied pretty universally in the sciences—obviously in biology, but it seems clear that in physics too the evolution of the universe was an unfolding of symmetry-breaking steps that rendered distinct the various particles and forces we observe today. Some physicists believe that even the basic laws of physics are emergent rather than somehow written into the fabric of reality.

For D'Arcy Thompson, there's a subtle dance of what biologist Jacques Monod called "chance and necessity"[1]—or what Monod's colleague François Jacob called "the possible and the actual"[2]—in the way forms emerge in nature. He ascribes a kind of primacy to "some lofty concepts, like space and number, [that] involve truths remote from the category of causation." (We tend to associate that view of geometry to Plato, but Thompson, ever the astute classicist, pointed out that Aristotle thought this way too.) Yet he adds that "natural history deals with the ephemeral and accidental, not eternal nor universal things." They come into being through and subject to the quirks of history.

Science is constantly pulled between these poles, as the phrases of Monod and Jacob testify. The physicist's dream of a "final theory" from which all else can be predicted looks increasingly like a fool's quest, and even in mathematics we must accept that the symmetries of an equation can be broken in its specific solutions. Contingency and a lack of predictability are intrinsic to both quantum mechanics (as far as we can currently tell) and chaos theory. But even in biology, which the evolutionary theorist Ernst Mayr distinguished from physical theory in being a predominantly historical and contingent science,[3] there's still a pull toward a kind of modern version of preformationism in the assertion that the genome is a blueprint for the organism. Thompson's perhaps apocryphal quote—and the far from apocryphal message of *On Growth and Form*—reminds us that real things arise from an unfolding of laws and plans in ways we can't necessarily foresee.

We can't (yet) escape our genes, but neither are we defined by them. It's easy to overlook the fact that questions about contingency and necessity in biology still haven't been resolved, and that some rather weighty matters reside in them. The notion of an organism as an individual is indispensable to natural selection, yet philosophers and biologists continue to debate what this individuality consists of. The question arises too in deciding the moral status of the human embryo, with all its rancorous ramifications. *On Growth and Form* might be seen to offer an appealing (to me) case that we acquire personhood and identity by degrees, in a process of growth that never stops until we do.

But of course the "form" that Thompson is mostly concerned with is one that we can perceive with our eyes: the arrangements of leaves and petals in flowers, the crack networks in mud and basalt, the shapes of cells and bubbles. You could see this in part as a (preemptive) rejoinder to Mayr: biology is not wholly distinct from physics and chemistry. Thompson's book is also a challenge to Theodosius Dobzhansky's famous assertion that "nothing in biology makes sense except in the light of evolution." Well, says *On Growth and Form*, at least *some* things do. Biological exceptionalism can be (and has been) taken too far.

Thompson's book is often, and rightly, regarded now as a precursor to studies of complexity, self-organization, and the emergence of order out of equilibrium. In this respect it was both before and behind its time. Concepts of phase transitions and symmetry-breaking, now so central to this field of inquiry, already existed, notably in the 1873 thesis of J. D. van der Waals, when the first edition was published, while Lars Onsager's theory of non-equilibrium thermodynamics in the 1930s could usefully have informed the revised edition of 1942—as could the work that had been done on embryo patterning and morphogenesis by Hans Spemann and others, not to mention all the understanding that had by then accumulated on genetics. Perhaps the most profound contribution to the theory of pattern formation after Thompson's death was Alan Turing's theory of reaction-diffusion dynamics in 1952;[4] but one can't have much confidence that Thompson would have taken much notice of that either, had it been formulated sooner.

This, though, shouldn't detract from the legacy. Turing, after all, *was* influenced by Thompson, whose magnum opus is one of just six references in Turing's remarkable paper. Mavericks who offer a fresh perspective in science are often as blinkered as they are visionary—maybe that's even a precondition of their role. For me, a large part of what makes *On Growth and Form* inspiring is that it is *sui generis*, indifferent almost to the point of contempt of the disciplinary boundaries that we seek to impose on scientific inquiry. What Thompson wants to do is not make a contribution to an academic field but gain some insight into why things are the way they are.

It's an added bonus—an equal part of the book's inspirational nature—that in the course of that quest he reminds us of the value of wonder and beauty, and of the importance of looking closely and seeing beyond what the prescriptions and proscriptions of our disciplines tell us to see. As an earlier polymath, John Herschel, said, "To the natural philosopher there is no natural object unimportant or trifling … a soap bubble … an apple … a pebble … He walks in the midst of wonders" (Figure 14.1).

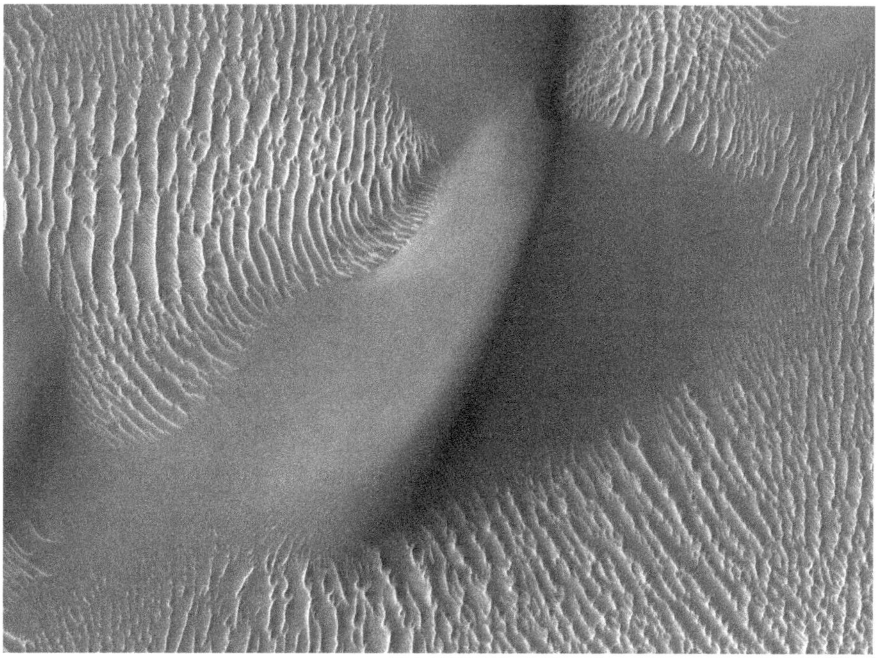

Figure 14.1 Photo from NASA of dunes on Mars. Courtesy of NASA/JPL-Caltech.

Image as Argument: D'Arcy Wentworth Thompson and Contemporary Scientific Discourse

Justine Kupferman

Every scientific publication is a story that is told to convince the world (or at least the scientific community) of the truth and importance of a new finding. And like many stories, visualization can help to tell the tale. For example, the tremendous recent progress in the field of neuroscience was spurred, in part, by new imaging techniques like functional magnetic resonance imaging (fMRI) and fluorescent microscopy. These methods have led to the creation of a profusion of new images and to subsequent new insights. A hundred years before this modern revolution in neuroscience, the renowned visual illustrations of two scientific giants, D'Arcy Thompson and Santiago Ramón y Cajal, helped convey their essential arguments and spread their ideas to a wide audience beyond their disciplines. The interplay between the textual scientific argument and the images in their works still resonates in contemporary neuroscience.

In the chapter Comparisons of Related Forms, from *On Growth and Form* (1917), Thompson shows that related forms can be quantitatively compared by mapping them in a Cartesian coordinate system. In Figures 761 and 762, Thompson used images of prehistoric mammalian skulls of the rhinoceros genus documented by Henry Fairfield Osborne, a scientist known for describing a new therapod (a dinosaur suborder characterized by hollow bones and three-toed limbs). Thompson applied his distortion grid method to *Hyrachyus*, which yielded *Aceratherium*, stating that "the essential difference between this skull [*Aceratherium*] and the former one may be summed up by saying that the long axis of the skull of *Aceratherium* has undergone a slight double curvature, while the upper parts of the skull have at the same time been subject to vertical expansion." These illustrations clearly demonstrate the utility of Cartesian coordinate transformations. Taken together, the text and the visual evidence are strongly compelling, as are similar analyses of leaf structure, copepods, and oceanic fish.

The neuroscientist Santiago Ramón y Cajal was a contemporary of Thompson and another polymath whose illustrations helped convey his scientific insights. As a young man Ramón y Cajal had wanted to be an artist and had worked as a medical illustrator before embarking on his medical training. Throughout his career as a doctor and neuroscientist, he made detailed observations of the many types of neurons in the brain and hypothesized, very often correctly, about their structure and function. To do this, he used a method developed by Camillo Golgi, with whom he shared the Nobel Prize, called the "black reaction" that stained individual neurons so that the cells of the brain could be seen in a way that had heretofore not been possible. Ramón y Cajal made detailed, microscopic observations of the different parts of the neuron—the axons that send signals to other neurons, the dendrites that receive information from other neurons, and the tiny gaps between connected neurons called synapses. One of his many important contributions that emerged from these observations was his discovery of dendritic spines, tiny protrusions that line the dendrites of neurons and are critical for proper synaptic communication between neurons. Accompanying his description of this discovery is an illustration of the dendrites of pyramidal neurons in the rabbit cerebral cortex. His drawing beautifully shows hundreds of these little spines dotting the neurons like flower buds on a Magnolia tree. Perhaps it is that we trust the powers of observation of someone who can create such an image or perhaps, as John Keats observed, it strikes us that something so beautiful must be true. Whichever the case, Ramón y Cajal's theory of the function of dendritic spines that emerged from his close observations of neurons gained wide acceptance quickly after its publication. Like his contemporary Thompson, it is impossible to separate the scientific arguments from the accompanying images or to underestimate the impact of his illustrations in gaining the acceptance of the scientific community.

The aesthetics of modern neuroscience seem in many ways to be vastly different than in the era of Thompson and Ramón y Cajal. Gone are days of the detailed hand drawing of microscopic observations. With fMRI machines, fluorescent microscopes, and computers to analyze it all, it may seem that contemporary scientific images create themselves, or perhaps the images seem to not be created at all but are instead objective snapshots of the world around us. Of course, this is not the least bit true. The scientist must make a thousand decisions in the process of creating such an image and many of those decisions have an aesthetic component. Just as in photography where exposure time, lighting, composition, saturation, focus, and brightness are all critical aspects of the final image, so too are each of these considerations critical in the production of a scientific image. All too often, too little care is taken in the crafting of a scientific image. The result is not just a less aesthetically pleasing or interesting image, but a missed opportunity to draw in the reader and persuade them unconsciously of the validity of the findings.

In my recent work, I studied the evolution of genes that are unique to humans. Many of these human-specific genes control aspects of brain development and function. The aim of this research is to better understand how dendritic spines are formed during brain development, and how human-specific genes have increased the density of dendritic spines on neurons of the cerebral cortex, thereby enhancing the cognitive abilities of ancient humans and allowing our species to thrive. From Thompson, I learned to think carefully about the physical structure of dendritic spines and ask what function is conferred to the neuron by this physical shape. To do this, I used modern molecular biology methods to label with a fluorescent tag the proteins that reside in dendritic spines, just as Ramón y Cajal used the then cutting-edge Golgi stain to label neurons. I keep Thompson and Ramón y Cajal in mind throughout the experimental process as I create the raw images that will be quantified and analyzed to test each hypothesis. Once the experiment is complete, a "representative image" is chosen to visually convey the main finding. For my studies, this might be a comparison between an unaltered mouse neuron grown in a petri dish and one that has a human-specific gene inserted into its genome. The "humanized" neuron has many more dendritic spines than normal mouse neurons, and I would select an image that visually reflects that finding (Figure 14.2). But this is not the only consideration in creating the image. My images are taken on a scanning laser confocal microscope, and I therefore consider factors such as magnification, focal

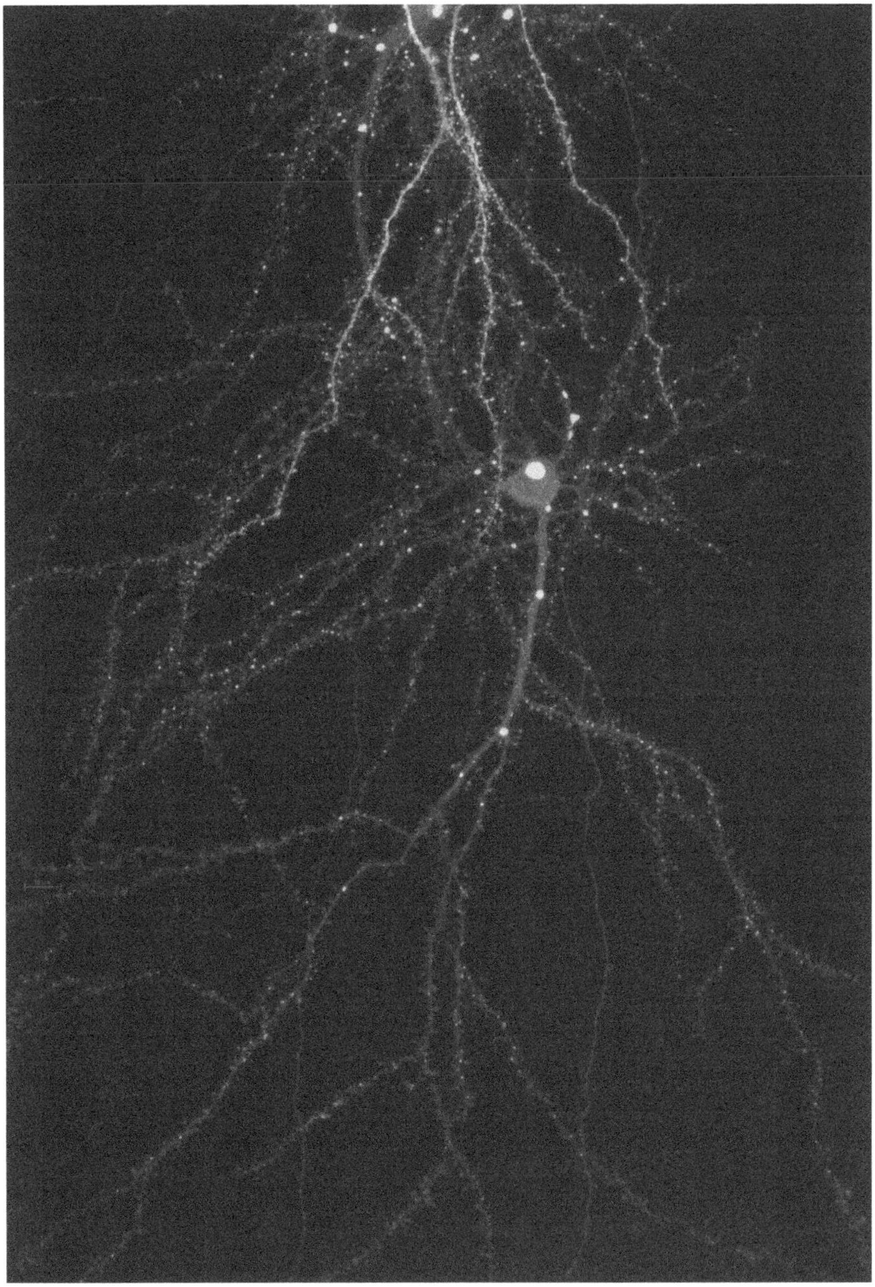

Figure 14.2 Justine Kupferman, *Humanized Mouse Neurons in a Dish*, 2017. Courtesy of Justine Kupferman.

plane, laser intensity, and, of course, the features of the neurons themselves. After the image is captured on the microscope, I must then choose how to add in color and light to best illustrate the features I wish to draw attention to. In the image shown grayscale, one neuron is colored green and contains one of these human-specific genes; its dendrites are dotted with hundreds of dendritic spines, each labeled by a tiny, purple dot. The neurons colored blue, by contrast, do not have any human genes and have fewer dendritic spines. In this image, I wish to draw attention to the humanized dendrite's dense thicket of dendritic spines. Learning from the examples of Thompson and Ramón y Cajal, I try to create an image that both accurately represent the findings of the experiment and is aesthetically appealing in the hope that the image will help persuade the neuroscience community to accept my findings and incorporate them into our growing understanding of the brain. Although these findings are quite humble in comparison to achievements of Thompson and Ramón y Cajal, I believe that thinking carefully about the visual qualities of image has strengthened the impact of these findings, and any scientist would do well to emulate these two scientific giants in the consideration of their images.

Reflections on Influence

Carolee Schneemann

From the window of my ancient house, I observe one leaf resisting the wind as it holds to the narrow branch of the hundred-year-old Black Locust tree. I regard this stubborn leaf as I heal from a four-and-a-half-hour femoral artery bypass surgery. Phylogeny recapitulates ontology—this concept persists as the leaf clings to its branch while I'm reviewing my early influences for this essay, the talismanic guides to my unpredicted future.

As a 17-year-old with a full scholarship to college, I was set free in the cultural prison of "Double-Knowledge." The mantra of the time was: "Don't set your heart on art, you're only a girl." Starting as a landscape painter, I visually concentrate on natural forms while physically feeling part of the landscape. I embrace my influences with a blind determination as I enter the forbidden territory of contemporary culture. For the most part, my influences are archaic. I'm aware that any female precedents are absent.

Freshman year I have found a fisherman's shack in South Harpswell Maine; and here I am painting seascape, landscape. I study the physical energies of sea shells, fish scales, cloud formations, the lines in my hand, the wings of butterflies, the patterns of my cat's fur ....

How, as a visual artist, am I to enter the physical dynamic of these vast and coherent permutations by which things are described and by which they realize their functioning? All this is deepened with my intense attachment on discovering D'Arcy Thompson's *On Growth and Form*.

> The medium is said to be permeable to the force, in greater or less degree than the standard medium, according as a variation of the density of the lines of force, from the standard case, under otherwise identical conditions, is in excess or defect.
>
> (Thompson, 171)

I want to explore art history and theories of visual perception. I'm ranging freely as a scavenger of culture—a crow finding the shine and gleam, from which I activate momentum from the painting surface to actual time and space. I discover a parallel influence in Focillon *The Life of Forms in Art*: "*The double problem of form in time: Its internal development as opposed to external circumstances*" (Focillon, 55).

These readings as well as the writings of Bachelard and the radical proposals of Antonin Artaud will enforce my experience of transposing forms. I absorb aspects of movement in nature and my body; an extension of kinetic visual properties. This mysterious process activates my eye and body—translating what is seen to what is marked upon a page. A myriad of influences will bring me into a confrontation with inherited cultural traditions.

Shared readings with my partner James Tenney address our research adventures in grasping the issues of our contemporary aesthetics. We search out challenging materials, such as Goethe's Color Theory, Von Helmholtz, Rudolf Steiner, Ludwig Wittgenstein, Erwin Schrodinger, Kurt Godel; as well as readings of Freud, and Jung. From Wilhem Reich we study the origins of fascism, sexual oppression; with Simone

De Beauvoir's *The Second Sex* (1949), all the gender distortions I experience are given historic context. A revelation! These sources merge to become the inspiration with which I situate myself as an external force: an explicit physical energy extending my visual premises into structured live actions. Beginning with the Kinetic Theater of the early 60s, then the visual morphologies and installations of the 70s and 80s.

In 1983 I undertook a sequential series of visual morphologies (Color Plate 13). *Fresh Blood: A Dream Morphology* presented two vivid dream elements. The connection of these two elements demanded an analysis of their shared "V" shape. My response to these elements connected once again to the early influence of evolutionary biology as I understood it in terms of patterns and visual components in nature, and the body as a wellspring of knowledge. I go back to Thompson:

> Moreover, the microscope seemed to substantiate the idea that there is no limit to the mechanical complexity which we may postulate in an organism, and no limit therefore to the hypotheses which we may rest thereon."
>
> (*On Growth and Form*, 159)

I organize a vocabulary of V shapes depicted in nature, in culture, in artifacts, rock formations, sacred erotic sculptures, and, insistently, in the human body. This simple unit would present me with ten years of visual permutations. The two V shapes were composed of a bouquet of dried leaves containing almost-human small blossom heads, and the companion dream element was an umbrella exhibiting a V "vector" shape—both opened and closed (Figure 14.3).

A subsequent morphological installation titled *Cycladic Imprints* (1989) was based on a vocabulary of double curves: Cycladic sculptures, violins, human torsos and genitals, animal skeletons and fronds. A projection system of 360 images activate 10' x 8' wall, on which eighteen motorized violins move in simple rhythms.

It is the basic conflagration of Thompson's poetics and mechanics of structure which assist my recognition of the power of Paleolithic art; much of it accredited to women artists. The intensive research of Marija Gimbutas finally ascribes the ancient history of Paleolithic artifacts in which sculpture and carvings establish a realm of creativity, long dismissed and misattributed. Works of menstrual time factoring, astrological systems, calendars, and references to lived experience had been translated as the growth of life and art took form.

It is here and with my most recent sculptural installation that I continue to experience the generative influence of D'Arcy Thompson and Focillon. This research provides me with permissions for melding theoretical concepts into a poetics of imagery and materials. My installation, *Flange 6 rpm* (2012), incorporates computer mechanisms, which determine the continuous motion of a grouping of extended vertical forms. Seven motorized sculptural units, containing hand-sculpted components. Each form is unique, cast in aluminum from a lost wax process. Sculptural units are mounted on a motorized base which moves them at 6 rpm—slowly, side to side, as well as forward and back—in a continuous motion so that the sculptural elements are almost touching, creating a sense of tension and unpredictability. A video projection of the fiery transformation is hugely enlarged and projected over or within the sculptural

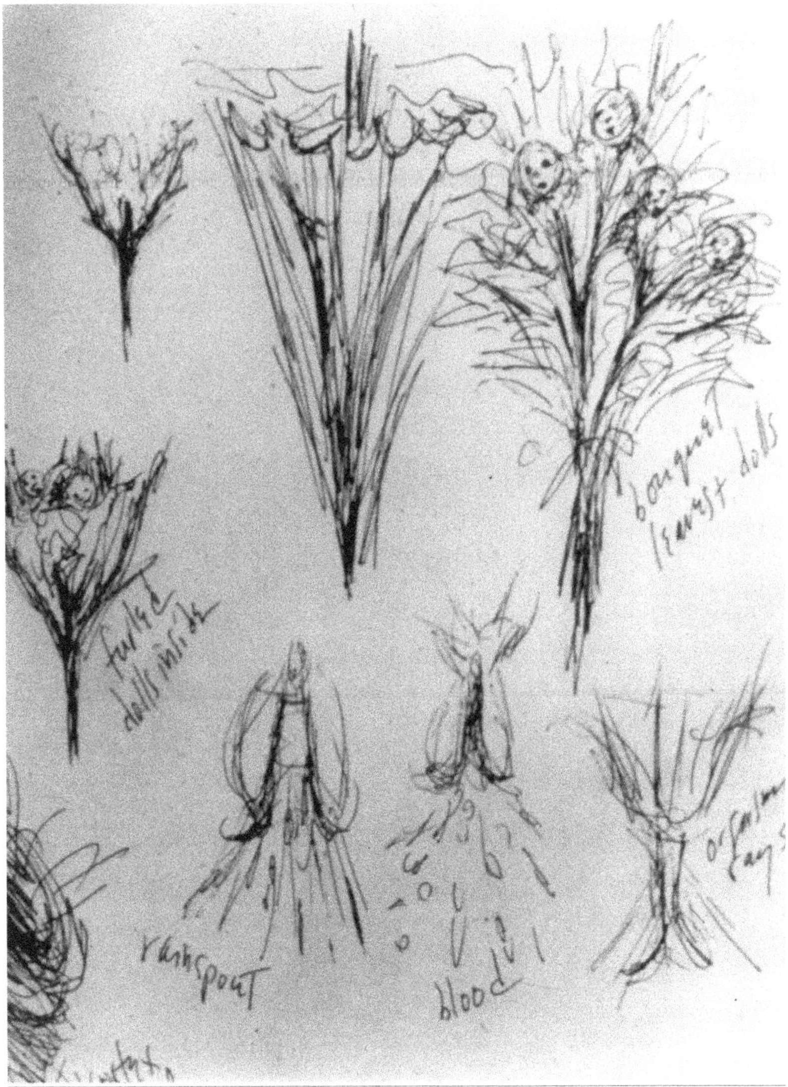

Figure 14.3 Carolee Schneemann, *Fresh Blood Drawing,* 1981 (c). Courtesy of Carolee Schneemann Foundation.

installation. The resulting image produces a luscious, flaming, blooming arena as the moving flanges open and close and shift in their threatening precarity.

> The quality of a medium filling the field of force may be uniform, or it may vary from point to point. In particular it may depend upon the magnitude of the field; and the quality of one medium may differ from that of another.
>
> (*On Growth and Form*, 171)

Labyrinth (1960) was my initial outdoor event. Living in Sidney Illinois with my partner, the composer, James Tenney—we were both on fellowships at the University of Illinois at Champagne Urbana, his in music, mine in painting. We had driven through an ice storm from Vermont to Illinois. Our old car was attached to a Uhaul, which contained a small upright piano, my paintings and materials, and boxes of our books. The car is so old that we have stuffed toilet paper rolls in the rusted-out interior so that Kitch's kittens will not fall out.

That spring, a tornado blew through our fragile cottage—crashing trees and raising rivulets of mud, dirt and rocks from the streambed. I invited friends to follow instructions on cards, instructing them to crawl, climb, and interact within the altered landscape. This was a dynamic I recognized as breaking the traditional "frame," and I would extend the dimensionality of my paintings with motorized elements, increasing the collage density with domestic objects; I had begun slicing through layers of my paintings to experience "the other side." In his chapter "The Structure of the Cell," Thompson writes:

> The quality of a medium filling the field of force may be uniform, or it may vary from point to point. In particular it may depend upon the magnitude of the field; and the quality of one medium may differ from that of another.
>
> (*On Growth and Form*, 171)

D'Arcy Thompson and Polycrystalline Pattern Formation

Bart Kahr

Chapter IX of *On Growth and Form* is "On Concretions, Spicules, and Spicular Skeletons," which examines, in the main, polycrystalline objects in the biological world. In this chapter, Thompson relied heavily on Ernst Haeckel's (1834–1919) famous *Challenger Monograph*[5] that vividly characterized radiolaria, protozoa with mineralized silaceous skeletons. Two more careful and influential students of pattern formation in nature than Thompson and Haeckel would be hard to find. But, while their eyes were directed in similar directions, they drew different inferences. Haeckel was an arch-Darwinist, while Thompson was a Darwinist for good reason only.

My exploration of Thompson's crystallization studies was a consequence of an accidental encounter with Haeckel. Browsing in a bookstore in Chicago more than ten years ago I came upon a large volume, *Visions of Nature: The Art and Science of Ernst Haeckel*[6] decorated on the front and back flyleaves with photographs of radial crystalline bodies, called spherulites. Moreover, the spherulites represented showed concentric optical rings. For this reason, they are called *banded spherulites*, fascinating polycrystalline objects of precisely the kind we have been studying in our laboratory at New York University. A banded spherulite of aspirin from the melt is shown in Figure 14.4. As the study of banded spherulites is an unusual research

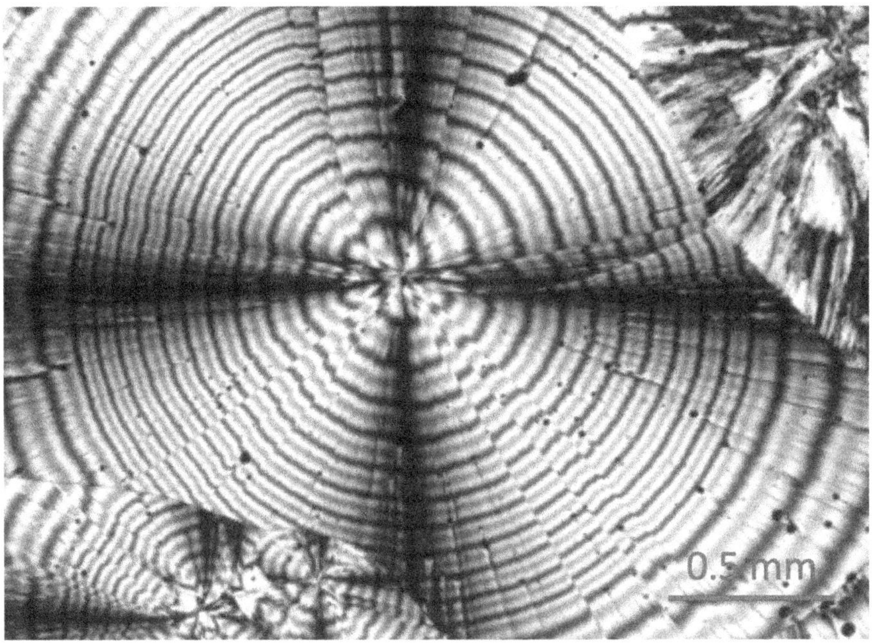

Figure 14.4 Banded spherulite of aspirin grown radially from the melt. Courtesy of AG Shtukenberg.

pursuit, imagine my surprise at finding our objects of interest in an art book, in place of the masterworks of graphic design by Haeckel that spill out of the oversized interior pages. Thompson's Chapter IX is filled with illustrations (2nd ed. for example pps. 655, 659, 660, 663, 666, and 672) of radial or spherulitic bodies, some with concentric markings or contrast as a consequence of rhythmic precipitation, but none that are precisely of the kind shown in Breidbach's *Visions of Nature*.

Breidbach told me that these samples on glass slides reside in Haeckel's lab, in Jena, preserved as a museum. The substances do not carry individual labels. Only "Krystallseelen" (Crystal souls) was written on the box containing them. However, he assured me[7] that they were small molecules obtained from Otto Lehmann (1855–1922).

Lehmann was a pioneer in the study of liquid crystals. He saw a microscopic resemblance of liquid crystal textures with biological patterns that prompted his *Flüssige Kristalle und Theorien des Lebens* (Liquid Crystals and Theories of Life) in 1908, in which he cautiously explored the connections between liquid crystals and living systems. Lehmann's banded spherulites—mostlikely the products of Lehmann's liquid crystals after they solidified—fascinated Haeckel, who believed that liquid crystals and self-patterning polycrystals were the "missing links," so to speak,[8] between the animate and inanimate, and they inspired Haeckel's book *Krystallseelen*[9] (*Crystal Souls*[10]), the label on the box. Lehmann drew the line as to whether liquid crystals are alive: assuredly not.[11] Haeckel was more hopeful.

Krystallseelen was published in 1917 as was the first edition of *On Growth and Form*. Therefore, Haeckel's greatest excesses were probably not available to Thompson at the outset of the latter's great program of morphogenesis. However, Thompson and Haeckel were already at odds. Of Haeckel's musings about biocrystalline form resulting from "the interaction of adaptation and heredity [that becomes] modified in form, and differentiated in a vast variety of ways, in the struggle for existence", Thompson said, "What Haeckel precisely meant by these words is not clear to me."[12] Thompson and Haeckel occupied antipodes on the philosophical spectrum of morphogeneticists.

Thompson's dismissiveness of Haeckel's outlook does not take away from the fact that there is a striking pattern in Figure 14.4, and in all banded spherulites. And, it is extraordinarily common pattern as we have discovered in our laboratories.[13] About one in three common substances that can be melted will form these concentric patterns after crystallization under some conditions including besides aspirin, DDT, malic acid, cholesterol, testosterone, coumarin, mannitol, and resorcinol among hundreds of others. The origin of this pattern was recognized by French scientists more than 100 years ago, but these scientists came from the world of mineralogy, a fertile ground for pattern formation by physical forces, but one which does not necessarily push back against the high tide of Darwinism, Thompson's aim.

The mystery of the banded spherulites is that as the crystallites grow radially, they twist to form helicoids. As the refractivity of the crystals rotates around the radial directions, optical contrast is created between optical polarizers giving concentric light and dark bands. It is as if one were looking at the threads of optical screws, arranged radially. But, helical crystals are strange crystals indeed. Crystals are by definition polyhedral. They are straight with sharp edges and flat faces. Helicoids have curvature and are more characteristic of forms found in the living world. Helicoids can't be single crystals (Figure 14.5).

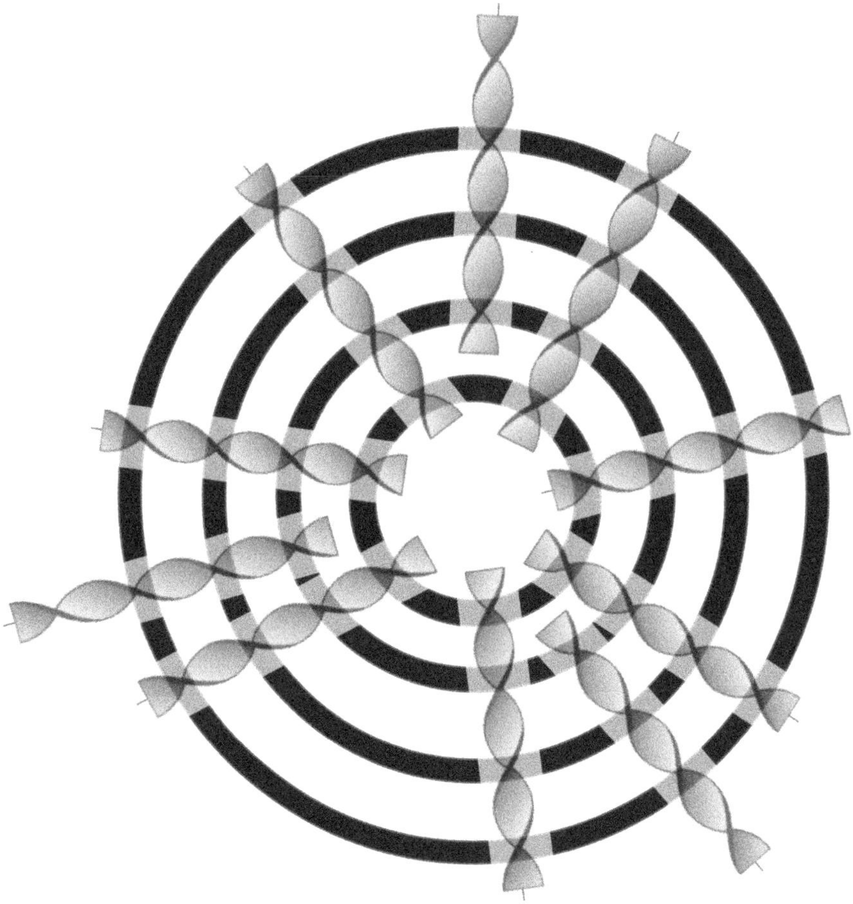

Figure 14.5 Architecture of banded spherulite composed of radial crystalline helicoids. Courtesy of Bart Kahr.

And so, the Haeckel/Thompson dynamic reemerges in curved crystals, and the very common pattern of the banded spherulite of so many simple substances still requires a mechanism of formation. Of course, there is no question today that this mechanism is purely one of physical forces, but it is devilishly difficult to identify with certainty how the necessary forces arise in the chemistry of crystal growth. Thompson's prescient speculations on these matters would be welcome, but this is one branch of pattern formation that did not fall into his wide purview.

Still, it would be all too easy, and less interesting, to take the side of Thompson and physics uncritically. There is a natural selection involved in spherulite formation. Some crystals grow, and die, starved for space and nutrient, whereas other nucleate and grow into daylight. This is a process whose outcome is based on a competition of individuals, albeit, one that doesn't depend on the gene as we know it. There is no evolution, just growth, however non-natural or contrary to the expectations for crystal forms that were well established in the 19th Century, long before *On Growth and Form*.

Five words with definitions:

1. Spherulite—aggregate of many needle or plank-shaped crystals growing radially
2. Banded spherulite—subset of spherulites that, when compressed in two dimensions, show in addition to a radial organization, concentric optical contrast
3. Rhythmic precipitation—oscillation of crystallization in time and space limited by diffusion of the crystallizing medium
4. Liquid crystal—a state of matter that flows like a liquid but with molecules having persistent orientations
5. Refractivity—deflection of a light (or sound) wave at an interface between two media governed by the speeds of light (or sound) in the respective substances

Conversations with Thompson

Ellen K. Levy

The world of things living, like the world of things inanimate, grows of itself, and pursues its ceaseless course of creative evolution. It has room, wide but not unbounded, for variety of living form and structure, as these tend towards their seemingly endless, but yet strictly limited, possibilities of permutation and degree ...[14]

D'Arcy Thompson

What is at stake in Thompson's world view? *On Growth and Form* is a clarion call to understand the nuts and bolts of how forms develop. The passage above encapsulates three important aspects of his outlook that recur throughout his work: analogy, self-organization, and constraints. The first issue, biological analogy, explains why disparate forms without common ancestry can be correlated through physical and mathematical laws. Thompson demonstrated that the occurrence of similar patterns in unrelated entities can reflect a parallel but independent structural response to a problem, often involving principles of self-organization ("creative evolution"), Thompson's second point. The third reference in Thompson's quote concerns constraints; he determined that only a finite number of designs are possible for organisms to assume as they develop.[15] Enrico Coen's *The Art of Genes* (2000) furthered my understanding of Thompson's points, while integrating later developments in molecular genetics. Coen and others pointed not only to Thompson's early grasp of complex systems in which spontaneous order could arise in both living and non-living systems, but to the importance Thompson placed on limits imposed by an organism's basic body plan during development.

All three issues figure greatly in the arts. Analogy, of course, pervades all creative acts, and Thompson looked to art for analogous solutions involving mathematics. Albrecht Dürer's *Four Books on Human Proportion* (1512–1528), for example, informed Thompson's well-known transformation grids.[16] Like Thompson, theorists Bruno Latour and Donna Haraway show that scientific meanings lie within broader cultural narratives and the reverse, that cultural narratives are situated at the heart of scientific terms. This anthology reflects that artists and architects sometimes expand on Thompson's principles of morphogenesis to provide political and military dimensions to their own works. I find such examples instructive; part of my self-assigned aim as an artist is to visualize paths of convergence between biology and culture. In this space, I briefly describe two installations, one past and one current, that engage analogy, self-organization, and constraints. I conceive these works not only as intimate conversations with a viewer, but with Thompson, in particular.

Past Conversations

Thompson's work paved the road to a new understanding of forces in evolution and to complex systems, a subject that held enormous interest for me prior to the millennium.

To explore complexity, I incorporated patent drawings and text in long scroll-like artworks to capture the dynamic feedback mechanisms of learning and economic profit that propel innovation. My work immediately assumed an evolutionary cast due to extensive commonalities between the myriad forms of organisms and the continuous spinoffs of artifacts.

All patents in the US patent office reference their antecedents, reflecting that learning has taken place. I built up genealogies of related inventions by tracking the listed references on the patents. The linkages reflect the ongoing adjustments and novelties inherent in patented inventions over a roughly 200-year span. Tracing these references suggests how seemingly obsolete technologies can regain relevance and become "exapted" (Stephen J. Gould's term) for new uses. One of my patent-based works, *Evolution*, was exhibited at the Field Museum of Chicago in 2006.[17] It is composed of drawings compiled from successive gene patents and digitally collaged with fragments of inventions, including Monsanto's infamous "terminator gene" and snippets from *Scientific American* and migration and evolutionary charts (Figure 14.6).

I later learned that Niles Eldredge, expert on trilobite fossils and co-author with Gould of the theory of punctuated equilibrium, had created a taxonomy of his cornet collection. By systemizing musical instruments rather than arthropods, Eldridge compared biological and cultural evolution, indicating occurrences of horizontal transfer to emphasize the linkages among cornets. He commented that lateral transfer of know-how between groups marks the development of cultural artifacts (akin to horizontal gene transfer in organisms) stating, "if you define evolution as the fate of transmissible information through time, you have a definition of evolution that can embrace both systems, biology and culture."[18] Eldredge connected the cornets in time laterally, with serial numbers attached and sorted by makers and models. He singled out the expansion of the curve inside a brass instrument arguing, "there is overall

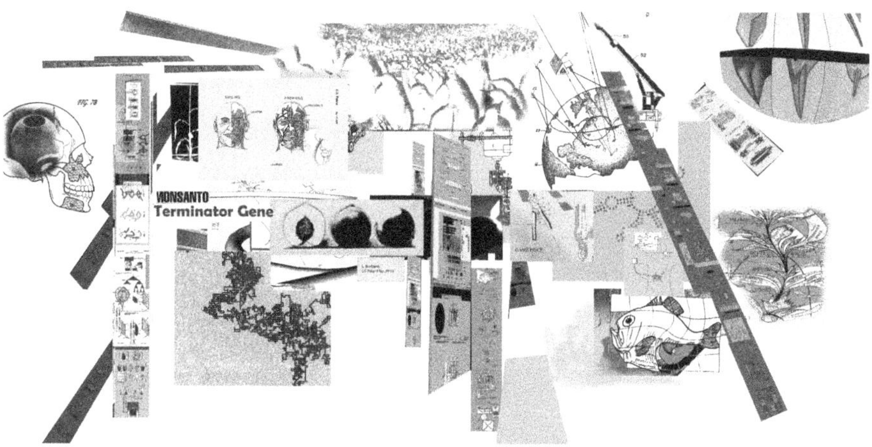

Figure 14.6 Ellen K. Levy, *Evolution*, exhibited at the Field Museum, Chicago, 2006. 32 × 64 inches, acrylic over archival print. Courtesy of Ellen K. Levy.

widening but not a simple monotonic function as in a gastropod. The shape of that tube can be described mathematically." He thus identified operative Thompson-like qualities in brass musical instruments through emphasizing mechanical features.

Present Conversations

Thompson was intrigued by experiments about biological liquid crystals that function in both biotic and abiotic settings.[19] For the German modern architect BrunoTaut, the form-generative qualities of crystals reflected both abstract concepts and political ideals. Taut saw in glass a hope of overcoming through transparency political and social divisions that had led to a world war.[20] Whereas Taut's evocative glass structures were intended to generate positive socio-political change and show the influence of environmental forces both on nature and architecture, in these dystopian days, my thoughts are on how J. G. Ballard used the process of crystallization to frame violent environmental catastrophe and the shattering of political norms. In *Crystal World*, Ballard steers Thompson's mathematical crystals via metamorphosis to a place where unfamiliar yet partially accurate distortions of the real become surreal. I aimed to portray this state in *Crossing Borders: Mexico*; it includes paint, prints, wood, metal, and crystals. I depict barriers over and under a desert landscape, creating a top-down/bottom-up dynamic. The landscape above includes grid-like images appropriated from the Internet in response to public calls for designs of walls to bar Mexican immigrants and refugees from the US. The lower part of the images shows geo-thermal self-organized formations beneath the surface, including the strange giant crystal, fence-like structures located below the earth's surface in Naica, Mexico. When installed, one views illusionistic sections of flooring that have seemingly been cut open to depict some of the giant crystals in Naica. (Naica formations are, of course, too inaccessible to be recruited as actual borders.) My "wall" pits enforced political ordering on top against the percolating conditions of bottom-up self-organization. The metaphor is augmented by wedges of materials (wood, metal) on which stills of a cellular automata program supply rules for simulating the growth of crystal-like formations. Actual crystals are on the floor (Color Plate 14).

On Growth and Form was a catalyst for me to locate potent artistic analogies in biology, history, ecology, and economics. Although organic metaphors typically suffuse my work and visually link it to Thompson, it is also the routine crossing of extra-artistic borders and a focus on evolution and self-organizing processes that mark me as a benefactor of his thinking.

The Vortex and D'Arcy Wentworth Thompson

Meredith Tromble

The grid is also a performative system, a physical technology for thinking through.
Art historian Dawna Schuld[21]

Tracing the echoes of *On Growth and Form* through my art is a submarine affair, they sound so deep and far from the shores of conscious memory. Was there a time when I could not picture D'Arcy Wentworth Thompson's stretchy fish? I cannot recall it, just as I cannot remember exactly where I first encountered them, perhaps in the *Whole Earth Catalog*, which praised *On Growth and Form* in 1969, or a slide lecture by my first painting teacher, Francis Sprout. But in 2011, when I walked into a CAVE (Computerized Automatic Viewing Environment) for the first time and saw images that one could stretch and move, accompanied by a Cartesian grid—which in the CAVE shone a rich fluorescent green—they seemed familiar. These projected images of data were relatives of Thompson's piscine transformations. What began as play for Thompson—his daughter Ruth reported that he amused his children with stretchy drawings made on a rubber sheet[22]—informed his ideas about form and forces. His methods seemed right at home in the CAVE with researchers at the University of California, Davis (UCD), who made their 3-D projections interactive with tracking and high-powered computation, and who had invited artists to play with their tools as part of generating new ideas. Following that first encounter, the geobiologist Dawn Sumner, a co-founder of UCD's KeckCAVES, became my primary collaborator on a still-growing body of artworks known as the *Vortex* series. Sumner investigates growth at vast time scales—from the beginnings of life on earth—and at molecular physical scales, describing growth and form in precisely articulated, measured ways that were inconceivable in Wentworth-Thompson's time. Sumner's concern, and one of the big puzzles of life's history, is how evolution sculpted the life-sustaining nanoscale structures of photosynthesis. Photosynthesis, the root support of growth for almost every living being, depends on form, on precise distances and shapes in the molecular respiratory chain that produces biological energy. The data Sumner brings to her questions spans aeons, from 3.5-billion-year-old rocks that may be fossilized cyanobacteria to contemporary microorganisms in Antarctic lakes; the quest to understand that data led to her work in interactive 3-D visualization. If Thompson could miraculously arrive in Sumner's lab today, he might need time to grasp her technique, but he could immediately appreciate her holistic approach to her science. As his daughter Ruth wrote in her biography of Thompson, "in his philosophy nothing could be isolated in a narrow groove; each and all combined to the ultimate purpose; everything must merge to the creation of the whole."[23]

Together Sumner and I developed an interactive image vortex filled with drawings based on dreams or stories collected from researchers. The vortex asserts the continuity between the physical and virtual worlds, following the transmigration of impulses and memories through images and texts: from the memory in a story-teller's mind, to the language of the story as it is told, to my drawing, to code text, to the visuals of the vortex and so into more minds as memories.

The vortex became the core of a rapidly expanding series of artworks including the *Dream Vortex*, programming for interactive head-mounted and flat screen 3-D displays; *Outside the Vortex* and *The Vortex*, two dance performances with the Los Angeles-based company Donna Sternberg & Dancers; two physical installations and a video installation, and numerous mixed-media drawings and digital prints. Scores of ancillary collaborators have spiraled into the project, including the physicist Jim Crutchfield who gave it a home at UCD's Complexity Sciences Center. Describing this voluminous series in any detail would quickly overflow the allotted bounds of this text, but what the works all have in common is the dynamic, spiraling image of a vortex that blurs many images into a potent, fertile whole (Color Plate 15). The vortex itself is four dimensional, active in time and present in space. The images that erupt from it are two-and-a-half dimensional, planar drawings that move in three-dimensional space, at the threshold of form. This dimensional flexibility arises from a spatial matrix, the

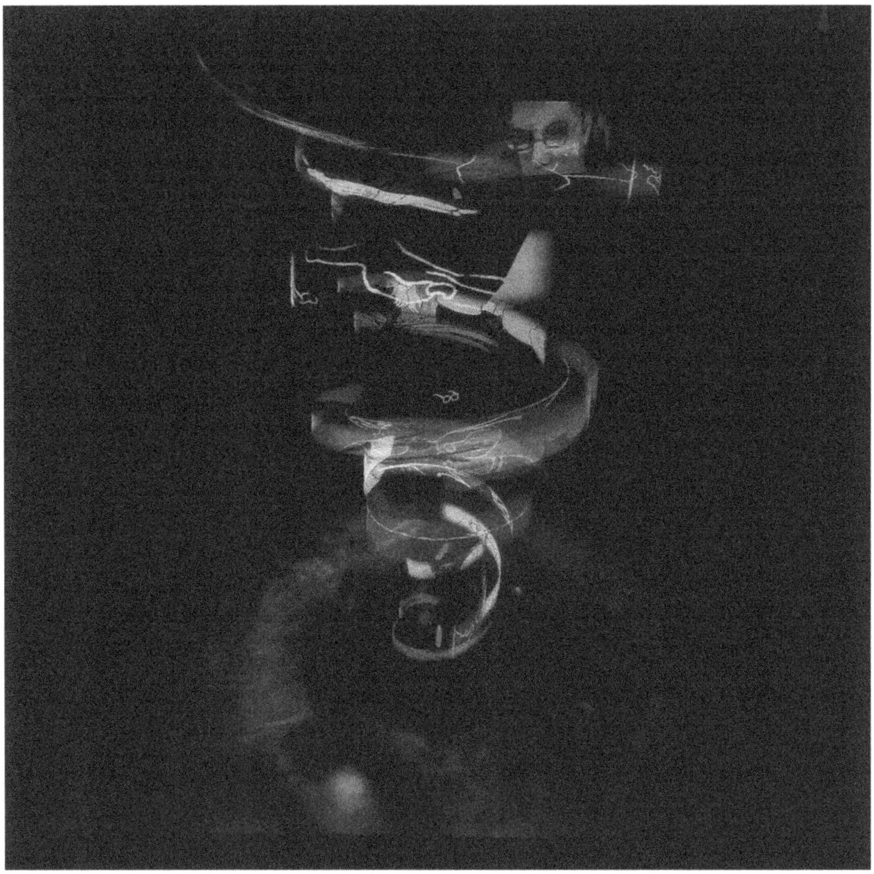

Figure 14.7 Meredith Tromble with Dawn Sumner, *Dream Vortex*, Virtual Reality Installation, *Physicist Dream*. Courtesy of Meredith Tromble with Dawn Sumner.

3-D version of Thompson's grid. Like the grid from which it arose, the vortex implies limitless extension. In principal, its growth spirals through time endlessly open to new iterations, with programming that ensures a degree of randomness and unpredictability within each occurrence (Figure 14.7).

This open condition is an artistic embrace of change and complexity, aligned with the study of nonlinear systems surrounding the project in the Complexity Sciences Center. In his essay on Thompson, the biologist Stephen J. Gould wrote that "Thompson was interested in the deformed net … as a diagram of forces ….The method of transformed coordinates is D'Arcy Thompson's provisional mathematics for complex structures."[24] Just as new visualization technologies have magnified the power of the grid since Thompson's time, contemporary physics can "see" mathematically nonlinear, dynamic, and multi-dimensional forces like those Thompson encapsulated in his prescient diagram.

Thompson's insight was so fundamental and generative that it courses through many different channels today. In an essay on the impact of Thompson's work, biologist P. B. Medawar described his influence as "diffused and widely pervasive."[25] As an artist whose practice has flowed toward Thompson through three channels, through fine art, interactive visualization tools, and physics, I celebrate his stretchy fish and the ocean of ideas around them.

Deployable and Other Structural Forms

Henry Petroski

On the morning I visited the Deployable Structures Laboratory at Cambridge University, Professor Sergio Pellegrino was occupied with students and technicians folding a newly hinged device into its compact transport configuration. Since he was preoccupied, I was shown around the lab by my host, Professor Chris Calladine, who recently had studied deployable engineering structures from a biological perspective. In his paper at the symposium on deployable structures held at Cambridge in 1998, Calladine noted that since "all biological structures *grow* and *develop*: in that sense they are *all* deployable." But he quickly added that studying growth and development in biological structures cannot necessarily be expected to provide useful lessons for deployable engineering structures, for the two processes are not analogous. He also noted that it is not fruitful to look to biological evolution for engineering design guidance, quoting the aphorism that "evolution is blind; technology is mind."

Where Calladine did see some potential inspiration from biology was at the level of molecular structures, which might "stimulate the constructive thought and imagination of our engineers." Other students of biological structures see inspiration in the wing folding of insects and the protective curling up of leaves in high winds. Steven Vogel, the late Duke University biologist, studied this latter phenomenon and wrote eloquently about it. Like Calladine, Vogel saw similarities between natural and human-made structures—which, after all, as D'Arcy Thompson noted, follow and are constrained by the same laws of physics and engineering—but did not believe that copying nature assured superior structures, deployable or otherwise. That is not to say that inspiration cannot be found in nature. Velcro is a familiar example.

The Deployable Structures Laboratory was full of models of engineered structures, many of which have the beauty of form and operation of a flowering plant. Among these are camera aperture-control devices that were seen to have applications for deployable roof structures to protect sports fans against rain or sunlight. Some are flat and appear to take the form of pressed flowers, while others are spherical and are suggestive of the intricate but graceful geometry of pinecones, sunflowers, and pineapples. Elsewhere in the laboratory were foldable tensegrity structures, those skeleton-like assemblages of wires in tension and struts in compression that date from the 1960s and that were given currency by Buckminster Fuller. Chuck Hoberman's sphere and other of his transformable toys and exhibits are obvious successors.

But how far can deployable structures be scaled up? D'Arcy Thompson explained the biological equivalent of what engineers call a size effect when he wrote in *On Growth and Form*, "An organism is so complex a thing, and growth so complex a phenomenon, that for growth to be so uniform and constant in all the parts as to keep the whole shape unchanged would indeed be an unlikely and an unusual circumstance. Rates vary, proportions change, and the whole configuration alters accordingly." Even a geometrically simple manmade object, like an obelisk, can be a complex thing structurally. It was not until Galileo pointed out that, in addition to geometry, the strength of the material had to be taken into account, that previously inexplicable failures were understood. Unlike politicians, engineers cannot simply declare something "too big to fail."

The forms of deployable structures seem virtually endless. Thomas Heatherwick's unique hydraulically operated Rolling Bridge at London's Grand Union Canal development is among the most creative. Even in common terrestrial applications they include a wide class of inflatable structures, ranging from air mattresses to the giant balloons used in Macy's Thanksgiving Day parade. Venetian blinds, Murphy beds, sleeper sofas, recliner chairs, and even bureau drawers and common doors are deployable domestic structures. As children we played with pogo sticks, Slinkys, and deployed the folding kickstands on our bicycles. In the kitchen, we use scissor-handle operated can openers, ingenious cork-extraction devices, sliding shelves, and lettuce baskets daily. In our office or study we use retractable ball-point pens, mechanical pencils, and staplers as a matter of course. Hotel rooms often have retractable clothes lines, collapsible ironing boards, and televisions on extendable and rotatable bases. Extension ladders are clearly deployable, as are retractable box-cutter knives, and a host of gripping and grasping tools. Many early cell phone designs relied on deployable structures for their casings, as do laptop computers. Our automobiles are equipped with airbags, storable cup holders, retractable seat belts, operable sunroofs, and a host of other deployable devices. And it is not only large things that are deployable. An ophthalmologist replacing a cataract with an implant makes an incision only large enough to fit a folded lens through. Once inserted behind the cornea, the plastic implant is allowed to deploy into a proper lens shape. Compact stents that deploy when in position in an artery also allow for minimally invasive surgery. Engineers working on similarly beneficial medical devices have been employing techniques inspired by the Japanese art of paper folding. These origami-like constructions deploy on arrival at a target location in the body. In fact, deployable structures are all around us and have become such common features in our daily lives that we hardly notice or feel them for what they are. Even a magazine might be said to be a deployable structure, as might the book in which this chapter is printed.

Properly curving or folding a structure—whether it be a spring-steel measuring tape or a slice of pepperoni pizza—always stiffens it. This is why cardboard boxes and tin roofs are corrugated, why tin ceilings are impressed with decorative patterns, and why unpressurized tin cans have circumferential ridges beneath the label. That automobile body panels often have a complex curvature serves not only to make them more stylish but also to impart a greater stiffness than a flat panel, thus enabling a thinner steel sheet to be used in their manufacture. Decorative ridges on computer and appliance housings similarly are designed to add stiffness as well as styling.

The folds in the robe of the Statue of Liberty help give the thin copper sheets of which the statue is composed a stiffness that enables it to maintain its shape in the high winds of Upper New York Bay. In 2012, when Hurricane Sandy ruined other structures on Liberty Island, the statue itself suffered no significant damage. Had the lady with the torch contained wide expanses of unfolded copper, its shape might change noticeably with the weather and make noises as the copper panels buckled in and out under changing temperatures and the varying pressure exerted by the wind. Similarly, if my retractable steel tape were perfectly flat, it would droop like a wet noodle whenever deployed more than a few inches. It would also have virtually no resistance to buckling when pushed against a baseboard. Folding and forming are staples of structural engineering design.

Notes

1 Monod, J., *Chance and Necessity* (London: Penguin, 1997).
2 Jacob, F., *The Logic of Life, and the Possible and the Actual* (London: Penguin, 1989), 3; Mayr, E., *What Makes Biology Unique?* (Cambridge: Cambridge University Press, 2007).
4 Turing, A. M., "The Chemical Basis of Morphogenesis," *Phil. Trans. R. Soc. Lond. B*, Vol. 237 (1952), 37–72.
5 Haeckel, E. Report on the Radioloaria Collected by H.M.S. Challenger during the Years 1873–1876, H.M.S. Stationary Office, Edinburgh, 1887.
6 Breidbach, O. *Visions of Nature: The Art and Science of Ernst Haeckel*, Prestel, Munich, 2006.
7 E-mail: O. Breidbach to B. Kahr, March 12, 2007.
8 Mackay, A. L., From "the Dialectics of Nature" to the Inorganic Gene, *Foundations of Chemistry*, 1999, 1, 43–56.
9 Haeckel, E. *Krystallseelen*, Leipzig, 1917.
10 Haeckel, E. *Crystal Souls* (trans. Mackay, A.) *Forma*, 1999, *14*, 1–204.
11 Sluckin, T. J.; Dunmur, D. A.; Stegemeyer, H., *Crystals That Flow*, Taylor and Francis, New York, 2004.
12 Thompson, D. *On Growth and Form*, Reprint of 2nd edn. (1942) Dover, New York, 1992; 691.
13 Shtukenberg, A. G.; Punin, Yu. O.; Gujral, A.; Kahr, B. Growth actuated bending and twisting of crystals, *Angew. Chem. Int. Ed.* 2014, *53*, 672–99.
14 Thompson, D. W., *On Growth and Form* (Cambridge: Cambridge University Press, 1917), 137.
15 Gould, S.J., "D'Arcy Thompson and the Science of Form," *New Literary History*, Vol. 2, No. 2 (Winter, 1971), 257.
16 Address delivered by D'Arcy Thompson at the opening of session 1903–1904, quoted in *The College* December 1903, 7–8; Jarron, M. "Editorial," *Interdisciplinary Science Reviews*, Vol. 38, No. 1 (2013), 1–11.
17 *Gregor Mendel: Planting the Seeds of Genetics* (cur C. Albano and M. Wallace, adv M. Kemp), orig. Field Museum, Chicago, IL, traveled to 7 other venues, including the National Museum of Health Medicine, Washington, DC, 2006.
18 Discussions with Niles Eldredge took place on Monday 6/12/2017 and 5/24, 2017.
19 Hyde, Stephen, "D'Arcy Thompson's Legacy in Contemporary Studies of Patterns and Morphology," *Interdisciplinary Science Reviews*, Vol. 38, No. 1 (March 2013), 17.
20 Whyte, Iain Boyd (trans. and ed.), *The Crystal Chain Letters: Architectural Fantasies by Bruno Taut and His Circle* (Cambridge, MA: MIT Press, 1985).
21 Schuld, Dawna, "Mind Matrix: Situating Cognition in the Sculptural Grid," in *The Routledge Companion to Biology in Art and Architecture*, edited by Sabine Flach and Jan Söffner 286 (London and New York: Routledge, 2017).
22 D'Arcy Thompson, Ruth, *D'Arcy Wentworth Thompson: The Scholar-Naturalist 1860–1948* (London: Oxford University Press, 1938), 178.

23 Ibid., 126.

24 Gould, Stephen Jay, "D'Arcy Thompson and the Science of Form," *New Literary History*, Vol. 2, No. 2 (Winter, 1971), *Form and Its Alternatives* (Baltimore, MD: The Johns Hopkins University Press, Winter, 1971), 245–6.

25 Medawar, P. B., "Postscript: Growth and Form," in *D'Arcy Wentworth Thompson: The Scholar-Naturalist 1860–1948*, edited by Ruth Thompson (London: Oxford University Press, 1938), 233.

Bibliography

McGeehan, Patrick (2013). "Statue of Liberty Is to Reopen, Fittingly, by the Fourth of July," *New York Times*, March 20, A21.

Pellegrino, S., (2001) ed. *Deployable Structures* (Vienna: Springer-Verlag).

Pellegrino, S. and S. D. Guest, (2000) eds. *IUTAM-IASS Symposium on Deployable Structures: Theory and Applications. Proceedings of the IUTAM Symposium held in Cambridge, U.K., 6–9 September 1998* (Dordrecht, The Netherlands: Kluwer).

Petroski, Henry (2004). "Deployable Structures," *American Scientist*, March–April, 122–126.

Thompson, D'Arcy W (1942). *On Growth and Form*, 2nd edn. (Cambridge: Cambridge University Press), 205.

Vogel, Steven (1998). *Cats' Paws and Catapults: Mechanical Worlds of Nature and People* (New York: Norton).

Index